Advances in Intelligent Systems and Computing

Volume 769

Series editor

Janusz Kacprzyk, Polish Academy of Sciences, Warsaw, Poland
e-mail: kacprzyk@ibspan.waw.pl

The series "Advances in Intelligent Systems and Computing" contains publications on theory, applications, and design methods of Intelligent Systems and Intelligent Computing. Virtually all disciplines such as engineering, natural sciences, computer and information science, ICT, economics, business, e-commerce, environment, healthcare, life science are covered. The list of topics spans all the areas of modern intelligent systems and computing such as: computational intelligence, soft computing including neural networks, fuzzy systems, evolutionary computing and the fusion of these paradigms, social intelligence, ambient intelligence, computational neuroscience, artificial life, virtual worlds and society, cognitive science and systems, Perception and Vision, DNA and immune based systems, self-organizing and adaptive systems, e-Learning and teaching, human-centered and human-centric computing, recommender systems, intelligent control, robotics and mechatronics including human-machine teaming, knowledge-based paradigms, learning paradigms, machine ethics, intelligent data analysis, knowledge management, intelligent agents, intelligent decision making and support, intelligent network security, trust management, interactive entertainment, Web intelligence and multimedia.

The publications within "Advances in Intelligent Systems and Computing" are primarily proceedings of important conferences, symposia and congresses. They cover significant recent developments in the field, both of a foundational and applicable character. An important characteristic feature of the series is the short publication time and world-wide distribution. This permits a rapid and broad dissemination of research results.

More information about this series at http://www.springer.com/series/11156

Herwig Unger · Sunantha Sodsee
Phayung Meesad
Editors

Recent Advances in Information and Communication Technology 2018

Proceedings of the 14th International Conference on Computing and Information Technology (IC2IT 2018)

Springer

Editors
Herwig Unger
LG Kommunikationsnetze
Fern Universität in Hagen
Hagen, Germany

Sunantha Sodsee
Faculty of Information Technology
King Mongkut's University of Technology
North Bangkok
Bangkok, Thailand

Phayung Meesad
Faculty of Information Technology
King Mongkut's University of Technology
North Bangkok
Bangkok, Thailand

ISSN 2194-5357 ISSN 2194-5365 (electronic)
Advances in Intelligent Systems and Computing
ISBN 978-3-319-93691-8 ISBN 978-3-319-93692-5 (eBook)
https://doi.org/10.1007/978-3-319-93692-5

Library of Congress Control Number: 2018947344

Printed on acid-free paper

This Springer imprint is published by the registered company Springer International Publishing AG part of Springer Nature
The registered company address is: Gewerbestrasse 11, 6330 Cham, Switzerland

Preface

This proceedings volume contains the papers of the 14th International Conference on Computing and Information Technology (IC2IT 2018) held on July 5–6, 2018, in Chiang Mai, Thailand. IC2IT is a platform for researchers to meet and exchange knowledge in the field of computer and information technology. The participants in IC2IT present their current research new findings and discuss with partners to seek new research directions and solutions as well as cooperation.

Springer has published the proceedings of IC2IT in its well-established and worldwide-distributed series on Advances in Intelligent and Soft Computing, Janusz Kacprzyk (Series Editor). This year there were total 88 submissions from 16 countries. Each submission was assigned to at least three program committee members, and at least two members must accept a paper in order to include in the proceedings. The committee accepted 33 papers for presenting at the conference and publishing the papers in the proceedings. The contents of the proceedings are divided into subfields: Data Mining, Machine Learning, Natural Language Processing, Image Processing, Network and Security, Software Engineering, and Information Technology.

We would like to thank all authors for their submissions. We would also like to thank all the program committee members for their support in reviewing assigned papers and in giving good comments back to the authors to revise their papers. In addition, we would like to thank all university partners in both Thailand and overseas for academic cooperation. Special thanks also go to the staff members of the Faculty of Information Technology at King Mongkut's University of Technology North Bangkok who have done many technical and organizational works. Without the painstaking work of Dr. Watchareewan Jitsakul, the proceedings could not have been completed in the needed form at the right time.

Finally yet importantly, we would like to thank all the speakers and audiences for their contributions and discussions at the conference that made the conference a success. We hope that the proceedings IC2IT will be a good source of research papers for future references with the state of the art.

April 2018

Herwig Unger
Sunantha Sodsee
Phayung Meesad

Organization

Program Committee

M. Aiello	Uni. Stuttgart, Germany
S. Auwatanamongkol	NIDA, Thailand
G. Azzopardi	Uni. Groningen, The Netherlands
T. Bernard	Syscom CReSTIC, France
C. Biemann	TU, Darmstadt
S. Boonkrong	SUT, Thailand
S. Butcharoen	TOT, Thailand
M. Caspar	Braunschweig, Germany
T. Chintakovid	KMUTNB, Thailand
K. Chochiang	PSU, Thailand
K. Dittakan	PSU, Thailand
T. Eggendorfer	HS Weingarten, Germany
M. Ghazali	UTM, Malaysia
M. Hagan	OSU, USA
W. Halang	FernUni, Germany
C. Haruechaiyasak	NECTEC, Thailand
U. Inyaem	RMUTT, Thailand
W. Jirachiefpattana	NIDA, Thailand
W. Jitsakul	KMUTNB, Thailand
S. Jungjit	TSU, Thailand
M. Kaenampornpan	MSU, Thailand
M. Komkhao	RMUTP, Thailand
A. Kongthon	NECTEC, Thailand
P. Kropf	UniNe, Switzerland
M. Kubek	FernUni, Germany
K. Kyamakya	AAU, Austria
U. Lechner	Uni. München, Germany
H. Lefmann	TU Chemnitz, Germany

P. Meesad	KMUTNB, Thailand
W. Nadee	PSU, Thailand
A. Nanthaamornphong	PSU, Thailand
K. Nimkerdphol	RMUTT, Thailand
M. Phongpaibul	TU, Thailand
J. Polpinij	MSU, Thailand
W. Pongphet	NIDA, Thailand
N. Porrawatpreyakorn	KMUTNB, Thailand
P. Prathombutr	DEPA, Thailand
P. Saengsiri	TIRTR, Thailand
P. Sanguansat	PIM, Thailand
T. Siriborvornratanakul	NIDA, Thailand
M. Sodanil	KMUTNB, Thailand
S. Sodsee	KMUTNB, Thailand
V. K. Solanki	CMRIT, India
W. Sriurai	UBU, Thailand
T. Sucontphunt	NIDA, Thailand
W. Tang	CityUni, Hong Kong
C. Thaenchaikun	PSU, Thailand
J. Thaenthong	PSU, Thailand
D. Thammasiri	NPRU, Thailand
N. Theeraampornpunt	PSU, Thailand
J. Thongkam	MSU, Thailand
N. Tongtep	PSU, Thailand
D. Tutsch	UNIWUP, Germany
H. Unger	FernUni, Germany
V. Wattanasoontorn	PSU, Thailand
M. Weiser	OSU, USA
W. Werapun	PSU, Thailand
N. Wisitpongphan	KMUTNB, Thailand
K. Woraratpanya	KMITL, Thailand
P. Wuttidittachotti	KMUTNB, Thailand

Organizing Partners

In Cooperation with
King Mongkut's University of Technology North Bangkok (KMUTNB)
FernUniversitaet in Hagen, Germany (FernUni)
Chemnitz University, Germany (CUT)
Oklahoma State University, USA (OSU)
Edith Cowan University, Western Australia (ECU)
Hanoi National University of Education, Vietnam (HNUE)

Gesellschaft für Informatik (GI)
Mahasarakham University (MSU)
Ubon Ratchathani University (UBU)
Kanchanaburi Rajabhat University (KRU)
Nakhon Pathom Rajabhat University (NPRU)
Phetchaburi Rajabhat University (PBRU)
Rajamangala University of Technology Krungthep (RMUTK)
Rajamangala University of Technology Thanyaburi (RMUTT)
Prince of Songkla University, Phuket Campus (PSU)
National Institute of Development Administration (NIDA)
Sisaket Rajabhat University (SSKRU)
Council of IT Deans of Thailand (CITT)
IEEE CIS Thailand

Contents

Software Engineering

Data Mining

Combining Multiple Features for Product Categorisation by Multiple Kernel Learning

Chanawee Chavaltada[1], Kitsuchart Pasupa[1(✉)], and David R. Hardoon[2]

[1] Faculty of Information Technology,
King Mongkut's Institute of Technology Ladkrabang, Bangkok 10520, Thailand
Kitsuchart@it.kmitl.ac.th
[2] PriceTrolley Pte. Ltd., Singapore 573969, Singapore

Abstract. E-commerce provides convenience and flexibility for consumers; for example, they can inquire about the availability of a desired product and get immediate response, hence they can seamlessly search for any desired products. Every day, e-commerce sites are updated with thousands of new images and their associated metadata (textual information), causing a problem of big data. Retail product categorisation involves cross-modal retrieval that shows the path of a category. In this study, we leveraged both image vectors of various aspects and textual metadata as features, then constructed a set of kernels. Multiple Kernel Learning (MKL) proposes to combine these kernels in order to achieve the best prediction accuracy. We compared the Support Vector Machine (SVM) prediction results between using an individual feature kernel and an MKL combined feature kernel to demonstrate the prediction improvement gained by MKL.

Keywords: Product classification · Multiple kernel learning
Feature combination

1 Introduction

Digital revolution, especially E-commerce, has had a tremendous influence on business. Online shops stock a lot of products and display their descriptions on website. This provides convenience and flexibility for customers as they shop and enquire about the products. Consequently, there is an enormous amount of data about products. Therefore, a method for effective query is needed. Search bar is a query tool that facilitates finding an item in a database. Using a search bar, a customer can search for an item by using specific keywords; the matched products are then listed. Moreover, an intelligent search can suggest relevant items or filter items by categories, making it easier for customers to see relevant data of their desired products. Therefore, effective classification of products for intelligent search is essential. However, manual product categorisation is too

H. Unger et al. (Eds.): IC2IT 2018, AISC 769, pp. 3–12, 2019.
https://doi.org/10.1007/978-3-319-93692-5_1

repetitive and labour-intensive when dealing with large data. It would be better to have an automated system to perform this task in order to reduce classification time and cost [1].

In this respect, machine learning has been applied to automation of product categorisation. Supervised learning is a typical type of machine learning that requires learning from training data so that the model will be trained to be efficient in classification before it is used for actual analysis. In a previous work, we used text data as a feature: we extracted product names and trained the model. Several classification techniques were tried out with this approach to determine the optimum technique [2]. A good model cannot be built by using text feature alone; a better model is expected when other kinds of features are added. Apart from product name, there are other potential features such as product description and picture of the product. Therefore, we considered using different aspects of an image of a product as features such as colour, shape, edge and interest point that can characterize an image [3].

Combining all of these features is a good way to improve performance; however, the diversity of features may cause improper weighting of each different feature. Therefore, we need some new method to combine the features and produce proper weights. Multiple kernel learning (MKL) is an algorithm that can combine several features by finding an optimal linear combination of kernels in order to solve this problem.

In this paper, we focus on demonstrating product categorisation with MKL which is based on Support Vector Machine (SVM) [4]. We collected a set of product names, product pictures and category paths from an online shopping website. The extracted features were text and image data. The text data was preprocessed and extracted with a term frequency-inverse document frequency (TF-IDF) method and the image data was extracted according to colour, shape, edge, and interest points. The image data was extracted into interest points with a Speeded-up Robust Features (SURF) method and clustered into a bag of visual words [5]. Category paths were transformed into vector labels and their structure was declared in a list of edges. Hence, this problem can be solved as a multi-label problem. In this work, we employed a Binary Relevance method to transform a multi-label problem into multiple binary problems without any learning across models [6,7], because at this stage-we aimed to evaluate whether combining features with MKL could really improve classification rate. To do so, we compared the prediction results of using individual features, concatenated features, and MKL features. The related theories are described in Sect. 2 followed by a description of collected data in Sect. 3. The experimental framework is discussed in Sect. 4. In Sect. 5, we present our results and finally conclude the paper in Sect. 6.

2 Methodology

2.1 Support Vector Machine

Classification is a data mining model that requires learning of statistical historical data (training data) to predict an outcome. SVM is one of powerful algorithms that can classify a wide range of data types.

The kernel function $\kappa(x_i, x_j)$ in SVM calculates the similarity between samples x_i and x_j. An SVM model can be defined as in Eq. (1).

$$f(x) = sign\Big(\sum_{i=1}^{m} \alpha_i y_i \kappa(x_i, x) + b\Big) \tag{1}$$

where α is the weight of each sample, b is the bias of the model, m is the number of samples, and $y_i \in \pm 1$ is a label of x_i which is a sample feature vector [8].

Let $\{\phi(x_m), y_m\}$ be input-output pairs where inputs $\phi(x)$ are in $\mathbb{R}^n$ feature space. The primal soft-margin SVM can be written as a following optimisation problem:

$$\begin{aligned} \min_{w\in\mathbb{R}^n, \varepsilon\in\mathbb{R}^m, b\in\mathbb{R}} \quad & \frac{1}{2}\|w\|_2^2 + C\|\varepsilon\|_1 \\ s.t. \quad & y_i(\langle w, \phi(x_i)\rangle + b) \geq 1 - \varepsilon_i \\ & \varepsilon_i \geq 0, i = 1, \cdots, m \end{aligned} \tag{2}$$

where ε is a slack variable, and $C \in \mathbb{R}^+$, is a penalty parameter. The original problem can be converted into a dual form by using the technique of Lagrange multipliers, see [9]. Normally, an SVM model gives one prediction for each sample; however, in this work, the SVM provided more than one predictions as it is a multi-label problem. We cast the problem as a Binary Relevance task [7]. It transformed a multi-label problem to several binary problems without any learning across models.

2.2 Multiple Kernel Learning

For heterogeneous data sources, combining the data with incorrect weighting is possible and the resulting model will be ineffective. MKL is an approach that can solve this problem by combining different data sources by finding an optimal linear combination of kernels. The combine kernel $\mathcal{K}_C$ was built by using a set of features that was mapped to a kernel function $\kappa_j(x, x')$ according to Eq. (3).

$$\mathcal{K}_C(\kappa_1, \cdots, \kappa_v) = \sum_{j=1}^{v} \lambda_j \kappa_j, \quad \lambda_j \geq 0 \tag{3}$$

Given a combination coefficient λ_j for each kernel κ_j, MKL solves λ_j in optimisation problem according to Eq. (4) below,

$$\begin{aligned} \min \quad & \frac{1}{2}\sum_{j=1}^{v}\frac{1}{\lambda_j}\|w_j\|_p^p + C\sum_{i=1}^{v}\varepsilon_i \\ w.r.t \quad & w \in \mathbb{R}^n, \lambda \in \mathbb{R}^+, \varepsilon \in \mathbb{R}^m, b \in \mathbb{R}, \\ s.t. \quad & y_i\Big(\sum_{j=1}^{v}\langle w_j, \phi_j(x_i)\rangle + b\Big) \geq 1 - \varepsilon_i \end{aligned} \tag{4}$$

where C is a predefined positive trade-off parameter, w_j is the weight of feature ϕ_j, ε_i is the vector of a slack variable and b is a bias term. Furthermore, p-norm regularisation is applied as a penalty function in the MKL, $p > 0$ for multiple kernel weights. The dual of this optimisation problem can be formulated as well, see [10] for more details.

3 Data Set

3.1 Data Collection

Data on 5,834 products were collected from an online shopping website in Singapore, including product names, pictures and category paths. Some categories had insufficient data; therefore, we set a condition to exclude categories that had the amount of data of less than one percent of the total data set. In the end, we obtained 36 categories in which their category paths could have two levels of maximum depth.

3.2 Feature Extraction

The product data set was extracted into different types of features which were text (ϕ_1), shape (ϕ_2), edge (ϕ_3), colour (ϕ_4), and SURF (ϕ_5) features, where $\phi_j \in \Phi$. We combined them by two approaches: (i) common concatenation that concatenated all individual features, $\phi_{con} = [\phi_1\ \phi_2\ \phi_3\ \phi_4\ \phi_5]$, and (ii) MKL technique that combined features by combining kernels, $\mathcal{K}_C(\kappa_1, \kappa_2, \kappa_3, \kappa_4, \kappa_5)$.

Text. Document data formatting is often done by converting the data into a useable compatible format for each respective process or text processing. There is a number of text preprocessing techniques such as punctuation removal, conversion of letters to lowercase, stop word removal and number removal. They are usually applied for information retrieval, information extraction and data mining. Text feature is commonly extracted to a vector format for several types of analysis. Term frequency-inverse document frequency (TF-IDF) is a text vectoriser that is able to find the appearance frequency of words in a document and measure the priority of words in order to construct a vector of the document. In addition, general text analysis requires a term of words to contain at least one word. A term of words could be more than one word. An n-gram model was used to construct terms of words. It is a linguistic probability model for predicting word objects in a sentence [11] for building a corpus. In this work, the set of product names was used as a feature; they were converted into lowercase letters in order to prepare them for text processing in which number, punctuation, and stop words were removed. Moreover, the product names were tokenised by an n-gram model with $n = \{1, 2\}$ [12] into a corpus. TF-IDF is one approach that can vectorise word sequences into vectors according to the frequency of the words, so the text vectoriser was built by using the corpus of a TF-IDF model. It was used to convert all product names into features for classification models.

Image. There are several approaches for extracting vector features from an image. In this experiment, we processed the set of images into four diverse feature types including colour, edge, shape and SURF feature. The colour of an image was separated into three channels–red, green, and blue–in matrix form. Each colour channel matrix was binarised into 64-bin histograms and concatenated into a feature for describing image colour. For edge and shape features, the image was converted to gray scale and a Canny edge detector technique was used for edge detection [13]. Then, the image would be filled with holes and small connected components in the image would be removed to detect the shape of the product. Moreover, an image had interest points that were selected at some distinctive locations in the image such as corners, blobs, and T-junctions. SURF method is a detector and descriptor of interest points that is fast and high-performance [5]. Each image of a product had a set of interest points, that were used as features. These sets of points were grouped by a clustering method, a k-mean algorithm, in order to build a bag-of-visual-words and map interest points to clusters [14], where $k = \sqrt{\frac{N}{2}}$ that is a rule of thumb [15] and N was the number of all interest points from all product pictures. The result of cluster mapping was a histogram of visual word counts to be uses as a feature.

Multiple Feature Combination. A combination of features can be built by a combination method. The feature set $\phi_j \in \Phi$. The common approach is to concatenate a set of features into a new feature ϕ_{con}. However, an individual feature alone was probably not discriminative enough to yield good performance; we need to improve the performance by combining different features [16]. MKL is a technique that combines kernels from several data types into a feature as described in Sect. 2.2. The set of features Φ was transformed into a set of kernels $\mathcal{K}$ by Eq. (4) which would be combined with weight λ.

3.3 Label Transformation

The set of category paths was used as label data and transformed into a list of binary vectors to be used in binary classifiers. After data categories or classes that were too small were rejected, the remaining data classes were processed in order to distinguish unique categories and unique paths. The unique paths were used to build a tree of product categories for describing path relations. For label vectors, each path of an item was encoded by mapping with categories. Table 1 shows the label for each item defined as binary vectors. A label $y_{i,c} \in \{\pm 1\}$ where 1 described the class which was in the path and −1 described the class that was not. The size of y was $N \times Q$ where N was the sample size and Q was the number of categories.

Table 1. Example of multi-label data set

Category path	Fresh	Vegetable	Meat	⋯	Snack
Fresh ⇒ Meat	1	−1	1	⋯	−1
Fresh ⇒ Vegetable	1	1	−1	⋯	−1
Fresh	1	−1	−1	⋯	−1
Snack	−1	−1	−1	⋯	1

4 Experiment Framework

We proposed a framework for evaluating five types of features mentioned in the previous section. These features were combined together by simple concatenation and by MKL. SVM was used to build models for each category as multiple binary learning models. The performance of the models needed to be evaluated as a multi-label problem. The results showed performance comparison between individual and combined sources. The data set was split into two sets, 90% for training and 10% for testing. The experiment was performed five times for finding statistical significance of the results.

Kernel functions used in this work were the following: linear kernel $\kappa(x,x')_{LIN} = \langle \phi_\kappa(x), \phi_\kappa(x') \rangle$; and radial basis function kernel (RBF) $\kappa(x,x')_{RBF} = \exp\big(-\gamma \|\phi_\kappa(x) - \phi_\kappa(x')\|_2^2\big)$, where γ is an adjusting parameter for RBF kernel [17]. The tuning parameters of the models with individual features and concatenated feature were the following: a penalty parameter of SVM which varied as $C = \{10^{-4}, 10^{-3}, \cdots, 10^3, 10^4\}$; a kernel function $\kappa = \{\kappa_{LIN}, \kappa_{RBF}\}$; and a γ in that κ_{RBF} that varied as $\{10^{-4}, 10^{-3}, \ldots, 10^3, 10^4\}$. Because MKL is a combination of kernels, each kernel was tuned as an individual feature. For the MKL method described in Sect. 2.2, an extra parameter needed to be varied, $p = \{0.5, 1, 1.5, 2\}$ for p-norm regularisation in MKL, and C value likewise needed to be tuned for the SVM model. The models with individual features and concatenated feature had 3 tuning parameters, while the models with MKL combined features had 12 tuning parameters. In the parameter tuning step, Particle Swarm Optimization (PSO) was used as the optimizing method. Details of PSO are described in [18]. For the PSO settings, we set the swarm size to 10 times the number of parameters and the maximum number of iteration to 50. Multi-label problem was evaluated by two types of evaluation: example-based and label-based. Example-based measurement could be complicated because a result could be completely correct, partially correct, or completely incorrect; therefore, we calculated the score with Eq. (5).

$$score(P_x) = \frac{Y_x \cap P_x}{Y_x \cup P_x} \tag{5}$$

Given Y_x was the set of true labels and P_x was the set of predicted labels for the testing data $x \in X$, then this score function was available in the accuracy function Eq. (6).

$$Accuracy_X = \frac{1}{N} \sum_{x \in X} score(P_x) \tag{6}$$

where N is the number of testing data X [19]. For label-based measurement, the calculation could be achieved by using two averaging operations: macro-averaging and micro-averaging. A binary evaluation measure $B(TP, TN, FP, FN)$ was calculated based on the number of true positive TP, true negative TN, false positive FP, and false negative FN. Let TP_c, TN_c, FP_c and FN_c be the measures for a class c, then the macro-averaging and micro-averaging operations were performed as follows:

$$B_{macro} = \frac{1}{Q} \sum_{c=1}^{Q} B(TP_c, TN_c, FP_c, FN_c) \tag{7}$$

$$B_{micro} = B\Big(\sum_{c=1}^{Q} TP_c, \sum_{c=1}^{Q} TN_c, \sum_{c=1}^{Q} FP_c, \sum_{c=1}^{Q} FN_c) \tag{8}$$

where Q is the number of classes [20]. For this paper, we used Area Under Receiver Operating Characteristic Curve (AUROC) that could deal with unbalanced data that used B_{macro} and B_{micro} for evaluation. Hence the optimised objective function by the PSO algorithm was AUROC with micro-averaging [21].

5 Experiment Results

We compared different types of features including individual features Φ, concatenated feature ϕ_{con} and MKL feature $\mathcal{K}_C$. Two types of evaluations were exhibited, i.e., example-based and label-based evaluations. Figure 1(a) shows the accuracy of seven feature types measured by example-based evaluation. It shows rather poor results with no result with higher than 50% accuracy; however, MKL was the best contender. Box plots of micro-AUROC and macro-AUROC are illustrated in Fig. 1(b) and (c), respectively, for label-based measurement. Both figures show a similar overall picture. Using text feature yielded the best results when only individual feature was considered. However, a combination of these features tended to improve the overall performance. It was clearly seen that MKL was able to achieve the best performance while concatenated features came in the second.

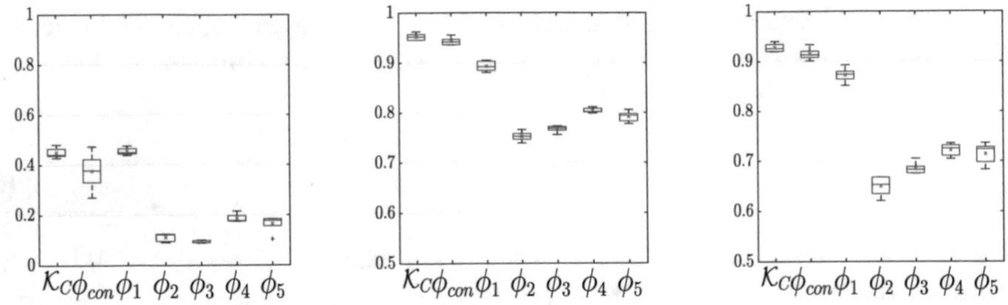

(a) Accuracy (example-based)

(b) micro-AUROC (label-based)

(c) macro-AUROC (label-based)

Fig. 1. Measurement from seven type of feature

We further verified the results by employing multiple comparisons with sign test as shown in Table 2. Using MKL and label-based measurement, a $p = 0.0625$ was achieved which was significantly better than using other features. However, MKL was not significantly better when using example-based measurement. The results were inconclusive for comparison between concatenated and text features but the mean difference of the results from MKL was better than those from the former two types of features. Moreover, the results from concatenated features tended to be better than those from individual features except for text feature that was not significantly different in accuracy comparison. However, for label-based measurement, the results from concatenated features were better than those from all individual features. Furthermore, the text feature in this experiment was significantly better than the image feature as evaluated by both example-based and label-based measurements.

Furthermore, we investigated the weight λ_j of each kernel ϕ_j computed by the MKL as shown in Fig. 2 (average across all runs). λ_4–the average weight of colour feature ϕ_4–was the highest weight among all others, while the minimum weight was λ_2–the average weight of shape feature ϕ_2. This means that the edge feature was the most essential feature in this task while the shape feature was the least discriminative feature for this data set. Although, using text feature alone to build a model yielded a better performance than other individual features, it was not the most discriminative features in MKL.

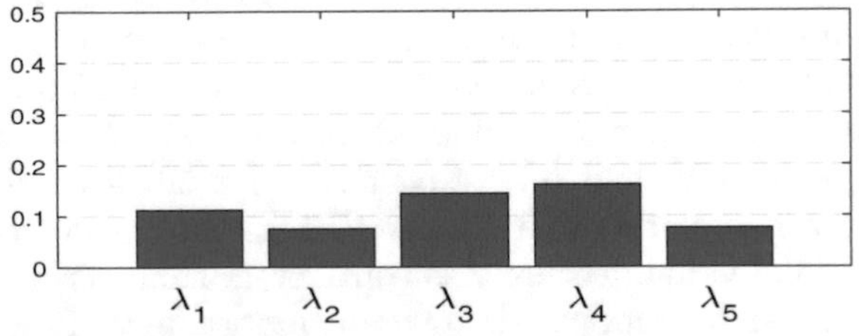

Fig. 2. The set of average lambda λ_j for each ϕ_j

Table 2. Multiple comparison with sign test–mean difference (MD) and its p-value. Bold face indicates statistical significance

Feature types		Accuracy		micro-AUROC		macro-AUROC	
		MD	p-value	MD	p-value	MD	p-value
κ_C	ϕ_{con}	0.074	0.375	**0.009**	**0.063**	**0.013**	**0.063**
κ_C	ϕ_1	−0.005	1.000	**0.058**	**0.063**	**0.055**	**0.063**
κ_C	ϕ_2	**0.339**	**0.063**	**0.200**	**0.063**	**0.278**	**0.063**
κ_C	ϕ_3	**0.357**	**0.063**	**0.185**	**0.063**	**0.244**	**0.063**
κ_C	ϕ_4	**0.264**	**0.063**	**0.147**	**0.063**	**0.206**	**0.063**
κ_C	ϕ_5	**0.286**	**0.063**	**0.161**	**0.063**	**0.214**	**0.063**
ϕ_{con}	ϕ_1	−0.079	0.375	**0.050**	**0.063**	**0.042**	**0.063**
ϕ_{con}	ϕ_2	**0.264**	**0.063**	**0.191**	**0.063**	**0.265**	**0.063**
ϕ_{con}	ϕ_3	**0.282**	**0.063**	**0.176**	**0.063**	**0.231**	**0.063**
ϕ_{con}	ϕ_4	**0.189**	**0.063**	**0.139**	**0.063**	**0.193**	**0.063**
ϕ_{con}	ϕ_5	**0.212**	**0.063**	**0.152**	**0.063**	**0.201**	**0.063**
ϕ_1	ϕ_2	**0.343**	**0.063**	**0.141**	**0.063**	**0.223**	**0.063**
ϕ_1	ϕ_3	**0.361**	**0.063**	**0.126**	**0.063**	**0.188**	**0.063**
ϕ_1	ϕ_4	**0.269**	**0.063**	**0.089**	**0.063**	**0.151**	**0.063**
ϕ_1	ϕ_5	**0.291**	**0.063**	**0.102**	**0.063**	**0.159**	**0.063**
ϕ_2	ϕ_3	**0.018**	**0.063**	**−0.015**	**0.063**	**−0.035**	**0.063**
ϕ_2	ϕ_4	**−0.075**	**0.063**	**−0.052**	**0.063**	**−0.072**	**0.063**
ϕ_2	ϕ_5	**−0.052**	**0.063**	**−0.039**	**0.063**	**−0.065**	**0.063**
ϕ_3	ϕ_4	**−0.093**	**0.063**	**−0.037**	**0.063**	**−0.037**	**0.063**
ϕ_3	ϕ_5	**−0.071**	**0.063**	**−0.024**	**0.063**	**−0.030**	**0.063**
ϕ_5	ϕ_4	0.022	1.000	0.013	0.375	0.007	1.000

6 Conclusion

This paper proposes a framework for automated product categorisation by using MKL to improve feature combination. We evaluated and compared seven types of features using multi-label measurement including example-based and label-based measurements. We have found that the performance of MKL combined features in label-based measurement was statistically and significantly better than those of individual features and concatenated features. However, when we used example-based measurement, some results were inconclusive, such as those between MKL and concatenated features, MKL and text features, and concatenated and text features. Nevertheless, three box plots of accuracy, micro-AUROC and macro-AUROC versus feature show that MKL was still the best overall. Consequently, we concluded that using the MKL method to combine kernels for improving performance of our model. In future work, using MKL, we aim to build a learning model framework that shares learning between models for each label in order to solve multiple label problems.

References

1. Simon, P.: Too Big to Ignore: The Business Case for Big Data, vol. 72. Wiley, Hoboken (2013)
2. Chavaltada, C., Pasupa, K., Hardoon, D.R.: A comparative study of machine learning techniques for automatic product categorisation. In: Cong, F., Leung, A., Wei, Q. (eds.) Advances in Neural Networks - ISNN 2017, pp. 10–17. Springer, Cham (2017)

3. Szummer, M., Picard, R.W.: Indoor-outdoor image classification. In: Proceedings of 1998 IEEE International Workshop on Content-Based Access of Image and Video Database, pp. 42–51 (1998)
4. Pasupa, K., Hussain, Z., Shawe-Taylor, J., Willett, P.: Drug screening with elastic-net multiple kernel learning. In: 13th IEEE International Conference on BioInformatics and BioEngineering, pp. 1–5 (2013)
5. Bay, H., Tuytelaars, T., Van Gool, L.: SURF: speeded up robust features. In: Leonardis, A., Bischof, H., Pinz, A. (eds.) Computer Vision - ECCV 2006, pp. 404–417. Springer, Heidelberg (2006)
6. Zhang, M.L., Zhou, Z.H.: A review on multi-label learning algorithms. IEEE Trans. Knowl. Data Eng. **26**(8), 1819–1837 (2014)
7. Li, Y.K., Zhang, M.L.: Enhancing binary relevance for multi-label learning with controlled label correlations exploitation. In: Pham, D.N., Park, S.B. (eds.) PRICAI 2014: Trends in Artificial Intelligence, pp. 91–103. Springer, Cham (2014)
8. Sonnenburg, S., Rätsch, G., Schäfer, C., Schölkopf, B.: Large scale multiple kernel learning. J. Mach. Learn. Res. **7**, 1531–1565 (2006)
9. Chapelle, O.: Training a support vector machine in the primal. Neural Comput. **19**(5), 1155–1178 (2007)
10. Hussain, Z., Shawe–Taylor, J.: Improved loss bounds for multiple kernel learning. In: Gordon, G., Dunson, D., Dudk, M. (eds.) Proceedings of the Fourteenth International Conference on Artificial Intelligence and Statistics. Proceedings of Machine Learning Research, vol. 15, pp. 370–377. PMLR, Fort Lauderdale (2011)
11. Jurafsky, D., Martin, J.H.: Speech and Language Processing, edn. 2. Prentice-Hall Inc. (2009)
12. Ryan, J.: Cinema information retrieval from Wikipedia. Technical report, Courant Institute of Mathematical Sciences (2014)
13. Canny, J.: A computational approach to edge detection. IEEE Trans. Pattern Anal. Mach. Intell. **8**(6), 679–698 (1986)
14. Tomasik, B., Thiha, P., Turnbull, D.: Tagging products using image classification. In: Proceedings of the 32nd International ACM SIGIR Conference on Research and Development in Information Retrieval, SIGIR 2009, pp. 792–793. ACM, New York (2009)
15. Kodinariya, T.M., Makwana, P.R.: Review on determining number of cluster in k-means clustering. Int. J. Adv. Res. Comput. Sci. Manag. Stud. **1**(6), 90–95 (2013)
16. Gehler, P., Nowozin, S.: On feature combination for multiclass object classification. In: 2009 IEEE 12th International Conference on Computer Vision, pp. 221–228 (2009)
17. Chang, Y.W., Hsieh, C.J., Chang, K.W., Ringgaard, M., Lin, C.J.: Training and testing low-degree polynomial data mappings via linear SVM. J. Mach. Learn. Res. **11**, 1471–1490 (2010)
18. Kennedy, J., Eberhart, R.: Particle swarm optimization. In: Proceedings of IEEE International Conference on Neural Networks, vol. 4, pp. 1942–1948 (1995)
19. Boutell, M.R., Luo, J., Shen, X., Brown, C.M.: Learning multi-label scene classification. Pattern Recogn. **37**(9), 1757–1771 (2004)
20. Tsoumakas, G., Katakis, I., Vlahavas, I.: Random k-labelsets for multilabel classification. IEEE Trans. Knowl. Data Eng. **23**(7), 1079–1089 (2011)
21. Wu, F., Han, Y., Tian, Q., Zhuang, Y.: Multi-label boosting for image annotation by structural grouping sparsity. In: Proceedings of the 18th ACM International Conference on Multimedia, pp. 15–24. ACM, New York (2010)

ARMFEG: Association Rule Mining by Frequency-Edge-Graph for Rare Items

Pramool Suksakaophong[1(✉)], Phayung Meesad[1], and Herwig Unger[2]

[1] Faculty of Information Technology,
King Mongkut's University of Technology North Bangkok, Bangkok, Thailand
pramool.s@email.kmutnb.ac.th, phayung.m@it.kmutnb.ac.th
[2] Chair of Communication Networks, Fern University of Hagen,
Universitätsstr. 27, Hagen, Germany
herwig.unger@fernuni-hagen.de

Abstract. Almost every global economy has now entered the digital era. Most stores are trading online and they are highly competitive. Traditional association rule mining is not suitable for online trading because information is dynamic and it needs a fast processing time. Sometimes products on online stores are out of stock or unavailable which needs to be manually addressed. Therefore, this work proposes a new algorithm called Association Rule Mining by Frequency-Edge-Graph (ARMFEG) that can convert transaction data to form a complete virtual graph and store items counting in the adjacency matrix. With the limitation of search space using the top weight which is automatically generated during frequency items generation, ARMFEG is very fast during the rule generation phase and can find association rules in all items from adjacency matrix which solves the rare item problem.

Keywords: Association rule mining · Rare items · Undirected graph
Adjacency matrix · ARMFEG

1 Introduction

Association Rule Mining has been widely used in business domains. It can be used for Market Basket data analysis which can increase sale volumes and help gain a high competitive edge. Association rule mining can find goods items that are often bought together. The Apriori algorithm is one of the most well-known association rules in mining [1]. Many applications are based on Apriori; however, it has a drawback with multiple scans over transaction data. Assigning minimum support is normally done by the user, but it can be hard to find the appropriate value. The association rules algorithms is based on a proper minimum support threshold. If it is set too low, the number of rules is large. However, if it is set too high, the number of rules is too small, which may not find interesting association rules. Processing time can be reduced by cutting less-frequent items. However, it can be impossible to find association rules for certain items.

H. Unger et al. (Eds.): IC2IT 2018, AISC 769, pp. 13–22, 2019.
https://doi.org/10.1007/978-3-319-93692-5_2

Such as items with less frequency but with high value. For example, wine and caviar may appear less in transaction data, but the marketing division still need it for sale promotions. FP-Growth is another widely used in association rule in mining [2]. It is faster than Apriori because of it count frequency on FP-Tree without generating candidate itemset. It only takes two passes when scanning transaction data. The limitation of FP-Growth is memory consumption during the rule mining process because of the duplicate node in the FP-Tree.

Therefore, in this paper, a new algorithm to find association rules for all items using fundamentals of a complete graph is proposed. This will be suitable for e-commerce or online stores that have a new product or when new items are introduced, and notify when frequently out of stock. The advantage of applying a complete graph is the weight of each item can be computed by a local search. Moreover, association rules for a rare item can be found easily from the resulted complete graph.

2 Literature Review

Association rule mining from market basket transactions has two main stages: frequent itemset generation and associated rule generation. Firstly, the frequent item generation produces a set of items that have support values of higher than or equal to the specified minimum support. Secondly, the associated rules generation yields rules based on the frequent itemset resulted from frequent itemset generation. The association rules must meet both minimum support and confidence criteria.

2.1 Primary Definition of Association Rule

Let $I = i_1, i_2, ..., i_d$ be the set of goods items in the market-basket transaction data and $T = t_1, t_2, ..., t_d$ be the set of transactions. Each transaction t_i contains a subset of items chosen from I.

The minimum support can be written as *minsup*. Finding support from Eq. (1) is achieved by calculating data that appears together with X and Y compared with proportion of all items N

$$Support, s(X \rightarrow Y) = \frac{\sigma(X \cup Y)}{N} \quad (1)$$

The confidence of a rule is called the $minconf$ which can be calculated from Eq. (2), items of X and Y that appear in transactions together are compared with items X

$$Confidence, c(X \rightarrow Y) = \frac{\sigma(X \cup Y)}{\sigma(X)} \quad (2)$$

The number of possible rules can be calculated from Eq. (3)

$$R = 3^d - 2^{d+1} + 1 \quad (3)$$

ARM is useful when discovering interesting relationships hidden in large datasets. However, the problem of Association Mining process is time-consumption. The letter d is the number of items; for example, if there are four items, the numbers of possible rules are 50 rules; if there are eight items, there exist 6,050 rules. If the number of items increase then the number of rules increases exponentially. Researchers in the areas of association rule mining have tried to propose algorithms that can reduce search space i.e., by setting *minsup* and *minconf*.

2.2 Set-Based Algorithm

The Apriori algorithm helps to reduce the number of candidate itemsets explored during frequent itemset generation. It would prune itemsets that do not pass the *minsup*. By anti-monotone property of support, the support of an itemset never exceeds the support of its subsets. However, a disadvantage of the Apriori algorithm is too many rounds of candidate generation.

2.3 Tree-Based Algorithm

The FP-Growth algorithm can reduce candidate itemsets by using the tree structure called FP-Tree [3]. It scans the database only twice when processing. A few techniques that mines out all the frequent itemsets without the generation of the conditional FP trees are used. However, a disadvantage of FP-Growth is the amount of memory units consumed. Therefore, to reduce the size of memory units, the BitTableFI method would convert data item into Bit to reduce the data size [4]. The other problem of association rule mining is to determine the *minsup* threshold. A few existing association rule mining algorithms determine both *minsup* and *minconf* by a user. Finding the best values of *minsup* and *minconf* is not an easy task. A few researchers use particle swarm optimization (PSO) to find the best values of *minsup* and *minconf* automatically [5]. Weighted Association Rule Mining (WARM) [6] is another technique that allows the user to specify multiple *minsup* to solve rare item problems [7]. Because a few items appear rarely in transactions but have high value, it needs to be promoted by a marketing department. WARM can also solve the problem by specifying a weight for each item.

2.4 Graph-Based Algorithm

Many types of a graph can be used for association rule mining. The Bipartite Graph [8] represents the relationship between items and transactions in two portions. FP-Growth-Graph [9] uses a graph instead of a tree in the frequent items generation phase of association rule mining. With enhancements of ARM to the graph called Ant Colony optimization [10], and weight graphs [11]. Graph-based Association Rule Mining (GARM) [12] converts each transaction into a multi-graph with colored edges. This becomes a weight graph and can be merged with a colored spanning tree of all colors to which specified vertex is connected.

The ARM may generate a lot of rules but it is not useful for sale departments because they are only interested in rules related to their products. Many ARM algorithms prune infrequent items before generating rules, so that infrequent items do not appear in the rules. To solve this rare items problem, implementation of a graph base is proposed. The advantages of a graph is simple operation, unique and independent on each node.

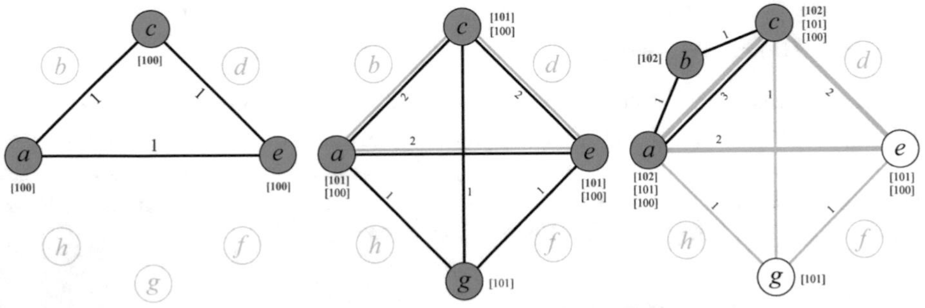

(a) Transaction 100 {*a*, *c*, *e*} (b) Transaction 101 {*a*, *c*, *e*, *g*} (c) Transaction 102 {*a*, *b*, *c*}

Fig. 1. The frequent itemset generation phase

Table 1. An example of transaction database

TID	Items
100	{Apple, Cake, Eggs}
101	{Apple, Cake, Eggs, Grapes}
102	{Apple, Beer, Cake}
103	{Apple, Cake, Diapers, Eggs}
104	{Apple, Cake, Eggs, Fish, Grapes}
105	{Diapers, Fish, Ham}
106	{Apple, Cake, Diapers, Eggs, Grapes}
107	{Beer, Diapers, Eggs, Grapes}
108	{Diapers, Fish, Ham}
109	{Diapers, Fish}
110	{Apple, Beer, Cake, Eggs, Grapes}
111	{Cake, Eggs, Fish, Grapes, Ham}
112	{Apple, Beer, Cake, Diapers}
113	{Fish, Grapes, Ham}
114	{Apple, Cake, Eggs, Ham}
115	{Grapes}

3 Methods and Algorithm

A graph data structure consists of two sets called Vertices or Nodes and Edges. This kind of graph is called an undirected completed graph. We assume that the characteristics of transactions from a shopping cart are not sequential. The weight graph represents by a adjacency matrix.

3.1 Frequent Itemset Generation

In the frequency itemset generation phase, this starts with counting unique items (**UI**) in a transaction database with red-black tree. Then construct graph ***FEG*** by adjacency matrix with the length of **UI**. Each transaction simulation is scanned as an undirected completed graph. Then counts each pair of the vertex to $\boldsymbol{FEG}_{i,j}$ and store HashMap Itemset (***HMI***) with a key as **ItemsID** and values as **TID** while each vertex is self-loop.

Pseudo code for Frequency Itemset Generation

```
UI = count Uniqe Items from transaction data (T)
Create Graph FEG(UI x UI)
for each transaction (t) in (T)
   NI = count Number of items in (t)
   while NI > 0
     create undirected completed graph each (t)
     Update +1 each pair of FEG(i,j)
     if FEG(i=j) HashMap HMI(Key=ItemsID, Value=TID)
   end while
end for
```

Table 1 is an example of the transaction database that consists of 8 items: a = Apple, b = Beer, c = Cake, d = Diapers, e = Eggs, f = Fish, g = Grapes, and h = Ham. Figure 1 shows the frequent itemset generation phase. The first transaction, TID100 is $\{a, c, e\}$ converted to a subgraph as a completed graph with tree nodes $\{a, c, e\}$. Each pair of nodes are counting in an adjacency matrix $\boldsymbol{FEG}_{i,j}$. In the case of i=j, it is a self-loop; the **TID** is stored in ***HMI*** with key = **ItemsID** and Value = **TID**. The second transaction, TID101 is $\{a, c, e, g\}$. We count items as a completed subgraph of four nodes. Then count in $\boldsymbol{FEG}_{i,j}$ plus 1 for each position in adjacency matrix and store **TID** in ***HMI*** in case of self-loop. The third transaction, TID102 is $\{a, b, c\}$. The items are counted as completed subgraph of tree nodes. Then count $\boldsymbol{FEG}_{i,j}$ plus 1 for each position in the adjacency matrix. All transactions were conducted the same. The result from frequency itemset generation shows in Fig. 2.

3.2 Define Top weight

After collecting item frequency counts in the adjacency matrix ***FEG***, one can define top weight by sorting the adjacency matrix ***FEG***. The data is then rearranged according to weight in descending order. Table 2 illustrates the list of top

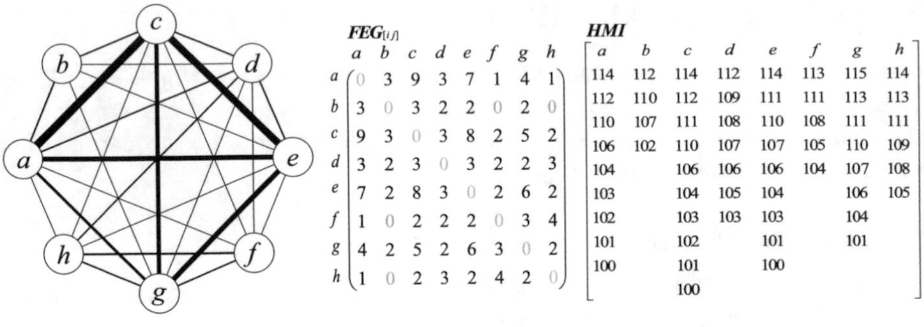

Fig. 2. Weight graph and array matrix of ***FEG*** and ***HMI***

weight, where the variable of top weight is $\boldsymbol{W}(9, 8, 7, 6, 5, 4, 3, 2, 1)$. If we set k=4, then the value 4th level of array $\boldsymbol{W}$ is 6. The top weight is used to prune the graph or set return from the Depth First Search (DFS) algorithm.

Table 2. List of top weight from 2-itemsets

W	2-Itemsets	Frequent	Support	Confidence
1	{a,c}	9	0.5625	1
2	{c,e}	8	0.5	0.8
3	{a,e}	7	0.4375	0.78
4	**{e,g}**	**6**	**0.375**	**0.67**
5	{c,g}	5	0.3125	0.5
6	{a,g}	4	0.25	0.44
7	{a,b}	3	0.1875	0.33
8	{b,d}	2	0.125	0.5
9	{a,f}	1	0.0625	0.1

3.3 Rule Generation

The rule generation phase starts from top-weight node shows in Fig. 3(a). The procedure starts with finding 1-itemsets from ***HMI*** that are greater than $\boldsymbol{W}$(k) and stored in a stack **NK**. The algorithm proceeds to find 2-itemsets from ***FEG*** which are higher than $\boldsymbol{W}$(k) using the DFS algorithm shows in Fig. 3(b). For each path of DFS is generate association rule by procedure RuleGen(). When searching for 3-itemsets until n-itemsets, the ARMFEG uses the weight by counting intersection from ***HMI***. Figure 3(c) illustrates the intersection $\{a, c, e\}$ equals 6 (100, 101, 103, 104, 106 and 114). The weight of intersection $\{a, c, e\}$ is higher than $\boldsymbol{W}$(k), as the consequent, e is taken into the generated rule set. The DFS procedure loops until process all path search that edge greater than $\boldsymbol{W}$(k).

Pseudo code for Rule Generation

```
k = defined level from Maximum weight
TopW = sort DESC uniqe weight from FEG
NK = Node list from HMI with weight >= k
   while Nk is not empty
   NK.push(v)
      DFS(FEG,v)    ( v is the vertex from NK )
      Stack S := {};    ( start with an empty stack )
      for each vertex u, set visited[u] := false;
      push S, v;
        while (S is not empty) do
        u := pop S;
          if (not visited[u] && weight >= k) then
             visited[u] := true;
             for each unvisited neighbour w of u
             push S, w;
             RuleGen(R,S)
          end if
        end while
      end DFS()
    end while
Save R to file
```

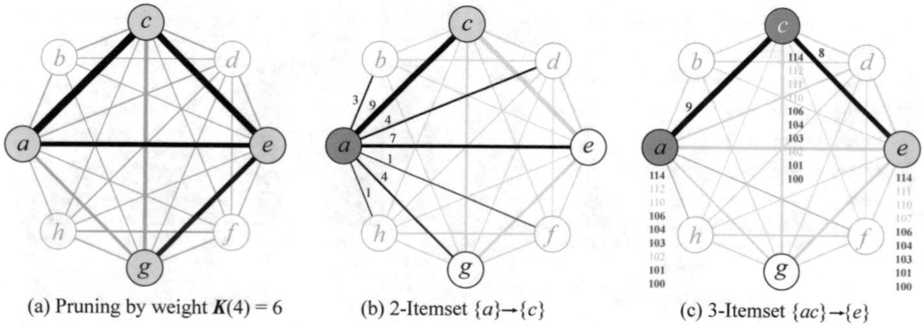
(a) Pruning by weight $K(4) = 6$
(b) 2-Itemset $\{a\}\rightarrow\{c\}$
(c) 3-Itemset $\{ac\}\rightarrow\{e\}$

Fig. 3. Rule generation

4 Results and Discussion

To illustrate the performance of the proposed method, a comparison study with existing algorithms including Apriori and FP-Growth techniques on four data sets: Chess, Connect, T10I4D100K, and T40I10D100K was conducted. Table 3 shows the data sets with characteristics consisting of transactions, the number of items, and an average of item per transaction. All algorithms in the experiments, i.e., Apriori, FP-Growth and ARMFEG, are coded in the JAVA programming language, developed by Eclipse IDE for Java Developers Version: Luna Service Release (4.4.2) ran on an Intel core i3 CPU @ 1.70 GHz, Windows 7 64 Bit OS and 4 GB of RAM.

Table 3. Characteristics of datasets use for experiment evaluation

Data	#*Items*	Avg. Length	#Transaction
Chess	75	37	3,196
Connect	129	43	67,557
T10I4D100K	870	11	100,000
T40I10D100K	942	40	100,000

4.1 Experimental Results

Figure 4(a) shows the processing and Fig. 4(b) illustrates the memory usage of each algorithm. The experimental results show that the efficiency of processing time for ARMFEG works faster than Apriori. However, FP-Growth had the fastest processing time. ARMFEG used less memory than FP-Growth in the T40I10D100K dataset, this is because the average number of items in each transaction was very high.

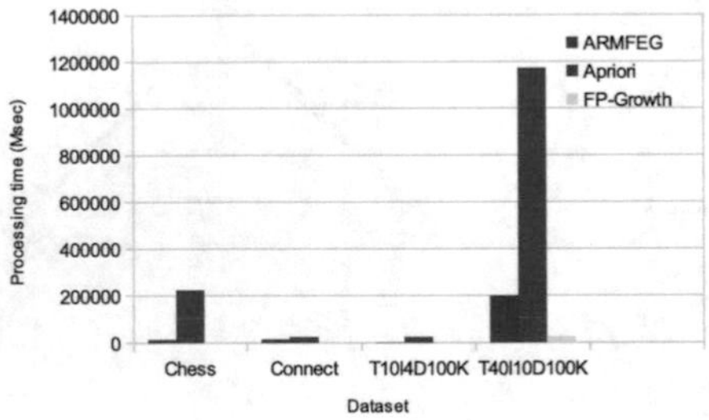

(a) Processing Time

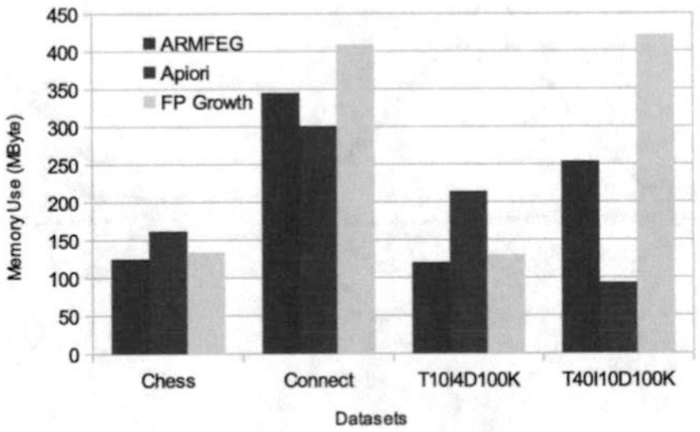

(b) Memory Use

Fig. 4. Processing time and memory use

4.2 Complexity

Complexity of the ARMFEG algorithm can be separated into two-phase. The first phase is frequent generation; unique items are counted by a red-black tree. It takes O(Log n), the complexity of search adjacency matrix is $O(|V|^2)$. The second phase is rule generation, which takes $O(|V|^2)$ and prevents exponential search space by defining the top-weight.

4.3 Rare Items

One of the benefits of graph data storage is that it can solve rare item problems. By employing the DFS algorithm, one can find the relationship of data by using the weight of each item from the adjacency matrix and generate association rules for any item. Figure 5(a) shown node *b* is rare item and we can generate rule from node *b*. The variable of weight b is w(3, 3, 2, 2, 2). If we set k = 4, then the value 4th level of array $\boldsymbol{W}$ is 2 (107 and 110). The variable k is used to prune the graph or set return from the Depth First Search (DFS) algorithm. Moreover, graph data storage can also reduce duplication of data and allows visualization of the relationship between items, even with fewer frequencies. The rare ARM allows merchants to conduct promotions to increase the chance of customers purchasing their goods.

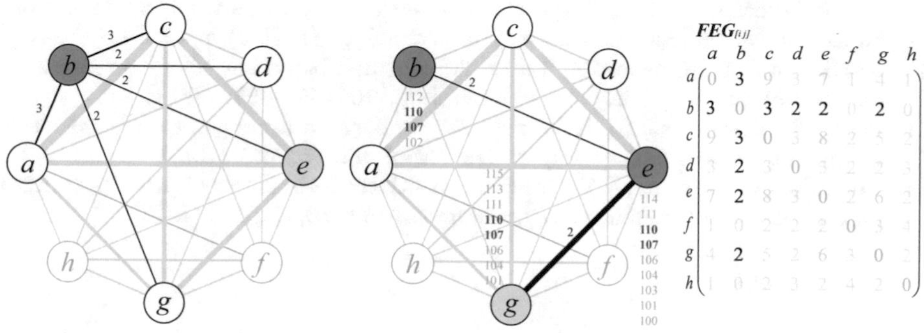

(a) Rare item node *b*, $\{b\}\rightarrow\{c\}$ (b) 3-Itemset $\{be\}\rightarrow\{g\}$ (c) Weight of node *b*

Fig. 5. Example of rule generation start from node *b*

5 Conclusions

ARMFEG is an association rule mining algorithm that is suitable for online shopping stores or dynamic data. Using a complete graph in a frequent itemset generation phase makes it possible to find association rules for all items, which solves the rare item problem. The complexity of search adjacency matrix is $O(|V|^2)$. The advantage of top weight is to prevent exponential search space and generate association rules with local search in a graph data.

References

1. Agrawal, R., Srikant, R.: Fast algorithms for mining association rules in large databases. In: Proceedings of the 20th International Conference on Very Large Data Bases, pp. 487–499. Morgan Kaufmann Publishers Inc., CA (1994)
2. Han, J., Pei, J., Yin, Y.: Mining frequent patterns without candidate generation. In: Proceedings of the 2000 ACM SIGMOD International Conference on Management of Data, pp. 1–12. ACM, Texas (2000)

3. Narvekar, M., Syed, S.F.: An optimized algorithm for association rule mining using FP tree. Procedia Comput. Sci. **45**, 101–110 (2015)
4. Dong, J., Han, M.: BitTableFI: an efficient mining frequent itemsets algorithm. Knowl. Based Syst. **20**, 329–335 (2007)
5. Asadi, A., Afzali, M., Shojaei, A., Sulaimani, S.: New binary PSO based method for finding best thresholds in association rule mining. Life Sci. J. **9**, 260–264 (2012)
6. Pears, R., Koh, Y.S., Dobbie, G., Yeap, W.: Weighted association rule mining via a graph based connectivity model. Inf. Sci. **218**, 61–84 (2013)
7. Liu, B., Hsu, W., Ma, Y.: Mining association rules with multiple minimum supports. In: The Proceedings of the Fifth ACM SIGKDD International Conference on Knowledge Discovery and Data Mining, California, pp. 337–341 (1999)
8. Duck, J., Long, J., Hwang, B., Keun-Ho, R.: Frequent pattern mining using bipartite graph. In: Proceedings of 18th International Workshop on Database and Expert Systems Applications, DEXA 2007, pp. 182–186 (2007)
9. Tiwari, V., Tiwari, V., Gupta, S., Tiwari, R.: Association rule mining: A graph based approach for mining frequent itemsets. In: Proceedings of International Conference on Networking and Information Technology (ICNIT), pp. 309–313 (2010)
10. Al-Dharhani, G., Othman, Z., Bakar, A.: A graph-based ant colony optimization for association rule mining. Arab. J. Sci. Eng. **39**, 4651–4665 (2014)
11. Pears, R., Pisalpanus, S., Koh, Y.S.: A graph based approach to inferring item weights for pattern mining. Expert Syst. Appl. **42**, 451–461 (2015)
12. Amal Dev, P., Shobana, N.V., Joseph, P.: GARM: A simple graph based algorithm for association rule mining. Int. J. Comput. Appl. **76**(16), 1–4 (2013)

Hybrid Filtering Approach for Linked Open Data Exploration Applied to a Scholar Object Linking System

Siraya Sitthisarn(✉) and Nariman Virasith

Knowledge Management and Semantic Web Technology Research Unit, Department of Computer and Information Technology, Faculty of Science, Thaksin University, Songkhla, Thailand
ssitthisarn@gmail.com, nvirasit17@gmail.com

Abstract. In this paper, a hybrid filtering approach for Linked Open Data (LOD) exploration is proposed. This approach is applied to a scholar object linking system. Our approach includes three techniques: cosine similarity, co-relation and voting by in-links. Cosine similarity and term co-relation techniques are used to compute the similarity and relationship between each pair of resources during graph expansion. Voting by in-links is adapted to identify the importance of resources within in the DBpedia graph. The evaluation study was conducted to assess the effectiveness of our approach. We found that the average precision of our filtering approach was 3 times higher than the exploration without filtering. Unfortunately, the average recall is slightly lower. A result of evaluation shows that our approach can reduce the information overload from LOD consumption.

Keywords: Linked data · Filtering · DBpedia · Scholar object linking

1 Introduction

Producing literature review is a significant task in research methodology. It enables researchers to acquire current knowledge, including substantive findings, as well as theoretical and methodological contributions to a particular topic based on the previous researches. To provide a literature review, researchers have to read a number of research papers. In the field of information technology, the papers include several technical terms such as computer languages, tools, approaches, algorithms and so on. Explanations of the technical terms are needed for newcomers in specific research areas. Unfortunately, if they do not have any background knowledge on the terms, they have to search information by using any knowledgeable resources. Nowadays, a search engine has become a primary tool to seek out associated web resources. However, these web pages may be inadequate because users need the flexibility in exploration to see more relevant information of the technical terms.

Linked Data [1] is a set of design principles for sharing machine-readable inter-linked data on the Web. Linked data becomes huge knowledge graphs referred to as a web of data. This enables semantic data from different sources to be connected and

H. Unger et al. (Eds.): IC2IT 2018, AISC 769, pp. 23–32, 2019.
https://doi.org/10.1007/978-3-319-93692-5_3

queried. The Linking Open Data project (LOD) is an example of the adoption of Linked data principles. The goal of the project is to "extend the web with data commons by publishing various open data sets as RDF on the Web and by setting RDF links between data items from different data sources." DBpedia [2] is one of LOD datasets constructed by extracting structured information from Wikipedia. It becomes the center dataset of LOD cloud because it plays an increasing important role, as a knowledge base, in enhancing the intelligent search and in supporting information integration.

Even though LOD is a potential technology for solving the above limitations, it is difficult for ordinary people to consume data. This is because the semantic query is the primary tool to access linked datasets and the query construction is beyond the ordinary users' skill. In addition, the pervasive spread of RDF data in LOD led to the information overload and this suddenly arises the challenge of how to retrieve and filter the content of interest from the huge data sources.

In this paper, we exploited LOD to support newcomers in their research by providing background knowledge about the technical terms, appearing in the research papers. The scholar object linking system was developed based on LOD graph exploration. To rule out the irrelevant information from the graph, a novel hybrid resource filtering approach was proposed by combining the benefit of three filtering techniques (i.e. cosine similarity, co-relation and voting by in-links). Cosine similarity and term co-relation was used to compute the similarity and relation between each pair of resource reached during the exploration. Voting by in-links based on social graph analysis was adapted to identify how important a resource is within a network of RDF in DBpedia. The result of the hybrid approach is the weight graph, where nodes represent the level of relevance between the two nodes. The hybrid approach can reduce the information overload due to LOD consumption.

2 Related Work

There are many applications in various domains used LOD as the huge knowledge space for locating related information [3, 4]. Unfortunately, result of LOD consuming leads to the information overload and this becomes the research challenge on how to retrieve and filter the content of interest from the huge data sources. As can be seen in Noia and colleagues' work [5], they proposed a content-based recommender system that exploits the movie data within DBpedia, Freebase and LinkMDB. The adaptation of vector space model is used to measure the similarity between movie resources within LOD graph and recommend to users. Apart from vector space model, categories of predicates are also used for detecting a similarity between two instances of movie. Similarly, Mirizzi and colleagues [6] proposed an approach to rank resources within DBpedia dataset. The resources are ranked according to the query, exploiting the structure of DBpedia graph. The approach computes a similarity value for each pair of resources during exploration. Co-relation between of labels of resources is computed by querying external information sources (e.g. Google, Bing and Delicious).

A filtering technique based on relationship in LOD graph is another way for resource filtering, Mulay and Kumar [7], proposed a framework for calculating the important of datasets and resources inside the datasets. Ranking score of resources is calculated using *owl:sameAs*. Three types of *owl:sameAs* are consider: outgoing link to a resource in a different dataset, incoming link from a resource from a different dataset, and incoming and outgoing links between a pair of resources. Besides, Vidal and colleagues [8] proposed an approach to semantic annotation with linked data in the domain of education. A context-based graph filtering algorithm is used to rule out irrelevant nodes in the subgraph. The algorithm is based on weight of relationship and term co-relation between resources.

In this research, we proposed the resource filtering approach during graph exploration. Our approach is based on primitive data filtering technique as [5, 6], Vector Space Model and term co-relation terms are adopted for measuring the similarity between resources. Voting by in-links, the novel technique is adapted to identify the important of resources based on a collective endorsement from other resources. With combination techniques, our approach computes potential subgraphs with relevant resources for users.

3 LOD Exploration

The LOD exploration is a module for exploring LOD datasets (i.e. DBPedia, DBLP and FOAF) in a scholar object linking system. The LOD exploration is to search resources and literals, associated with the technical terms. Basic exploration includes search operations in forward and backward directions. The framework of LOD exploration is depicted in Fig. 1. The framework starts with a tokenization component. The tokenization technique is based on a dictionary and computer lexicons. In this research, we acquired computer technical terms from ACM (Computing Classification System) and Encyclopedia of Computer Science. The tokens that match vocabulary from the computer lexicons are contained in the list.

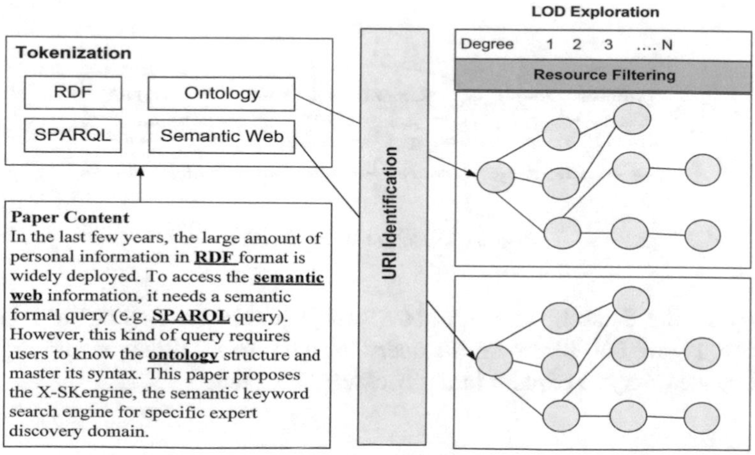

Fig. 1. LOD exploration framework

After a user selects the technical term from a list, the component of LOD exploration will map that term with the corresponding URI, then the URI are placed in the SPARQL query. Next, the module submits the query to DBPedia SPARQL endpoint. The exploration can be expanded for more information by using the N-degree of exploration, where *n* is the search distance from the original searched term. During expansion process, the resources unrelated with the technical term are ruled out, using our resource filtering approach. Only relevant and contextualized resources are included in the subgraph.

3.1 Basic Exploration

In this stage, resources with labels that enclose the technical term" are queried from the DBpedia dataset using property *rdfs:label*. The resource with namespace <http://dbpedia.org/resource/> is firstly selected. However, if there is no resource with that namespace, or there are more than one candidate resources, we will calculate degree of similarity of the technical term with label of the returned resources. This mapping returns a ranked list of relevant resources with hit scores. A resource with the highest hit score is the representative.

Basically, exploration is conducted in two directions: forward and backward. For the forward exploration, the resource's URI is defined as a subject. Then we look for other resources and literals that are objects of the SPARQL query statements. The literals (e.g. an abstract) will be directly displayed to users, while the resources can be expanded to related information. In the other way (i.e. backward exploration), the resource's URI is identified as an object of the SPARQL query statements. After execution, the resources that related to the object are retrieved. Also, it can be explored for more information in higher level of search degrees.

For example, when a user selects the term "Semantic Web", the system will locate the resource's URI associated to the terms, that is *dbr:Semantic_web*. After that, the resource "*dbr:Semantic_web*" is defined as subject of a SPARQL query for forward exploration and as object for backward exploration as depicted in Fig. 2.

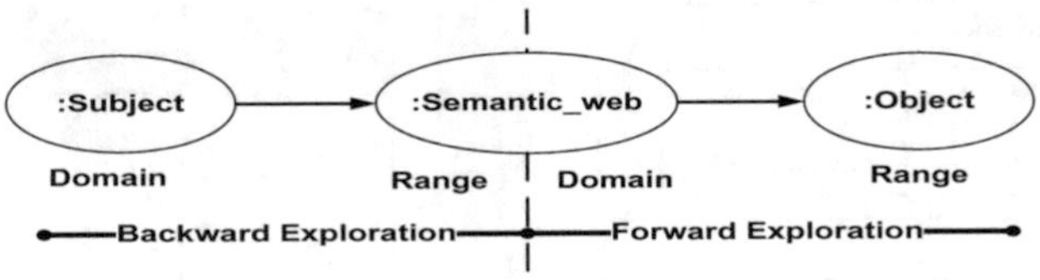

Fig. 2. Direction of exploration

To access the dataset, we construct SPARQL query for forward and backward exploration. Figure 3(a) illustrates the query for the forward direction and Fig. 3(b) is depicted as the query to explore in the backward direction.

```
SELECT ?predicate ?object
WHERE {
   dbr:Semantic_web ?predicate ?object.
}
```

a. Forward Direction

```
SELECT ?subject ?predicate
WHERE {
   ?subject ?predicate dbr:Semantic_web.
}
```

b. Backward direction

Fig. 3. SPARQL query corresponded to directions of exploration.

3.2 N-degree of Exploration

For n-degree of exploration, a user can identify level of search-degree for accessing more related information beyond the result from basic exploration. Given the corresponding resource, identified as a root node, this is explored up to a predefined distance (*N-degree*). A user can configure in the initial setting. Mirizza and colleagues [6] studied that setting $N = 2$ is the best distance because the retrieved resources are still highly correlated to the root node. If $N = 1$, many relevant resources are lost. Conversely, if user set $N > 2$ the exploration returns a number of non-relevant resources. Even though the higher search degree provides more information to users, some resources and relations are not significant (e.g. duplicate resources, URI is not active and so on) causing users to consume overloaded information. Definitely, this requires data cleaning and data filtering computation.

In this research, the resource filtering process is conducted during graph exploration. Given the root node, the exploration module browses the DBpedia graph from the root node to a number of traversing degrees equal to *N* (i.e. Max depth that a user defines). As depicted in Fig. 4, initially, for each node *v* discovered during exploration, the resource filtering module computes to eliminate the nodes that have low level of relevance to the root node or parent nodes. Result of filtering is a set of node u which is defined as the set of parent nodes for iterative expansion in next degrees.

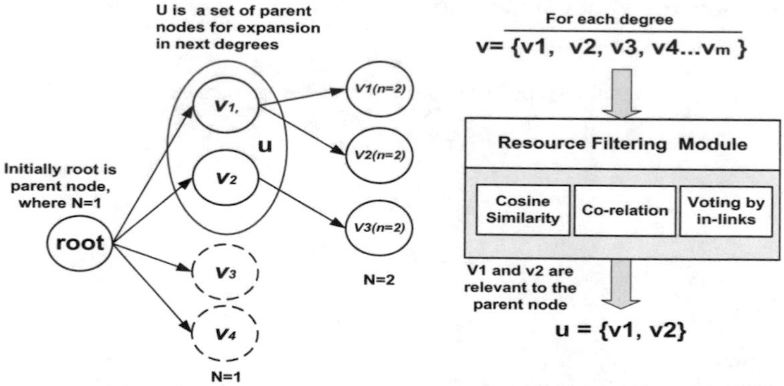

Fig. 4. Resource filtering during graph exploration

3.3 Filtering Resource for LOD Exploration

Filtering is the significant process when accessing to the large space of information. This is because users want to see only relevant information. This section explains the guiding techniques of filtering for the LOD exploration including (1) cosine similarity between resources, (2) co-relation between resources and (3) voting by in-links. As depicted in Fig. 4 given two resources *u* and *v* in the DBpedia graph, *u* is member of a set of parent nodes and *v* is a member of exploring node related to each *u*. These techniques are conducted during graph exploration to filter the returned resources.

Cosine Similarity Between Resources
Cosine similarity is used to compare how much *u* and *v* are similar with each other, exploiting their abstracts. Figure 5 illustrates the example of SPARQL query for finding abstract of resource *<dbr:Semantic_web>*.

```
PREFIX dbont: <http://dbpedia.org/ontology/>
PREFIX dbr: <http://dbpedia.org/resource/>
        SELECT DISTINCT  ?object
        WHERE {dbr dbont:abstract  ?object.
        FILTER(!isLiteral(?object) || lang(?object) = '' || langMatches(lang(?object), 'en'))}
```

Fig. 5. SPARQL query for finding abstract of resource

The vector space model is used to compute the similarity of resource's abstracts. The vector space model [9] is accomplished by assigning weights to index terms in abstracts. Each abstract is represented as vectors in n-dimensional space. Degree of similarity between a pair of vector *abstract:u* (d_u) and vector *abstract:v* (d_v) is computed in Eq. (1). The term frequency–inverse document frequency (*tf-idf*) is combined in the cosine similarity as a weighting factor as depict in Eqs. (2) and (3). Finally, a ranking of similarity values are returned.

$$sim(d_u, d_v) = \frac{\sum_{i=1}^{t} w_{i,u} \times w_{i,v}}{\sqrt{\sum_{i=1}^{t} w_{i,u}^2} \times \sqrt{\sum_{i=1}^{t} w_{i,v}^2}} \tag{1}$$

where

$$w_{i,u} = tf_{i,u} \times idf_{i,u} \tag{2}$$

$$w_{i,v} = tf_{i,v} \times idf_{i,v} \tag{3}$$

The similarity values are used to filter out insignificant resources in each list. Each resource that has a score lower than a pre-defined threshold (e.g. 0.25) will not be considered. Consequently, we have a list of candidate resources that will be filtered by other two techniques (i.e. voting by in-links and correlation between resources)

Co-relation Between Resources

Co-relation technique aims to evaluate the tie strength between two DBpedia resources using publication information from other LOD datasets (i.e. DBLP and IEEE publication).

Co(u,v) indicates the term co-relation between resource u and resource v. To derive the co-relation of resources, we consider a list of research articles that contain term t_u ($list(A_u)$), as well as a list of articles that contain term t_v, ($list(A_v)$). Where t_u, t_v are associates to label of resource u and v respectively. The weight of *Co(u,v)* can be calculated as Eq. (4).

$$Co(u, v) = \frac{list(A_u) \cap list(A_v)}{list(A_u) \cup list(A_v)} \tag{4}$$

Voting by In-links

In-links or in-degree centrality measures how popular a node is and its value shows prestige. In this paper, voting by in-links is considered as voting from the set of parent nodes to the child nodes as depicted in Eq. (5). If a node receives many in-links from the parent nodes, then it is received a kind of collective endorsement [10].

For expansion in the 1st degree, there are only direct links from the root node so that the root node should not be considered as the parent set. To address this problem, we defined a set of parent nodes of the 1st degree as the resources that have relation *dct:subject* with the root node. While, for the next degree of expansion, the set of filtered resources in the previous degree are defined as the parent nodes.

$$C_d^{in}(v_i) = \frac{d_i}{n-1} \tag{5}$$

Where d_i is the in-links (number of adjacent edges) of node v_i, n is number of parent nodes in the previous degree.

In above Eqs. (4) and (5), we now define the overall association score using voting in-links and co-relation criteria as depicted in Eq. (6).

$$R_{score} = C_d^{in}(v_i) + Co(u, v) \tag{6}$$

In Eq. (6), the resources that have associate score higher than the pre-defined threshold is assumed as relevant resources and they are defined as the parent nodes for expansion further. However, there is a potential to adapt weights for each criterion in future work.

Exploration Walk-Through

This section, we explain the example of LOD exploration by illustrating example of exploration of the term "Semantic Web". As regards Fig. 6, if a user clicks the technical term "Semantic Web". The system will map this term with the corresponding resource. For mapping, it is found resource *dbr:Semantic_web* that has namespace <http://dbpedia.org/resource/> therefore it is determined as the corresponding resource and also defined as the root node. Given the root node, the exploration module browses

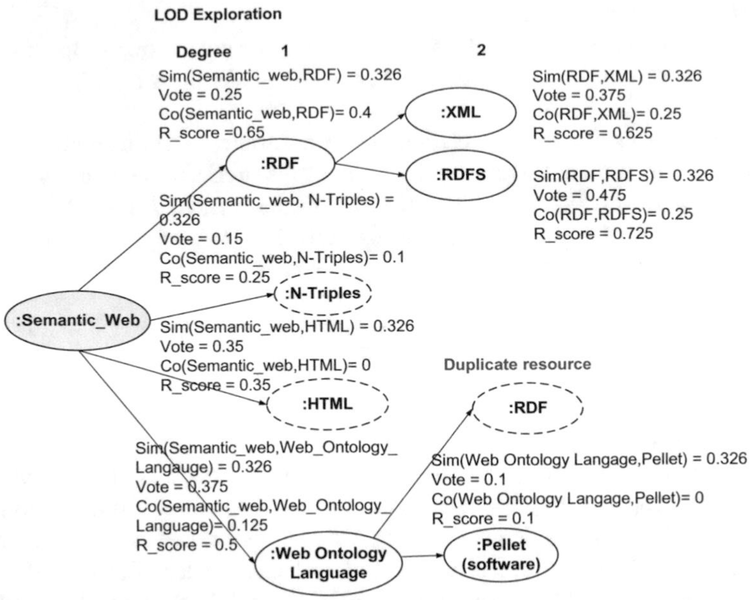

Fig. 6. Example of LOD exploration with the technical term "Semantic Web"

the DBpedia graph from the root node to a number of traversing degrees equal to N. This example, we set N = 2.

For the 1st degree of exploration, cosine similarity is used for comparing the similarity between the abstract of the root and abstract of each exploring nodes. The nodes with higher than or equal to S_THREADHOLD (i.e. 0.25) is chosen as a set of candidate nodes (*D*). Given set *D*, the voting by in-links technique and co-relation between resources are calculated. Finally, the overall association rank is conducted to determine a set of potential relevant resources. As depicted in Fig. 6 the resources with R_score, higher than or equal to R_THREADHOLD (i.e. 0.5) is determined as relevant resources. This consists of: *dbr:RDF* and *dbr:Web_Ontology_Language*. These resources are parent nodes for expansion in the next degree.

4 Evaluation

4.1 Experiment Setup

An experiment was conducted to measure the effectiveness of our LOD graph exploration and filtering techniques comparing to the exploration without the filtering module. Precision and recall were used as metrics of the evaluation. Precision = $\frac{|Ra|}{|A|}$ and Recall = $\frac{|Ra|}{|R|}$, where |R| is the number of relevant resources related to the experts' recommendation; |A| is the number of resources that system provides; and |Ra| is the number of the returned resources from system that are relevant to the recommendation of experts.

The participants in this experiment were three lecturers and two master degree students from Thaksin University, who do searches in the field of semantic web. We provided 20 technical terms from 5 research papers. Then each participant was requested to write the related articles to each term. Next, we uploaded the papers to the scholar object linking system and selected a term from the list. The experiment started with 1 degree of exploration and we recorded the related resources that system provided. These resource's articles were compared to the recommended articles from each participant. After that, the average precision and recall were calculated. The iterative processes were conducted for 2–3 degrees of exploration. The experiment was repeated by using the same system without filtering module.

4.2 Result and Analysis

For average precision, Table 1 shows that the LOD exploration with filtering module provided the average precision values at 76.6%, at the 1st degree. While, the exploration without filtering module provided average precision at 22.84 at the same level. In addition, the result shows the trend of average precision decreased. This can be explained that, if we expand the DBpedia graph further, the number of return resources will be higher but the number of recommended articles from each participant is stable. However, the average precision of our filtering approach was 3 times higher than which one without the module.

Table 1. The average precision for the filtering technique

LOD exploration	Average precision		
	1st degree	2nd degree	3rd degree
No filtering module	22.84	14.36	9.83
With filtering module	76.6	70.5	65.5

In contrast, Table 2 shows that LOD exploration without filtering module provided higher recall than the exploration with our filtering approach. This is because LOD exploration without filtering module retrieval all resources. While the LOD exploration with filtering module filter the resources that have high potential to be relevant with the root. This can make some relevant resources lose during filtering process. However, we found the highest recall value of our approach is 90% at the 3rd degree.

Table 2. The average recall for the filtering technique

LOD exploration	Average recall		
	1st degree	2nd degree	3rd degree
No filtering module	56.86	90.35	95.3
With filtering module	50.51	85.4	90

According to precision and recall values, we conclude that the system can reduce the information overload due to LOD consumption.

5 Conclusion and Future Work

In this study, we discussed the LOD exploration with resource filtering approach to reduce the information overload due to LOD consumption. Our filtering approach is a combination of primitive data and relation filtering techniques. It includes vector space model, co-relation between terms, and voting by in-links. As regards integration of these techniques, our approach computes potential subgraphs with relevant resources for users. Our future work will study the potential of association rule to filter the relevance information in LOD.

References

1. Bizer, C., Heath, T., Berners-Lee, T.: Linked data - the story so far. Int. J. Semant. Web Inf. Syst. **5**(3), 1–22 (2009)
2. Auer, S., Bizer, C., Kobilarov, G., Lehmann, J., Cyganiak, R., Ives, Z.: DBpedia: a nucleus for a web of open data. In: Aberer, K. et al. (eds.) The Semantic Web. LNCS, vol. 4825, pp. 722–735. Springer, Heidelberg (2007)
3. Attwood, T.K., Kell, D.B., McDerMott, P., Marsh, J., Pettifer, S.R.: Utopia documents: linking scholar literature with research data. Bioinformatics **26**(18), 568–574 (2010)
4. Fafalios, P., Papadakos, P.: Theophrastus: on demand and real-time automatic annotation and exploration of (web) document using open linked data. In: Web Semantics: Science, Services and Agents on the World Wide Web (2014)
5. Noia, T.D., Mirizzi, R., Ostuni, V.C., Romito, D., Zanker, M.: Linked open data to support content-based recommender systems. In: The 8th International Conference on Semantic Systems, pp. 1–8. ACM, New York (2012)
6. Mirizzi, R., Ragone, A., Noia, T.D., Sciascio, E.D.: Ranking the linked data: the case of DBpedia. In: Benatallah, B., et al. (eds.) ICWE 2010. LNCS, vol. 6189, pp. 337–354. Springer, Heidelberg (2010)
7. Mulay, K., Kumar, P.S.: SPRING: ranking the results of SPARQL queries on linked data. In: The 17th International Conference on Management of Data (COMAD 2011), pp. 1–10. Computer Society of India, Mumbai (2011)
8. Vidal, J.C., Lama, M., Otero-Garcia, E., Bugarin, A.: Graph-based semantic annotation for enriching education content with linked data. Knowl. Based Syst. **55**, 29–42 (2014)
9. Baeza-Yates, R., Ribeiro-Neto, B.: Modern Information Retrieval. Addison Wesley, New York (1999)
10. Easley, D., Kleinberg, J.: Networks, Crowds and Markets: Reasoning about a Highly Connected World. Cambridge University Press, Cambridge (2010)

Sugarcane Yield Grade Prediction Using Random Forest with Forward Feature Selection and Hyper-parameter Tuning

Phusanisa Charoen-Ung(✉) and Pradit Mittrapiyanuruk

Department of Computer Science, Faculty of Sciences, Srinakharinwirot University, Bangkok, Thailand
{Phusanisa.cha, praditm}@g.swu.ac.th

Abstract. This paper presents a Random Forest (RF) based method for predicting the sugarcane yield grade of a farmer plot. The dataset used in this work is obtained from a set of sugarcane plots around a sugar mill in Thailand. The number of records in the train dataset and the test dataset are 8,765 records and 3,756 records, respectively.

We propose a forward feature selection in conjunction with hyper-parameter tuning for training the random forest classifier. The accuracy of our method is 71.88%. We compare the accuracy of our method with two non-machine-learning baselines. The first baseline is to use the actual yield of the last year as the prediction. The second baseline is that the target yield of each plot is manually predicted by human expert. The accuracies of these baselines are 51.52% and 65.50%, respectively. The results on accuracy indicate that our proposed method can be used for aiding the decision making of sugar mill operation planning.

Keywords: Yield grade prediction · Machine learning · Random forest
Forward feature selection · Hyper-parameter tuning

1 Introduction

In each harvest season, sugar mills need to know the estimated yield of sugarcane in each harvest season for the purpose of operation planning. For each production year, the field surveying staffs of sugar mill will collect the data of each plot. Conventionally, the field experts of sugar mills will use their experience to estimate the sugarcane yield of each plot based on the surveying data, the historical yield profile of each plot and each farmer. The main disadvantage of using this human based yield estimation method is that there is a large discrepancy between estimated yields and actual yields.

In this work, we propose a machine learning based method for sugarcane yield grade prediction. The yield grade of each plot is assigned to low-yield-volume, medium-yield-volume or high-yield-volume. Then our goal is to predict the yield grade of a plot. The features used in the prediction are the plot characteristics, the sugarcane characteristics, the plot cultivation scheme and the rain volume.

Our proposed method is based on Random Forest (RF) technique [7]. In training phase, we propose to a RF model training procedure by using a forward feature

H. Unger et al. (Eds.): IC2IT 2018, AISC 769, pp. 33–42, 2019.
https://doi.org/10.1007/978-3-319-93692-5_4

selection [8] in conjunction with hyper-parameter tuning. We implement our method by using the Python Anaconda. In particular, we use the RF implementation of Scikit-Learn [1]. The contribution of this work is that we develop a machine learning based sugarcane yield grade prediction method for the data obtained from sugarcane plots around a sugar mill in Thailand. Our method outperforms the prediction provided by human-expert.

The remainder of the paper are listed as follows. The related works are reviewed in Sect. 2. In Sect. 3, we explain about the data used in the prediction as well as the details of our proposed methods. Then, we report the prediction accuracy and give a discussion in Sect. 4. Finally, the conclusion is drawn in Sect. 5.

2 Related Work

The literatures about sugarcane yield prediction are presented in [2–5]. These works are directly related to our work. In [2], the authors propose a sugarcane yield prediction method using a random forest algorithm. The features used in this work include: (i) biomass index computed from Agricultural Production Systems sIMulator (APSIM), (ii) yields from two previous years, (iii) cumulative rainfalls, radiations and daily temperature ranges in two different time periods, (iv) Southern Oscillation Index (SOI), and (v) 3-month running average sea surface temperature in the Niño 3.4 region. In [3], the authors propose a method to predict the sugar content from sugarcane harvests in the year of 2011–2012 where each observation is referred to one block in the farms. The authors use 53 features in the prediction. These features belong to four groups: (i) soil physics and soil chemistry, (ii) weather, (iii) agricultural practices, and (iv) crop-related information. Three different machine learning techniques are used in the prediction including Support Vector Regression, Random Forest, and Regression Trees. Also, the RRelief algorithm [6] is used for feature selection. The authors report that the prediction accuracy of their best method in term of Mean-Absolute-Error is 2.02 kg/Mg. Also, they report that their method can predict 90% of the observation within a precision of 5.40 kg/Mg.

In [4], the authors propose the study of the effects of hyper-parameter tuning, feature engineering and feature selection in sugarcane yield prediction techniques. For feature engineering, a set of features derived from the original features are calculated and used in the prediction, e.g., the rate of fertilization for each nutrient (N, P and K), the weather description for four different periods, etc. The feature selection is performed by using RRelief algorithm [6]. The machine learning techniques used in the study include Support Vector Machine (SVM), Random Forest (RF), Regression Tree (RT), Neural Network (NN), and Boosted Regression Trees (BRT). The hyper-parameter tuning is accomplished by grid search with 10-fold cross validation where the Mean-Absolute-Error is used as the metric. The authors evaluate 66 combinations of six machine learning techniques, tuning, feature selection, and feature engineering. The authors report that the BRT, SVM, and RF give the better performance among the others where the RF is the best. In [5], the authors propose the method to solve the harvest scheduling problem for a group of sugarcane growers that supply sugarcane to a mill in Thailand. A neural network is applied to predict the sugarcane yields to be used in the harvest scheduling. The features used in the prediction include crop class, farming skill, cultivars, soil type, the use of irrigation, the age of the cane in days from cultivation to harvest, the average value of the daily minimum/maximum temperature, the average daily rainfall (mm.), and the accumulated daily rainfall since germination (mm.).

With regard to the feature selection for crop yield prediction, the works in [8, 9] are presented. In [8], the authors propose a forward feature selection method for Wheat Yield Prediction. Their problem is recast as regression problem. The authors use two different predictive methods, i.e., Regression Tree and Support Vector Regression in their work. Inspired by this work, we adopt their idea into our proposed method presented in Sect. 3. In [9], the authors evaluate several common predictive modeling techniques applied to crop yield prediction using a method to define the best feature subset for each model. The predictive modeling methods studied in [9] include Multiple linear regression, stepwise linear regression, M5- regression trees, and artificial neural networks (ANN).

3 Data and Method

3.1 Data

Data Collection Process. The sugarcane production data used in the work are provided by a sugar mill in Thailand. The main sources of data are obtained by collecting the data at the farmer plots (e.g. cane class/type, soil type, area, fertilizer etc.) and collecting the data (e.g., actual yield) at the mill when the farmers deliver their sugarcane to the mill.

Two different application programs are used in the data collection. These programs use MS-SQL Server as the database backend. The first application program, referred to as "GIS-System", is used to record the data collected from the farmer plots. These data are associated with the GPS-based spatial information of plot location. The second application program, referred to as "Cane-Accounting", is used to record the actual yield data for each plot at the time the farmers deliver their sugarcanes to the mill.

The rainfall volume data are collected from about 100 mill-owned rain stations located around the farmer plots. Then, the rainfall volume data of each plot is derived from the closest rain station to the plot. The closest rain station to a plot is determined by using the GPS spatial data that are in the database mentioned earlier.

From the database of both application programs, we write a SQL script to aggregate the data and export the dataset to CSV file used for the predictive modeling (explained in the next section). Note that, the variable "YieldGrade" is the target of our prediction. The value is used to justify the "YieldGrade" of each plot. Totally, we have 12,520 records in the dataset that are derived from the production data in year 2558/2559 (7,450 records) and in year 2559/2560 (5,070 records). Finally, the yield grade of each plot is assigned according to the value of yield amount by using the criteria considered the crop cutting and the criteria set by mill operation management into 3 different grades: (i) Low-yield-volume (Grade = 1), (ii) Medium-yield-volume (Grade = 2), and

(iii) High-yield-volume (Grade = 3). The criteria used to assign the yield grade can be explained as follows. First, the yield grade is defined as **"Low"** or **Grade = 1** when the sugarcane yield is less than 7 Tons/Rais. Second, the yield grade is defined as **"Medium"** or **Grade = 2** when the yield is in the range 7 to 12 Tons/Rais. Finally, the yield grade is defined as **"High"** or **Grade = 3** when the yield is above 12 Tons/Rais. The list of variables (features) in the dataset can be summarized in Table 1.

Data Preprocessing. First, we perform an exploratory data analysis (EDA). There are some records of missing value (NULL). That is, there are 41 records, 45 records, and, 43 records that have no value in the fields "Type_Cane", "FertilizerType" and "Fertilizer" respectively. Also, there are some records that correspond to the outliers. A record is specified as outlier if there is the value in a categorical variable which the count of this variable for the whole data is very small comparing to the most frequency (largest count). The source of these missing data and outliers could happen during the data entry process. We opt to fill in these missing values and replace the values of outliers with the mode (most frequency) of the corresponding fields. After the data preprocessing, we apply the one-hot encoding (dummy variables) to the categorical variables. Finally, we randomly split the dataset into the train data and the test data with 70:30 (Train:Test) ratio. The number of records in the train and the test datasets are 8,765 records and 3,756 records, respectively.

In the training set, the number of records that are Grade 1, Grade 2, and Grade 3 are 3,933, 3,759, and 1,073, respectively. And, the number of records that are Grade 1, Grade 2, and Grade 3, in the testing set are 1,712, 1,607, and 437, respectively. The distributions of yield grade in the training and testing datasets are shown in Fig. 1. Clearly, the distributions of target variables of the train and test datasets are similar.

3.2 Yield Grade Prediction Using Random Forest with Forward Feature Selection

In this section, we present our Random Forest based method for sugarcane yield grade prediction. As an overview, the input to the system is a record of data with a preselected set of variables (listed in Table 1) and the output is the predicted yield grade of the corresponding record. At the model training step, we propose to use a forward feature selection in conjunction with hyper-parameter tuning to improve the accuracy of prediction.

As a review, Random Forest (RF) is a supervise machine learning algorithm that can be used for both regression and classification tasks. The RF model is an ensemble of decision trees. Each decision tree (DT) is trained with random subset of training data where the sampling is performed with replacement. Furthermore, at each step in the DT construction, the best feature of split is chosen from a random subset of features. To make the prediction of a test data instance, first we make the prediction for each DT. Then the predictions from all DTs in the model are aggregated either by hard voting or soft voting. Moreover, as the random subset of training data are sampled during RF model training, there will be about 37% of training instances that are not used in each DT construction. These samples are called Out-of-bag (OOB) instances. Alternatively,

Table 1. List of variables in the dataset.

Name	Type	Detail
Class_cane	Category	Cane class which has 3 different types: 1st rattoon cane, 2nd rattoon cane and 3rd rattoon cane
Type_Cane	Category	Cane type which has 4 different types: LK92-11, K84-200, K99-72 and Khonkean3
WaterType	Category	Irrigation sources which has 3 different types: Rain, Groundwater and Natural canal
ActionWater	Category	Irrigation action type which has 2 different types: Water pour and Rain
Epidemic	Category	Epidemic control method which has 2 different types: Preimergent and Herbicide
FertilizerType	Category	Fertilizer types which has 2 different types: Chemical and organic
Fertilizer	Category	Fertilizer formula which has 4 different types: 46-0-0, 15-15-15, 16-16-16 and 25-7-7
TypeSoil	Category	Soil type which has 4 different types: Loam, Silty clay and Ferus soil
GrooveWide	Category	Groove wide of the plot which has 4 different types: 120, 130, 140, and 150 cm
YieldOldGrade	Category	Yield Grade of the plot from the previous season which has 3 different types: Grade 1, 2, and 3
TargetGrade	Category	Target yield grade that is provided by human expert (Grade 1, 2, and 3)
TargetOldGrade	Category	Target yield grade from the previous season (Grade 1, 2, and 3)
FarmerContractGrade	Category	The farmer grade provided by the financial department according to his/her profile of sugarcane yields in previous years and his/her profile of financial debts with the mill in case that the financial aid is provided to the farmer which has 3 different types: A: Good, B: Fair, and C: Poor
Area_Remain	Continuous	The actual plot area that is used in the sugarcane cultivation (rais)
Distance	Continuous	The distance from plot to sugarcane mill (km)
Rain_Vol	Continuous	The rain volume in area of plot (mm)
ContractsArea	Continuous	The amount of sugarcanes that the farmer commits to deliver to the mill (Tons/Rais)
YieldGrade	Category	Yield grade (Grade 1, 2, and 3)

we can also use these OOB instances to evaluate the predictive accuracy of RF model by averaging the evaluations of OOB instances of DTs of the RF model.

In this work, we use Random Forest model in Scikit-Learn [1]. The input features are listed in Table 2 but not include YieldGrade, which will be the target of prediction. First, we try to build the RF predictive model by using the default parameters set up by

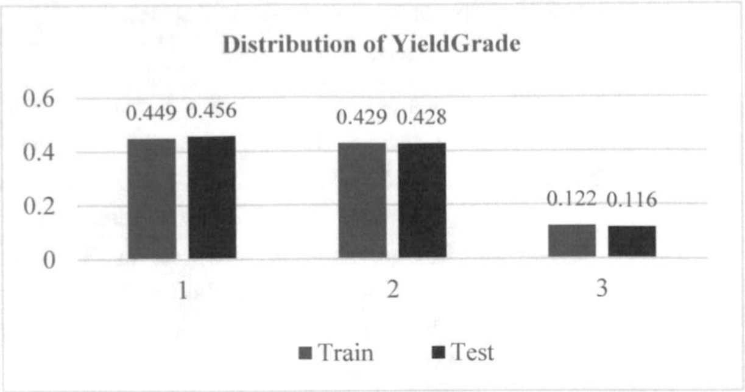

Fig. 1. The histograms show the distributions of yield grades in the training set (blue) and testing dataset (red).

Scikit-Learn. We found that the accuracy of yield grade classification on both the training set and the testing set are 97.96% and 68.29%, respectively. The OOB accuracy score is 64.61%. As the accuracies on the training set and the testing set are largely different, this indicates that the problem of model overfitting.

Algorithm 1: RF model training with Forward Feature Selection and Hyper-parameter tuning
Input: Training data: `X`=features, `Y`=class label, `F`=set of features
Output: Trained model: `best_clf` and best hyper-parameters: `best_params`

```
1:  clf_1, params_1=GridSearchCV(RandomForestClassifier(),X[F],Y)
2:  impVal = clf_1.feature_importance()
3:  F=sort_descending(F, impVal)
4:  S_0={}
5:  for i=1 to |F|
6:        S_i = S_i-1 ∪ F[i]
7:        clf = RandomForestClassifier(params_1)
8:        ValScore[i]=cross_validation_score(clf,X[S_i],Y)
9:  end for
10: k = argmax_i ValScore[i]
11: best_clf,best_params=GridSearchCV(RandomForestClassifier(),X[S_k],Y)
```

To improve the generalization of our model, we propose the RF model construction using a forward feature selection with hyper-parameter tuning that can be shown as Pseudo-code in Algorithm 1 with the naming adapted from Scikit-Learn library. Our proposed feature selection method is inspired by the idea presented in [8]. The details can be explained step-by-step as follows:

First, as shown in Line 1, we train a random forest classifier with grid search and cross validation. In this step of model training, all features (denoted by `F`) of the training data are used. The trained model and the best parameters are obtained in the variables `clf_1` and `params_1`, respectively. Then, in Line 2, we compute the feature importance values (denoted by `impVal`) of all features according to the above trained model by using the function `feature_importance`. Note that, for the purpose of saving the processing time of model training, we actually implement these 2 steps by separately tuning the hyper-parameters: n_estimators, min_samples_leaf, max_features, and max_depth. After that, we train a random forest classifier with the best values of these hyper-parameters found in the separate tuning. Then, the feature importance value of each feature is computed from this trained RF model.

Next, in Line 3, we sort the set of features F according to their importance values in descending order. Then, in Line 5–10, we iteratively add the feature in the order of importance value one-by-one into the feature list denoted by S_i. Note that the list of features in the ith iteration is constructed from the list of features in the previous iteration $(i-1)$ by adding the new feature whose importance score is decreasing. This step corresponds to Line 6. Then, in Line 7–8, we train a random forest classifier on the training data but we use only the above list of features (S_i). We denote the training data with this set of features as $X[S_i]$ in Line 8. Specifically, in this step of model training, we use the same hyper-parameters found in Line 2 (params_1) as the model is instantiated in Line 7. Then, the cross validation is applied to evaluate the accuracy of model. The mean of validation scores when model training using the set of features S_i will be calculated and recorded accordingly into ValScore[i].

After all iterations are done, we search for the set of features S_k that gives the maximum mean validation score. This step corresponds to Line 10. Finally, a random forest model is trained on the training data using only the best set of features S_k. In this step of training, the grid search with cross validation is used to fine-tune the best hyper-parameters. This step corresponds to Line 11. The trained model obtained from this step is the output of our proposed training method. We can use the model to predict the yield grade of a data instance.

4 Result and Discussion

We evaluate our yield grade classification method by measuring the accuracy of prediction on the test dataset. The prediction of a plot is considered as the correct classification only if the predicted yield grade is the same as the actual yield grade in the dataset. We compare the prediction of our RF based method with two different non-machine-learning baselines. The first baseline, referred to as Baseline-1, is to predict the yield grade of each plot based on the actual yield grade from the last year. This baseline is based on the intuition idea that if we do not have any machine learning system, we can use the information of actual yield from the last year as the prediction of yield grade of the same plot in the incoming year. From the dataset, the accuracy of Baseline-1 is 51.52% and is shown in Table 3. The second baseline, referred to as Baseline-2, is based on the prediction by using the target of yield grade where we let a human expert to manually give a target grade of each plot based on their experience. Then the accuracy of Baseline-2 is 65.50% and is shown in Table 3.

For the result of our proposed model, we report the intermediate results and the final result corresponding to some major steps in Algorithm 1 as follows.

The implementation that corresponds to Line 1–2 in Algorithm 1 is achieved by separately tuning the hyper-parameters: n_estimators, min_samples_leaf, max_features, and max_depth. The tuning range of n_estimators is 1 to

Table 2. Feature importance values of the RF based model and the mean validation score at each iteration of forward feature selection.

Feature name	Feature importance value	Mean validation score (after adding the feature)
TargetGrade	0.60273	0.66374
FarmerContractGrade	0.22070	0.69577
ContractsArea	0.05302	0.70264
Area_Remain	0.03336	0.70208
Distance	0.02211	0.69649
Rain_Vol	0.01935	0.69864
YieldOldGrade	0.01931	0.70232
Fertilizer	0.00702	0.70367
WaterType	0.00586	0.70511
Class_Cane	0.00353	0.70503
TargetOldGrade	0.00341	0.70511
GrooveWide	0.00285	**0.70591**
EpidemicType	0.00194	0.70487
Type_Cane	0.00144	0.70495
ActionWaterType	0.00139	0.70487
FertilizerType	0.00134	0.70511
TypeSoil	0.00065	0.70407

Table 3. The accuracies of the baselines and our propose method

Method	Baseline-1	Baseline-2	KNN	SVM	Naïve Bayes	AdaBoost	Our method
Accuracy	51.52%	65.50%	48.43%	49.79%	58.87%	65.50%	**71.88%**

1000 with step size of 100 (i.e., 1, 100, 200, …, 1000). The tuning range of `max_features` is 1 to 7 with step size of 1. The tuning range of `max_features` is 1, 2, 3, …, 17. Note that we have 17 features in the dataset. The tuning range of `min_samples_leaf` consists of 20 different values that are equally spaced between 2 to 40. And, the tuning range of `max_depth` consists of 10 values that are equally spaced between 5 to 50. After separate tuning, we obtain the best hyper-parameters as follows: `n_estimators = 300`, `min_samples_leaf = 14`, `max_features = 4` and `max_depth = 10`. When we re-train the RF model by using all features in the training data with, the above hyper-parameters, we obtain the accuracy on both the training set and the testing set are 75.24% and 71.86%, and the OOB accuracy score us 71.65%.

By using this trained RF model, we calculate the feature importances by using the attribute "`feature_importances_`" of the "`RandomForestClassifier`" class in Scikit-Learn. The features are sorted in the descending order according to the feature importance value. The sorted feature list will be used in the iteration of forward feature selection in Line 5–9. The feature importance values obtained in the dataset are listed in Table 2. As the feature "TargetGrade" is the most important feature, it seems that the predictions provided by human expert have an impact to our proposed method. We will discuss this issue by comparing our method with the baseline that uses this feature as the prediction.

For each iteration of forward feature selection in Line 6–8, the mean validation score when adding the feature to the model is shown in Table 2 (right side). Corresponding to Line 10 of Algorithm 1, the maximum value of validation score of all feature sets is *0.70591* where the number of features is *12*. That is, the algorithm selects to use 12 features as follows: TargetGrade, FarmerContractGrade, ContractsArea, Area_Remain, Distance, Rain_Vol, YieldOldGrade, Fertilizer, WaterType, Class_-Cane, TargetOldGrade and GrooveWide.

Corresponding to Line 11 of Algorithm 1, we perform the hyper-parameter tuning by applying the grid-search with the 5-fold cross-validation on the training set. The ranges of parameters used in the grid search are listed as follows:

```
params = {'n_estimators': [100,200,300, 400, 500],
          'min_samples_leaf': [10,12,14,16,20],
          'max_depth': [5,10],
          'max_features': [None, 4,5,6]}
```

We obtain the best score of 5-fold cross validation equal to *71.83%*, where the best parameters are listed as `max_depth: 10`, `max_features: 6`, `min_samples_leaf: 14` and `n_estimators: 100`. By using the trained RF model with the best parameters, the accuracy on both the training set and the testing set are 75.83% and 71.88%, respectively. This value of accuracy is the final result of our model.

We also compare our RF based method with other machine learning methods. Specifically, we use K-Nearest Neighbors (KNN), Support Vector Machine (SVM), Naïve Bayes, and AdaBoost. Each of these methods are trained with the default parameters of Scikit-Learn. The accuracies of these methods on the test dataset are shown in Table 3. From the accuracies reported in Table 3, we found that our method outperform the first baseline (51.52%) that predicts the yield grade of each plot based on the actual yield of the last year. This indicates that we can improve the accuracy of prediction about 20% by using one of our proposed methods instead of making the predictions without having any machine learning system. The accuracy of our method is also better than the ones of KNN, SVM, Naïve-Bayes, and AdaBoost.

With respect to the second baseline, we found that our proposed method outperforms the Baseline-2 about 6% of accuracy. At we explain earlier that the feature "TargetGrade" is the most important feature in the RF based method. Especially, from Table 2, the feature importance of "TargetGrade" is 60.27%. This means that if we incorporate a machine learning technique (i.e., RF in our case) with other features to the target yield grade ("TargetGrade") provided by human expert, we can improve the accuracy of prediction about 6%. From the standpoint of sugar mill operation, this could be a big gain as a decision making aid tool.

5 Conclusion

In this work, we propose methods for classifying the sugarcane yield grade at the plot level from the information about plot characteristics. Our method is based on Random Forest algorithm where our implementation is based on the random forest of Scikit-Learn [1]. The data used in this work is acquired from a set of sugarcane plots around a sugar mill in Thailand. The data consists of 12,521 records from two production years. We split the data into the training set and the testing set. The training dataset is used to build the models to predict the yield grade in the testing dataset. The classification accuracies of our random forest based method on the testing dataset are 75.83% and 71.88%, respectively. Our proposed method outperforms the baseline that uses the actual yield of the last year as the prediction (51.52%) and the baseline that the target yield of each plot is manually predicted by human expert (65.50%).

Some possible topics for future work can be listed as follows. First we can investigate the improvement in the prediction accuracy in the case that daily the temperature information in the local area and more information about the soil characteristics of each plot are available as in some related works. Second, a comprehensive study about the effects of hyper-parameter tuning, feature engineering and feature selection on the dataset could be a potential work. Finally, a model stacking technique can be applied to improve the prediction accuracy.

References

1. Pedregosa, F., et al.: Scikit-learn: machine learning in Python. J. Mach. Learn. Res. **12**, 2825–2830 (2011)
2. Everingham, Y., et al.: Accurate prediction of sugarcane yield using a random forest algorithm. Agron. Sustain. Dev. **36**(27), 1–9 (2016)
3. de Oliveira, M.P.G., Bocca, F.F., de Rodrigues, L.H.A.: From spreadsheets to sugar content modeling: a data mining approach. Comput. Electron. Agric. **132**, 14–20 (2017)
4. Bocca, F.F., Rodrigues, L.H.A.: The effect of tuning, feature engineering, and feature selection in data mining applied to rained sugarcane yield modelling. Comput. Electron. Agric. **128**, 67–76 (2016)
5. Thuankaewsing, S., Khamjan, S., Piewthongngam, K., Pathumnakul, S.: Harvest scheduling algorithm to equalize supplier benefits: a case study from the Thai sugar cane industry. Comput. Electron. Agric. **110**, 42–55 (2015)
6. Robnik-Šikonja, M., Kononenko, I., Konokenko, I.: An adaptation of Relief for attribute estimation in regression. In: Proceedings of the Fourteenth International Conference on Machine Learning, pp. 296–304. Morgan Kaufmann Publishers Inc., San Francisco (1997)
7. Breiman, L.: Random forests. Mach. Learn. **45**, 5–32 (2001)
8. Ruß, G., Kruse, R.: Feature selection for wheat yield prediction. In: Bramer, M., Ellis, R., Petridis, M. (eds.) Research and Development in Intelligent Systems XXVI, pp. 465–478. Springer, London (2010)
9. Gonzalez-Sanchez, A., Frausto-Solis, J., Ojeda-Bustamante, W.: Attribute selection impact on linear and nonlinear regression models for crop yield prediction. Sci. World J. (2014)

The Efficacy of Using Virtual Reality for Job Interviews and Its Effects on Mitigating Discrimination

Rico A. Beti(✉), Faris Al-Khatib, and David M. Cook

Edith Cowan University, Joondalup, WA 6027, Australia
rico.beti@gmail.com, faris.khatib@hotmail.com, d.cook@ecu.edu.au

Abstract. Virtual reality (VR) is an emerging technology that has already found successful application in a variety of different fields, including simulation, training, education, and gaming. While VR technologies have been considered for use in recruitment practices, available research on the topic is limited. In all stages of the recruitment process, social categorization of job applicants based on ethnicity, skin color, and gender, as well as other forms of discrimination are contemporary issues. This study examined the efficacy of using virtual reality technology as part of job interview strategies and evaluated its potential to mitigate personal bias towards job applicants. The use of VR technology as a job interview concept has been shown to have a high rate of acceptance and initial barriers for first-time users are low. The method provides benefits for job applicants in terms of social equity and performance.

Keywords: Diversity · Employment · Recruitment
Applicant identity · Applicant equity · Equality · Positive action
Virtual reality job interviews · Job discrimination
Perceived discrimination · HTC Vive

1 Introduction

Despite rigorous anti-discrimination laws present in many countries, job applicants often remain judged by their name, gender, physiology, ethnicity, or place of origin, and subsequently experience more difficulty in applying for a position than individuals from ethnic majorities. This issue is particularly prevalent in the early stages of the application process where the vulnerability to personal bias and social categorization by recruiters is the highest [1]. While these disadvantages in the beginning of the application process, specifically during the résumé screening phase, can partially be alleviated via the means of anonymous job applications, it is still possible that the discrimination may simply be postponed and surface during later stages, such as the personal face-to-face interview [2]. Nonetheless, given that anonymous job applications have gained in popularity and have shown promising results [3], a key factor in combating personal

H. Unger et al. (Eds.): IC2IT 2018, AISC 769, pp. 43–52, 2019.
https://doi.org/10.1007/978-3-319-93692-5_5

discrimination during the interview stage seems to be the reduction of social cues received by the recruiters to a minimum. A potential way to achieve this goal, and ultimately to reach a higher degree of equity between job applicants, may be the employment of emerging technologies in the hiring processes, such as virtual reality environments.

This paper discusses the efficacy of conducting job interviews in a VR environment and subsequently examines the possibility for a potential increase in social equity amongst applicants.

2 Background

While specific available literature in regards to VR in job interviews is limited due to the technology's novelty and due to the subjective nature of personal discrimination, related studies show that VR technology seems promising in reducing conceptual bias and offers opportunities for an increase in equity for disadvantaged people [4]. VR and similar digital environments have been used successfully for training purposes, specifically for effective preparation for job interviews [5]. VR has furthermore been found to potentially offer benefits in terms of stress management and stress reduction [6], as well as enhancing the perception of presence during stressful situations such as job interviews [7]. Virtual reality environments have also seen application in healthcare where several studies have shown positive impacts on individuals with autism [8] and psychiatric disabilities [9]. This by itself can be regarded as a major advantage in terms of equity when compared to traditional interviewing methods. Combined with algorithmic approaches [4] and video-based communication [10], personal bias towards interviewees may further be reduced.

In terms of personal discrimination during job applications and interviews, existing studies seem to agree that social categorization definitely is a problem, but to an extent that could not have been clearly quantified yet [1,2]. Lloyds Banking Group is one of the early adopters of VR technology for the purpose of job interviews and have already started integrating VR environments into their standard hiring practices, arguing that VR allows testing situations that "would otherwise be unfeasible in the conventional assessment process" [11].

3 Viability Study

To test the efficacy of using VR technology for the purpose of job interviews, a study was undertaken involving simulated job interviews under controlled conditions. The study examined a total of 24 university students that were evenly divided into two groups of 12 students each based on their course type. *Group A* consisted of 12 students enrolled in courses directly related to information and communication technologies (ICT), such as Computer Science and Cyber Security. *Group B* consisted of 12 students enrolled in courses that are not directly related to ICT, such as Business, Accounting, and Nursing. Participants were recruited from Edith Cowan University campuses (Perth, Western Australia) and were randomly selected based on random time intervals between participant approaches.

This study hypothesized that (H_1) VR technology is a viable alternative for interviewing job applicants effectively and that (H_2) the absence of direct face-to-face contact with interviewees positively affects applicant equity.

4 Method

Each of the 24 students were invited to participate in a simulated virtual reality job interview for the fictitious position of *Student Ambassador* at their university. The virtual interview consisted of two main stages: *(1)* the formal oral interview and *(2)* the completion of interactive tasks within the VR environment. The simulated job interviews were conducted with each participant individually and anonymously. Participants were communicating with the recruiter exclusively via the VR headset's built-in microphone and a pair of audio headphones, and neither party could see the other. The recruiter, however, was able to observe the scene from the perspective of the participants within the virtual environment as the gameplay was streamed to the recruiter's computer. Before the commencement of the interview process, the participants were given a brief introduction of the HTC Vive headset (Fig. 3) and its hand-held controllers (*"wands"*) (Fig. 3), followed by an explanation of how to navigate through the virtual environment by the means of physical movement and *"teleportation"*[1]. Only a single controller was required for the simulation. The side buttons were used to initiate teleportation and the trigger allowed direct interaction with certain objects in the scene. The participants were then assisted in putting on and properly adjusting the VR headset and the audio headphones. From this point on, the recruiter would guide the participants through the simulated job interview process within the virtual environment. After completing the interview and its interactive parts, participants were assisted in removing the VR headset and audio headphones and then asked to complete a questionnaire.

The questionnaire consisted of a set of six initial demographic questions followed by two separate parts. The first part contained questions in regard to perceived discrimination during previous face-to-face interviews in terms of ethnicity, skin color, and gender. The second part contained questions about the participant's experiences during the VR interview, including comfort and focus, ease of use, and potential difficulties related to navigation and interaction.

4.1 Room Setup

The virtual environment in which the job interview was conducted consisted of three adjacent rooms (Fig. 1), each serving a specific purpose. The first room is the main office featuring the recruiter's desk, a chair on which the participants

[1] Moving instantly to a different location within the virtual environment without physical movement.

would be physically seated initially, a logic puzzle in the form of a poster on the wall, and an interactive whiteboard. The second room is larger than the main office and contains a coffee table that is placed in the center. The third room is the corridor which connects the main office with the second room. Two paintings were placed on the wall. All rooms were furnished, decorated, and used standard daytime office lighting, with the intent to provide a realistic and immersive virtual reality experience.

The interview took place within a physical area of approximately 2.5 m × 3.0 m, defining the boundaries within the virtual environment (Chaperone boundaries) accordingly. The main office was slightly larger, requiring the use of teleportation in the early stages of the interview.

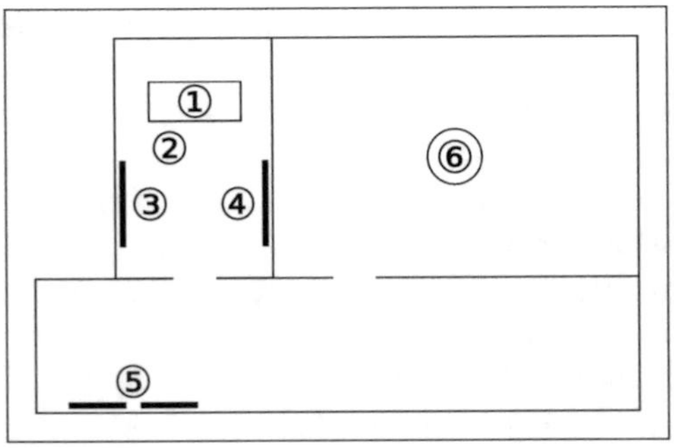

Fig. 1. Virtual environment room setup; (1) Desk, (2) Participant's chair, (3) Logic puzzle, (4) Whiteboard, (5) Paintings, (6) Coffee able

4.2 Interview Process

For the first stage of the interview, the participants were seated on a physical chair positioned in such a way that it aligned with a virtual chair in the scene facing the recruiter's virtual desk. The recruiter asked several common interview questions regarding the course in which the participants is currently enrolled, their previous work experience, and skills required for the position. After approximately five minutes, the participants were asked to stand up and the chair was subsequently removed by a researcher, allowing the participants to make use of the entire VR area.

The second stage of the interview consisted of three distinct tasks. Firstly, the participants were asked to solve a puzzle requiring abstract reasoning (Fig. 2) and to draw the solution onto the virtual whiteboard. This cognitive task was designed to test the participant's abilities to recognize visual clues and interact with virtual objects to convey their thought process. Secondly, participants were asked to leave the main office room by the means of teleportation and look at two paintings in the hallway, followed by a short discussion with the recruiter to test their attention to detail. Thirdly, the participants were asked to step into the second room adjacent to the main office. The center of the room featured a coffee table, on top of which a number of small-sized objects were placed. In order to be able to clearly identify the objects, participants were required to kneel and get in close proximity, testing the ability to execute precise movements around an area of interest within the VR environment.

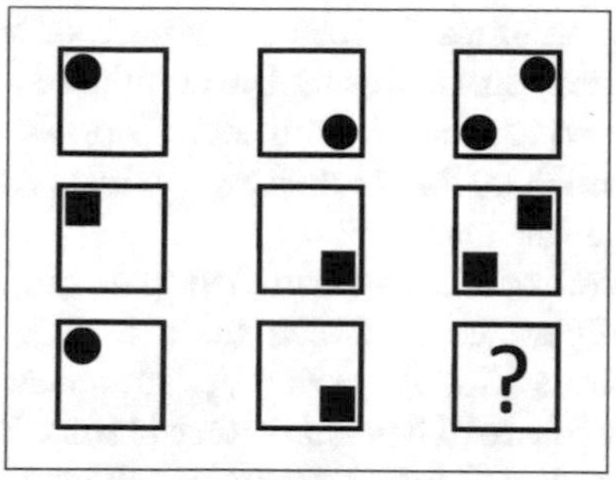

Fig. 2. Abstract reasoning task used to test interaction with the whiteboard

4.3 Equipment Specifications and Software

Several hardware and software components are required to set up a VR job interview. For the creation of the virtual reality experience, the HTC Vive VR System has been used, which consists of a VR headset (Fig. 3) to display the 3D scene, a pair of controllers (*"wands"*) (Fig. 3) for navigation and interaction, and two "lighthouse base stations" used for tracking the participant's position and movement. The headset's 5 m long tether was connected to an ASUS Rog Strix GL 700 VM GC 175T Gaming Notebook, featuring an Intel I7 Quad-Core processor (2.8 GHz) with 16 GB of RAM and an NVIDIA GeForce GTX 1060 graphics card. This setup is considered VIVE Ready [12].

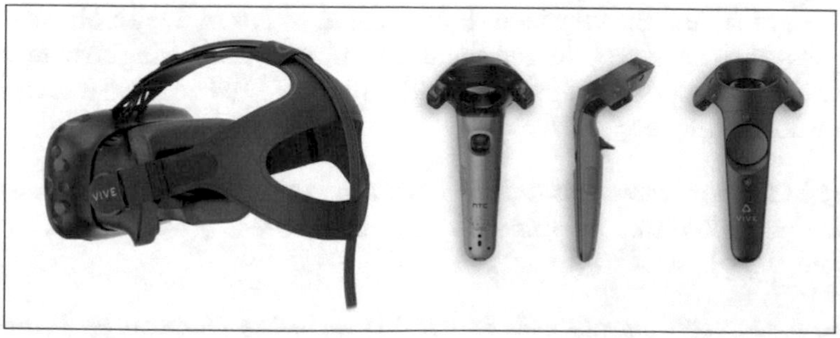

Fig. 3. HTC Vive headset (l.), Vive controllers ("Wands", ambidextrous) (r.) [13].

The virtual environment was created in the *VR Home Sandbox Game* [14]. Gameplay and audio were streamed in real-time to the recruiter's computer using TeamViewer 12 [15].

5 Results

The results of the questionnaires completed by each of the 24 participants were combined with observations made during the simulated virtual reality interview and subsequently analyzed through a qualitative process to determine the viability of employing VR technology for conducting job interviews and its implications on perceived equity amongst applicants.

Data points show similarities between Group A (ICT) and Group B (Non-ICT) and significant differences between male and female participants. The majority of all participants (61.5%) prefer VR interviews over traditional face-to-face interviews in the future. The results were similar between Group A (63%) and Group B (60%) overall but vary by gender, where 75% of male participants prefer VR interviews compared to 50% of female participants. Out of the 24 participants interviewed, 7 (29%) indicated that VR was somewhat helping them and 6 (25%) felt that VR greatly improved their ability to stay more focused during the interview. In contrast, 9 (37%) indicated that VR had little effect on their ability to concentrate and only 1 participant (4%) thought that VR was not helpful at all.

> *"My feelings about conducting an interview via VR was rest-assured once I felt more comfortable when we started. To my surprise, the graphical nature (game-like) aspect of the experience made me less conscious about how I was being perceived and therefore allowed me to focus on my own thoughts more carefully." (ICT Student 04, ♂)*

> *"VR is good and let me feel relaxed." (Non-ICT Student 06, ♀)*

The majority of participants (71%) experienced no difficulties in getting accustomed to being inside the virtual environment and generally found it "easy" or "very easy" to navigate through the scene and to interact with objects within. 14% expressed to have experienced slight difficulties while trying to adapt to VR. Some participants stated that they would have preferred to have a longer introduction prior to the interview.

> *"May be better if there was a little bit of time before the interview starts to allow the participants to walk around and get used to the control." (ICT Student 02, ♀)*

> *"I think because I am not used to VR I struggled to focus on the interview questions and was more interested in looking at the environment." (Non-ICT Student 09, ♀)*

Out of the 24 participants tested, 13 (54%) stated that they have never been in a VR environment before. From those 13 first-time users, 11 (85%) found it "easy" or "very easy" on average to navigate through the virtual office and to interact with objects, while 2 (15%) experienced minor difficulties.

Three questions were asked regarding perceived discrimination during previous face-to-face job interviews in terms of ethnicity, skin color, and gender. 4 participants (2♂, 2♀) have not had a face-to-face job interview or chose not to address these questions, leaving the answers of 20 participants (10♂, 10♀) to be considered.

6 participants (30%) indicated that their ethnicity, skin color, or gender "often" or "very often" affected their chances of getting the job. The perception of being discriminated was significantly higher among female participants (Table 2) when compared to male participants (Table 1).

Table 1. Perceived discrimination among males in previous face-to-face interviews.

$n_{♂} = 10$	Ethnicity	Skin color	Gender
Often	2	0	2
Very often	0	0	0

Table 2. Perceived discrimination among females in previous face-to-face interviews.

$n_{♀} = 10$	Ethnicity	Skin color	Gender
Often	4	3	1
Very often	0	0	2

When asked about the differences in terms of perceived discrimination between the VR interview and previous face-to-face interviews, 7 out of the 20 participants (35%) that had experienced at least one face-to-face job interview indicated that they felt more comfortable and less judged during the VR interview.

A final data point can be drawn from the interview, during which participants were asked to solve a logic puzzle (Fig. 2) and draw the correct answer onto the virtual whiteboard. The puzzle has two possible solutions where the circle is placed in the top right-hand corner and the square in the bottom left-hand corner, or vice-versa (clockwise and anticlockwise rotation), both of which were accepted. Among Groups A and B, 10 out of 24 participants (42%) were able to derive the correct answer. Group A had a slightly higher correctness rate of 50% compared to Group B's 33%.

6 Discussion

The results of this study show that the concept of VR job interviews enjoys a high level of acceptance within the tested demographic of which the majority prefers future interviews to be conducted in VR as opposed to traditional face-to-face conversations. The degree of acceptance was particularly high amongst male participants, and may be partially attributed to the "wow"-effect the technology induced in some participants. It can be concluded that the simulated interview has struck a balance between interaction, immersion, and imagination, which is essential for achieving a positive user experience [16]. Results also indicate that there are no significant difficulties associated with VR navigation and object interaction for most participants, including first-time users of the technology. However, depending on the complexity of the tasks involved in the interview, a longer and more detailed introduction may be required prior to commencing the interview. Some participants experienced a blurry vision, which may explain the low correctness rate concerning the logic puzzle, as it was a common error to mistake the circle for a square. Significantly, the data suggests that VR technology can help individuals to feel more focused and relaxed as they are immersed in the virtual experience and isolated from external audiovisual cues not related to the interview. Lau and Lee [17] state that virtual reality environments encourage individuals to suggest, explore, and evaluate ideas, which is likely to be desirable for applicant evaluation in a formal setting.

With 30% of the participants having perceived discrimination due to their ethnicity, skin color, or gender in previous face-to-face job interviews, the results of this study indicate that perceived social categorization is a significant issue, particularly among female participants. Social cues, which are known to influence the recruiters in their decision making process [1], can be limited with the use of VR technology due to the absence of direct face-to-face contact. This reduction may have been experienced by the 35% of the participants who indicated that they felt more comfortable and less judged during a VR interview when compared to their previous face-to-face interviews. To some extent, the promising results of anonymous job applications [3] can be carried over into the interview process by the means of using VR technologies.

Some participants that prefer traditional face-to-face interviews argued that the lack of non-verbal feedback makes it more difficult to discuss with the recruiter.

> *"I also prefer face-to-face interviews so you can gauge the interviewer's non-verbal reactions." (Non-ICT Student 09, ♀)*

This drawback, however, also exists when using other approaches avoiding direct face-to-face contact are employed, such as phone screenings [18] and video chat to some extent. In comparison, VR job interviews offer more opportunities in terms of interaction and conveying information visually, making them a more powerful alternative.

Despite the positive reception of the VR approach, the method requires a non-trivial setup and has substantial hardware requirements. Sensors must be placed in an adequately spaced area, and the headset needs to be tethered to a high-end computer with a CPU/GPU capable of rendering the VR scene in real-time using specialized software. This may be a barrier of entry into mainstream use of the technology for the near future.

7 Recommendations

Based on the results of this study, it is recommended that further research be conducted into different approaches of employing VR technology for job interviews to unlock its full potential. The capabilities of VR reach far beyond static simulations of real environments and the inclusion of 3D animations, physics engines, and advanced interactions with objects substantially distinguishes virtual interviews from traditional face-to-face assessments. Furthermore, the viability of using VR in the hiring process is likely to vary between different industries and business requirements, and shall be subject of further analysis.

At present, there is no software package available that is designed specifically for virtual reality job interviews. While the applications used were of adequate quality to accomplish all goals, their shortcomings introduced unnecessary difficulties for the participants. For example, imprecise teleportation often resulted in participants clipping through virtual walls, which was a major source of disorientation when navigating. Moreover, some participants indicated that they prefer a "multiplayer-mode" which would allow them to see and interact with the recruiter in the virtual environment. Most of these issues could be alleviated with purpose-built software.

8 Conclusion

The use of VR technology is a viable alternative to traditional face-to-face job interviews and has found general acceptance within the tested demographical segment. It has been shown that conducting a job interview within a virtual environment offers benefits in the form of stress reduction and an increased perceived level of presence when answering questions and completing interactive tasks. Furthermore, the use of VR technology for job interviews can reduce the number of social cues received by the recruiter and thus increase equity amongst the job applicants. While most participants experienced little difficulty during navigation and interaction, some initial training is required prior to the interview.

It is important to remember that VR is still an emerging technology and as such further research in the field of using VR as a recruitment tool is required. To address the current shortcomings, particularly the issue with clipping through walls and the participant's tendencies to walk out of the set VR boundaries, specialized software packages are needed. The refinement of technology and process will ultimately allow a larger audience to take advantage of the benefits VR job interviews offer.

References

1. Derous, E., Pepermans, R., Ryan, A.M.: Ethnic discrimination during résumé screening: interactive effects of applicants' ethnic salience with job context. Hum. Relat. **70**(7), 860–882 (2017)

2. Rinne, U.: Anonymous job applications and hiring discrimination. IZA World of Labor (2014)
3. Krause, A., Rinne, U., Zimmermann, K.F.: Anonymous job applications in Europe. IZA J. Eur. Labor Stud. **1**(1), 5 (2012)
4. Savage, D.D., Bales, R.A.: Video games in job interviews: using algorithms to minimize discrimination and unconscious bias (2016)
5. Chang, B., Lee, J.T., Chen, Y.Y., Yu, F.Y.: Applying role reversal strategy to conduct the virtual job interview: a practice in second life immersive environment. In: 2012 IEEE Fourth International Conference on Digital Game and Intelligent Toy Enhanced Learning (DIGITEL), pp. 177–181. IEEE (2012)
6. Villani, D., Riva, G.: Does interactive media enhance the management of stress? Suggestions from a controlled study. Cyberpsychol. Behav. Soc. Netw. **15**(1), 24–30 (2012)
7. Villani, D., Repetto, C., Cipresso, P., Riva, G.: May I experience more presence in doing the same thing in virtual reality than in reality? An answer from a simulated job interview. Interact. Comput. **24**(4), 265–272 (2012)
8. Smith, M.J., Ginger, E.J., Wright, K., Wright, M.A., Taylor, J.L., Humm, L.B., Olsen, D.E., Bell, M.D., Fleming, M.F.: Virtual reality job interview training in adults with autism spectrum disorder. J. Autism Dev. Disord. **44**(10), 2450–2463 (2014)
9. Bell, M.D., Weinstein, A.: Simulated job interview skill training for people with psychiatric disability: feasibility and tolerability of virtual reality training. Schizophr. Bull. **37**(suppl–2), S91–S97 (2011)
10. Kroll, E., Ziegler, M.: Discrimination due to ethnicity and gender: how susceptible are video-based job interviews? Int. J. Sel. Assess. **24**(2), 161–171 (2016)
11. Lloyds Banking Group: Introducing virtual reality to attract the best digital and it talent (2016). http://lloydsbankinggroupdigital.com/introducing-virtual-reality-to-attract-the-best-digital-talent
12. HTC Vive: VIVE Ready Computers (2017). https://www.vive.com/ready/
13. HTC Vive: HTC Vive Pressroom - Vive Photos (2016). https://www.htc.com/us/about/newsroom/htc-vive-press-kit/
14. Dandover: VR Home (2017). http://store.steampowered.com/app/575430/VR_Home/
15. TeamViewer: TeamViewer for Windows (2017). https://www.teamviewer.com/en/download/windows/
16. Rebelo, F., Noriega, P., Duarte, E., Soares, M.: Using virtual reality to assess user experience. Hum. Factors **54**(6), 964–982 (2012)
17. Lau, K.W., Lee, P.Y.: The use of virtual reality for creating unusual environmental stimulation to motivate students to explore creative ideas. Interact. Learn. Environ. **23**(1), 3–18 (2015)
18. Block, E.S., Erskine, L.: Interviewing by telephone: specific considerations, opportunities, and challenges. Int. J. Qual. Methods **11**(4), 428–445 (2012)

Natural Language Processing

Short Text Topic Model with Word Embeddings and Context Information

Xianchao Zhang, Ran Feng, and Wenxin Liang(✉)

Dalian University of Technology,
Dalian 116024, Liaoning, People's Republic of China
{xczhang,wxliang}@dlut.edu.cn, tomasvon2014@outlook.com

Abstract. Due to the length limitation of short texts, classical topic models based on document-level word co-occurrence information fail to distill semantically coherent topics from short text collections. Word embeddings are trained from large corpus and inherently encoded with general word semantic information, hence they can be supplemental knowledge to guide topic modeling. **G**eneral **P**ólya **U**rn **D**irichlet **M**ultinomial **M**ixture (GPU-DMM) model is the first attempt which leverages word embeddings as external knowledge to enhance topic coherence for short text. However, word embeddings are usually trained on large corpus, the encoded semantic information is not necessarily suitable for training dataset. In this work, we improve the GPU-DMM model by leveraging both context information and word embeddings to distill semantic relatedness of word pairs, which can be further leveraged in model inference process to improve topic coherence. Experimental results of two tasks on two real world short text collections show that our model gains comparable or better performance than GPU-DMM model and other state-of-art short text topic models.

Keywords: Topic modeling · Short texts · Topic coherence

1 Introduction

Short text such as tweets and micro-blogs has become a prevalent information source on the Internet and mining topical information from short text collections is vital for many applications. Compared to long text documents, short text documents are characterized with short document length and vast of vocabulary, which cause sparse word co-occurrence patterns on document level. Therefore, classical long text topic models based on word co-occurrence information such as LDA [1] and PLSA [2] are usually impeded to distill semantically coherent topics from short text collections.

This work was partially supported by National High Technology Research and Development Program (863 Program) of China (No. 2015AA015403) and National Science Foundation of China (No. 61632019).

H. Unger et al. (Eds.): IC2IT 2018, AISC 769, pp. 55–64, 2019.
https://doi.org/10.1007/978-3-319-93692-5_6

To alleviate the sparsity of word co-occurrence patterns for short text collections, several heuristic strategies have been proposed. One of the straightforward strategies is aggregating short text snippets into long pseudo-documents according to auxiliary information. For example, in the context of Twitter dataset, metadata such as authorship, hashtag, timestamp and location can be leveraged to aggregate tweets before applying classical topic models [3]. However, this strategy largely depends on training dataset and has limited scalability because such auxiliary information is not always available. Another strategy is to impose strong assumption for generative process of short text documents. For example, Dirichlet Multinomial Mixture (DMM) model [4] assumes that each document is associated with only one topic. Biterm Topic Model (BTM) [5] directly models the generative process of word pairs in corpus. Dual-Sparse Topic Model [6] employs the "spike and slab" prior for LDA to capture sparsity characteristic of topic-word distribution and document-topic distribution. Pseudo-document-based Topic Model (PTM) and Sparse-PTM (SPTM) in literature [3] assume that each short document is sampled from a long pseudo-document to obtain additional document level word co-occurrence information. Usually these models do not need external auxiliary information but are accompanied with complex inference process.

On the other hand, word embeddings [7,8] have been successfully applied in many natural language processing tasks, and semantic information encoded in word embeddings can be regarded as external knowledge to guide topic modeling for short text collections. Based on DMM model, Li et al. [9] propose the GPU-DMM model, it leverages the General Pólya Urn (GPU) model to incorporate the general word-pair semantic relatedness knowledge obtained from word embeddings to enhance topic coherence. However, word embeddings are usually trained on large corpus, the encoded semantic information of word embeddings is not necessarily suitable for training dataset.

In this work, we improve the GPU-DMM model in two aspects. Firstly, we distill semantic relatedness of word pairs from word embeddings and filter the semantic relatedness information by Point Mutual Information (PMI) between two words from training dataset. In model inference process, this information can be further leveraged to improve topic coherence. Secondly, instead of using DMM model for generative process, we take the assumption same as Twitter-LDA [10], i.e., each short document is associated with only one topic, but words in this document are not necessarily semantically related with the topic. Hence in this work, we assume that each word in a document is generated from either a normal topic or a background topic.

2 Related Work

Word embedding is a way of distributed representation for words, each word can be represented as a real-valued vector. There are many prevalent implementations of word embedding such as Continuous Skip-gram model, Continuous Bag-of-words model and Glove model [7,8,11]. Word embeddings are usually trained on large corpus such as Wikipedia and inherently encoded with general semantic and syntactic information of words, which can be revealed by algebraic calculation. For example, $vec(king) - vec(man) \approx vec(queen) - vec(woman)$, where $vec(w)$ is the word embedding of word w.

Li et al. [9] propose the GPU-DMM model which improves the DMM model by exploiting word embeddings and the GPU model in inference process to obtain coherent topics. Specifically, there are two steps for this model. Firstly, semantically relatedness of word pairs can be acquired from word embeddings by cosine similarity. Secondly, probabilities of co-occurrence of semantically related words under same topics can be increased by the sampling scheme of the GPU model. Hence semantic coherence of topics can be enhanced via the two steps.

3 The Proposed Model

In this section, we present the proposed model for short text topic modeling. There are four steps in this model, firstly, we train word embeddings from a large corpus, and calculate semantic similarity between two words by cosine similarity of corresponding word embeddings. Secondly, we construct semantically related word set (SRWS) for each word in vocabulary of training corpus. Then we leverage Point Mutual Information (PMI) to filter inappropriate semantic relatedness of word pairs for training dataset. Finally, we employ the GPU model in the inference process to enhance co-occurrence patterns of semantic related words under same topics to obtain coherent topics.

3.1 Generative Process of the Model

In this paper, we change the generative process of the GPU-DMM model for short text by incorporating background topic information. Specifically, each short document is associated with only one topic, and each word in short document can be generated from either the topic or a global background topic. Because not every word in a short document is semantically related to the corresponding topic, so our generative process is more rational for short documents. For each word in a document, there is a switch variable indicating whether the word is sampled from a normal topic or from the background topic. The detail of the generative process of this model can be referred as follow:

(a) Sample topics distribution for document collection: $\theta \sim Dirichlet(\alpha)$
(b) Sample words distribution for the background topic: $\phi_B \sim Dirichlet(\beta)$
(c) Sample switch variables distribution: $\psi \sim Beta(\gamma)$
(d) For each topic $k \in 1, \ldots, K$
  - Sample words distribution: $\phi_k \sim Dirichlet(\beta)$
(e) For each document $d \in 1, \ldots, D$
  - Sample a topic: $z_d \sim Multinomial(\theta)$
  - For each word $i \in 1, \ldots, N_d$
    - Sample a switch variable: $y_{di} \sim Bernoulli(\psi)$
    - If $y_{di} = 0$, sample w_{di} from $Multinomial(\phi_{z_d})$, else sample w_{di} from $Multinomial(\phi_B)$

The corresponding graphic model of the generative process is showed in Fig. 1.

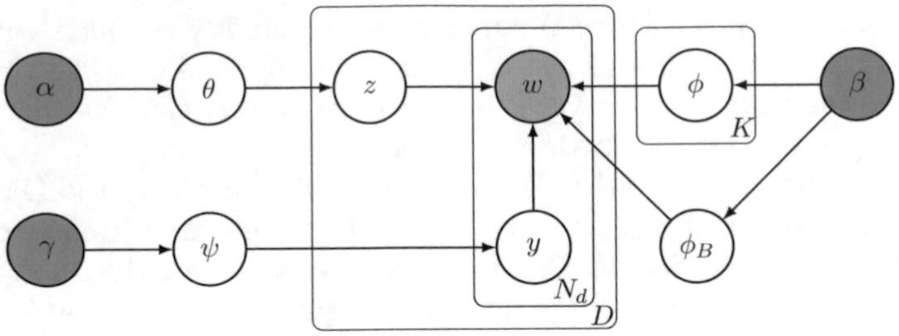

Fig. 1. BDMM

3.2 Obtain Word Pair Semantic Relatedness

We leverage word embeddings to obtain semantic relatedness between two words in this subsection. Specifically, for word w_1 and w_2, the corresponding word embeddings are $\boldsymbol{v_1}$ and $\boldsymbol{v_2}$, the semantic relatedness (SR) of w_1 and w_2 is defined as $SR(w_1, w_2) = cos(\boldsymbol{v_1}, \boldsymbol{v_2}) = \frac{\boldsymbol{v_1} \cdot \boldsymbol{v_2}}{\|\boldsymbol{v_1}\| \|\boldsymbol{v_2}\|}$, where the $cos(\boldsymbol{v_1}, \boldsymbol{v_2})$ is cosine similarity between $\boldsymbol{v_1}$ and $\boldsymbol{v_2}$. Then we construct a SRWS for each word in vocabulary of training corpus. For word w, the SRWS of w is constructed as $\mathcal{S}(w) = \{w_o | w_o \in V, SR(w, w_o) > \epsilon\}$, where V is the vocabulary list.

However, word embeddings are encoded with general semantic and syntactic information which may be not suitable for training dataset. For example, "bachelor" and "undergraduate" are not semantically related when it comes to "family" topic, so we use the PMI between two words to filter this unreasonable information. PMI is a prevalent method to measure the association of two words in the context of text mining. Given two words, w_1 and w_2, the PMI between word w_1 and w_2 is defined as follow:

$$PMI(w_1, w_2) = log \frac{P(w_1, w_2)}{P(w_1)P(w_2)} \tag{1}$$

where $P(w)$ denotes the probability of word w appearing in documents, which is $P(w) = \frac{|D_w|}{|D|}$, D_w is a set of training documents containing word w, D is training document collection, and $|D|$ is the number of documents in collection D. And $P(w_1, w_2) = \frac{|D_{w_1, w_2}|}{|D|}$ represents the probability of co-occurrence of w_1 and w_2 in same documents. We redefine the SRWS for word w, which is $\mathcal{S}(w) = \{w_o | w_o \in V, SE(w, w_o) > \epsilon, PMI(w, w_o) >= \eta\}$. The PMI value between two words is obtained from training dataset, and word embeddings are trained from external corpus, so we define the SRWS by both global and local context information.

Since words in a SRWS are supposed to share similar semantic and syntactic information, when we see one of them in a topic, it's reasonable to see other words under this topic with high probability.

GPU model is a popular model in statistics. In GPU model, there is a urn originally contains some colorful balls, when a ball is randomly drawn from the urn, its color is recorded, and two balls of that color are put back to the urn along with a certain number of balls with similar colors. In the context of topic model, topic and word can be regarded as urn and ball respectively. So the proportion of balls in a urn can be analogous to word distribution under a topic. The process of GPU model is just like the Gibbs sampling procedure for topic models. To incorporate semantic relatedness between words into inference process, we resort to the GPU model. Specifically, given word w and its SRWS $\mathcal{S}(w)$, the amount of promotion for $w_o \in \mathcal{S}(w)$ when we see word w under a topic is given by Eq. 2:

$$\lambda_{w,w_o} = \begin{cases} 1 & w = w_o \\ SR(w, w_o) & w_o \neq w, w_o \in \mathcal{S}(w) \\ 0 & \text{else.} \end{cases} \tag{2}$$

However, according to the generative process of this model, words in a document are divided into topical words and background words. Considering background words are not semantically related, so we barely apply the GPU model to topical words in each document.

3.3 Model Inference

We employ the collapsed Gibbs sampling for parameter estimation, however, the GPU model is nonexchangeable which means the joint probability of words under a specific topic is not invariable to permutation of word sequences, which causes intractable inference. For calculation simplicity, we take the assumption which regards every word as the last word and ignores the subsequent words and their topic assignments as [12]. The latent variables need to be sampled are $\boldsymbol{y}$ and $\boldsymbol{z}$, and the parameters $\boldsymbol{\theta}, \boldsymbol{\phi}, \boldsymbol{\phi_B}$ and $\boldsymbol{\psi}$ can be estimated via maximum a posteriori estimation (MAP) from samples. For document d, Gibbs sampling equation for its topic assignment $\boldsymbol{z}_d$ is given by Eq. 3:

$$P(\boldsymbol{z}_d = k | \boldsymbol{z}_{-d}, \boldsymbol{w}_d, \boldsymbol{y}_d) \propto (n_{k,-d} + \alpha) * \frac{\prod_{w \in d, y_w = 0} \prod_{i=0}^{n_d^w - 1} (\tilde{n}_{k,-d}^w + \beta + i)}{\prod_{j=0}^{N_d - 1} (\tilde{n}_{k,-d} + V\beta + j)} \tag{3}$$

where $\boldsymbol{z}$ is topic assignments for documents, and n_k is number of documents associated with topic k, n_d^w is number of appearance of word w in document d, $\tilde{n}_k^w$ is statistics of word w associated with topic k, and $\tilde{n}_k$ is statistics of words associated with topic k in corpus. The subscript $-d$ indicates document d is excluded when we calculate corresponding statistics. For i-th word w in document d, the Gibbs sampling equation for $\boldsymbol{y}_{d,i}$ is given by Eqs. 4 and 5:

$$P(\boldsymbol{y}_{d,i} = 1 | \boldsymbol{z}_d, \boldsymbol{w}, \boldsymbol{y}_{-d,i}) \propto (n_{-d,i}^{y=1} + \gamma) * \frac{n_{B,-d,i}^w + \beta}{n_{B,-d,i} + V\beta} \tag{4}$$

$$P(\boldsymbol{y}_{d,i} = 0 | \boldsymbol{z}_d, \boldsymbol{w}, \boldsymbol{y}_{-d,i}) \propto (n_{-d,i}^{y=0} + \gamma) * \frac{\tilde{n}_{z_d,-d,i}^w + \beta}{\tilde{n}_{z_d,-d,i} + V\beta} \tag{5}$$

where $n_d^{y=1}$ is number of background words in document d, n_B^w is number of word w associated with background topic, n_B is number of background words in corpus, and the subscript $-d, i$ means that i-th word of document d is excluded when we calculate these variables. Given sufficient samples, the parameter ϕ can be estimated by MAP, which is represented as:

$$\phi_{k,w} = p(w|k) = \frac{\tilde{n}_{k,w} + \beta}{\tilde{n}_k + V\beta} \tag{6}$$

The Gibbs sampling procedure of the proposed model is showed in Algorithm 1. SRWS of each word in vocabulary is calculated before the process of Gibbs sampling, and time complexity of this step is $O(|V|^2)$, where the $|V|$ is the size of vocabulary list. From the Algorithm 1, we can read out the time complexity of Gibbs sampling process of the model, which is $O(|D|lK)$, where $|D|$ is the number of documents in collection D, l is the average length of each document in D, and K is the number of topics. So the total time complexity of the model is $O(|V|^2 + MaxIteration|D|lK)$.

Algorithm 1. Gibbs Sampling Process

Input: Hyper-parameters: $\alpha, \beta, \gamma, K, \mathcal{S}$

Output: Posterior topic-word distribution of K topics: $\{\phi_1, \ldots, \phi_K\}$

Initialize the statistics before Gibbs sampling
for $iteration \leftarrow 1$ **to** $MaxIteration$ **do**
  for $d \leftarrow 1$ **to** $|\boldsymbol{D}|$ **do**
    $z_{old} \leftarrow \boldsymbol{z}_d$; $n_{z_{old}} \leftarrow n_{z_{old}} - 1$;
    for each $w \in \boldsymbol{D}_d$ **do**
      if $y_{d,w} == 0$ **then**
        $\tilde{n}^{w}_{z_{old}} \leftarrow \tilde{n}^{w}_{z_{old}} - 1$; $\tilde{n}_{z_{old}} \leftarrow \tilde{n}_{z_{old}} - 1$;
        for each $w_o \in \mathcal{S}(w)$ **do**
          $\tilde{n}^{w_o}_{z_{old}} \leftarrow \tilde{n}^{w_o}_{z_{old}} - \lambda_{w,w_o}$; $\tilde{n}_{z_{old}} \leftarrow \tilde{n}_{z_{old}} - \lambda_{w,w_o}$;
    Sample z_{new} according to Eq. 3 for document d;
    $\boldsymbol{z}_d \leftarrow z_{new}$; $n_{z_{new}} \leftarrow n_{z_{new}} + 1$;
    for each $w \in \boldsymbol{D}_d$ **do**
      if $y_{d,w} == 0$ **then**
        $\tilde{n}^{w}_{z_{new}} \leftarrow \tilde{n}^{w}_{z_{new}} + 1$; $\tilde{n}_{z_{new}} \leftarrow \tilde{n}_{z_{new}} + 1$;
        for each $w_o \in \mathcal{S}(w)$ **do**
          $\tilde{n}^{w_o}_{z_{new}} \leftarrow \tilde{n}^{w_o}_{z_{new}} + \lambda_{w,w_o}$; $\tilde{n}_{z_{new}} \leftarrow \tilde{n}_{z_{new}} + \lambda_{w,w_o}$;
    Sample variable y for each word in each document according to Eqs. 4 and 5

4 Experiments and Results

4.1 Experimental Setup

Dataset. We utilize the **Web Snippet** [13] and **Amazon Reviews** [14] to train topic models. **Web Snippet** contains 12340 web search snippets classified into 8 categories priorly and we treat each snippet as a short document. And the **Amazon Reviews** is a set of product reviews from Amazon spanning May 1996 to July 2014. Instead using all texts, we randomly sample 20000 short reviews categorized into 7 clusters from the original dataset. Grammar and vocabulary usage in reviews are usually informal which increases difficulty of learning coherent topics. Before applying topic models, we preprocess the two datasets in four steps: (1) lowercase all words in document collections; (2) remove stop words and non-alphabetic characters; (3) remove words appearing less than 5 documents; (4) remove documents containing less than 3 words. After preprocessing, for **Web Snippet** dataset, the average length of each document is 14.7. And for **Amazon Reviews** dataset, we get 19980 short reviews, and the average length of each review is 17.5, so documents in both collections are sparse and short enough.

Word Embedding. We leverage Wikipedia corpus[1] as training dataset, and Google word2vec toolkit with Continuous Skip-gram model[2] to train word embeddings. The dimension of the word embeddings is clamped to 300 just as the setting in literature [9]. To get semantically related word set $\mathcal{S}(w)$ for word w, we set the threshold ϵ to be 0.7 and η to be 0 for both datasets, because the effective of ϵ can be reconciled by the PMI, so $\mathcal{S}(w)$ is not too sensitive to the value of ϵ. For each word w in vocabulary lists of both datasets, we get $\mathcal{S}(w) = \{w_o | SR(w, w_o) > 0.7, PMI(w, w_o) >= 0\}$.

Baseline Methods and Parameter Setting. We compare our model with five state-of-art short text topic models, which are DMM, BTM, PTM, SPTM and GPU-DMM. We set common parameters for all models uniformly, such as $\alpha = 50/K$, $\beta = 0.01$ and $maxIteration = 1500$. We set the hyper-parameters of PTM and SPTM as the literature [3] suggests, and the number of long pseudo-document for our datasets is set to be 100 for the average length of each short document. For the proposed model, we set the hyper-parameter $\gamma = 0.5$.

4.2 Semantic Coherence Evaluation

To verify the effectiveness of our model, firstly we manage to measure the semantic coherence of topics learned from two datasets by topic models. We utilize the *topic coherence* proposed in the literature [15] as the criterion for this task. Higher *topic coherence* value indicates higher semantic coherence of topics. Experimental results in literature [15] have shown that performance of *topic coherence* is highly consistent with manual annotation. For topic k, the definition of the *topic coherence* is given by Eq. 7:

$$C(k; V^k) = \sum_{m=2}^{M} \sum_{l=1}^{m-1} log \frac{|D_{w_m^k, w_l^k}| + 1}{|D_{w_l^k}|} \tag{7}$$

where $V^k = [w_1^k, \ldots, w_M^k]$ is the top M words under the topic k sorted by its probability in descending order. And $|D_{w_1, w_2}|$ is the number of documents containing both w_1 and w_2. In this task, we set the number of topics,

[1] https://dumps.wikimedia.org.
[2] https://code.google.com/archive/p/word2vec/.

Table 1. Topic coherence

Topic number	PTM	SPTM	BTM	DMM	GPU-DMM	Our Model
Web Snippet dataset						
20	−958.44	−1021.19	−907.17	−955.60	−936.98	**−856.43**
40	−930.30	−1005.04	−872.89	−916.01	−890.25	**−870.32**
60	−919.83	−1047.04	−859.49	−900.55	−869.96	**−824.83**
80	−907.55	−1083.30	−833.57	−879.75	−861.69	**−829.77**
Amazon Reviews dataset						
20	−983.81	−922.89	−843.77	−859.98	**−828.98**	−838.87
40	−964.73	−1006.0	−849.54	−912.06	−866.95	**−844.30**
60	−935.39	−1000.22	−882.78	−911.96	−899.10	**−856.57**
80	−920.82	−913.76	−883.37	−913.76	−910.54	**−867.14**

$K = \{20, 40, 60, 80\}$. The *topic coherence* results of topic models are showed in Table 1. We can see that our model gains a clear edge compared to other benchmark models on both datasets, which shows high semantic coherence of topics learned by the proposed model. Surprisingly PTM and SPTM perform poorly on both datasets, and the reason might be the origin assumption adopted by PTM and SPTM which assigns each short document more than one topic and suffers from the sparsity of training dataset to cause the bad performance. We also observe that GPU-DMM can get superior results compare to DMM as the results in literature [9]. The BTM models the word co-occurrence patterns straightly, so it shows robustness in different datasets. But high time complexity of the Gibbs sampling process for BTM prevents this model to be applied in large corpus.

4.3 Classification Evaluation

To evaluate the semantic distinctiveness between topics, we conduct classification experiments. Topics learned by topic model can be regarded as a low-dimensional representation for documents. The more discriminative topics are, the higher performance classifiers can obtain. As mentioned in literature [9], the topic distribution conditional document can be inferred in two different ways:

$$p(z = k|d) \propto p(z = k) \prod_{w \in D_d} p(w|z = k) \tag{8}$$

$$p(z = k|d) \propto \sum_{w \in D_d} p(z = k|w) p(w|d) \tag{9}$$

where $p(w|d)$ can be estimated by word frequency int document d, and $p(z = k|w)$ can be obtained by bayesian rules $p(z = k|w) \propto p(z = k)p(w|z = k)$. In this work, we adopt the second strategy to obtain topic distributions under documents. We employ the *Random Forest* (RF) implemented in sklearn[3] as classifier. For parameters of the RF, we set $n\_estimators = 80$ and $max\_depth = 15$, which is tradeoff between efficiency and accuracy. We utilize the accuracy and f1-measure to quantify the results of classification in Fig. 2.

[3] http://scikit-learn.org/.

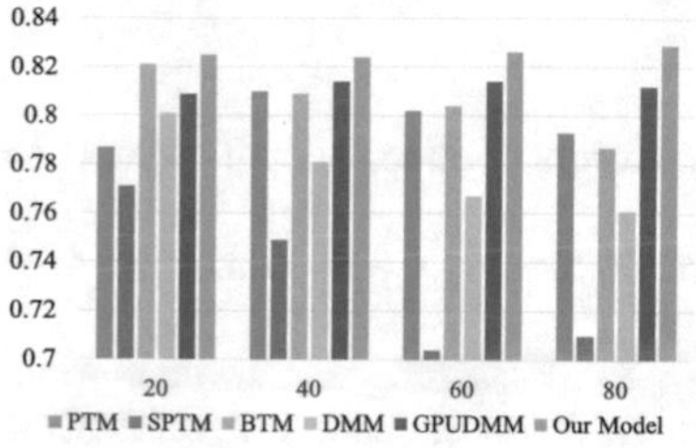

(a) Accuracy of Amazon Reviews

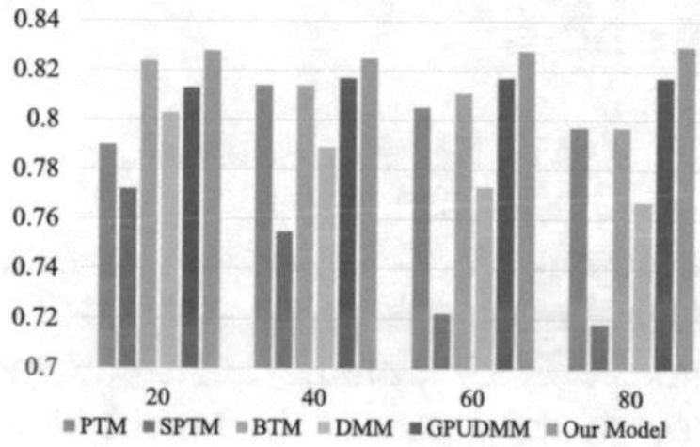

(b) F1 measure of Amazon Reviews

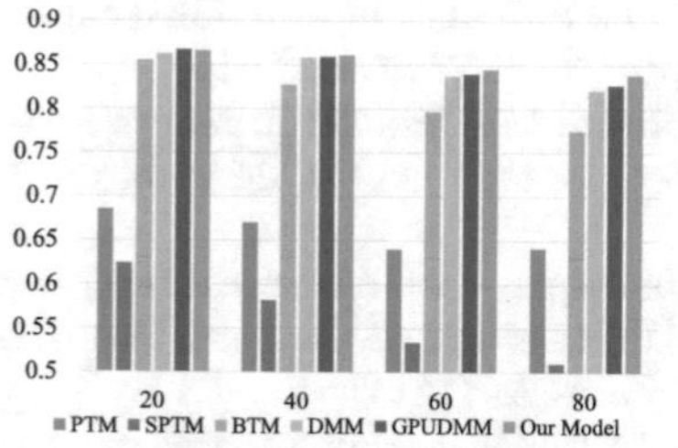

(c) Accuracy of Web Snippet

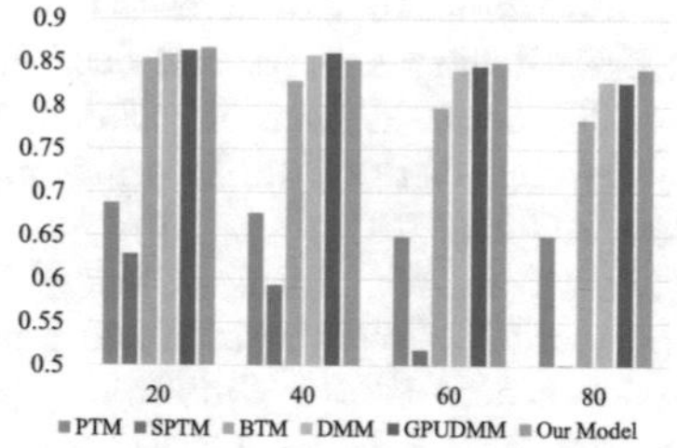

(d) F1 measure of Web Snippet

Fig. 2. Classification experiments on datasets

From results of classification, we observe that our model gains superiority to GPU-DMM model and other models in almost cases especially in **Amazon Reviews** dataset. Both GPU-DMM model and our model outperform DMM model which validates that word semantic relatedness information can truly help enforce semantic coherence of topics. With the increase of topic number, our model shows robustness compared to other models, which demonstrates that it can distill discriminative topics under different settings of topic granularity. BTM and DMM model gain comparable performance even without word embedding information in **Web Snippet** dataset, but fail on **Amazon Reviews**. We infer that effectiveness of these models are heavily influenced by the quality of training dataset. The PTM and SPTM still perform poorly on both dataset, but we observe the results on **Amazon Reviews** is much better than **Web Snippet**, so the scale of the dataset is a influential factor for the PTM and SPTM.

5 Conclusion

In this work, we improve the GPU-DMM model to propose a new topic model which leverages both word embeddings and context information to obtain coherent topics from short text collections. However there are still several problems we should pay attention to, for example, large training collections such as Twitter could be considered to valid the generalization ability of our model, And we can also test different implementations of word embeddings for the proposed model.

References

1. Blei, D.M., Ng, A.Y., Jordan, M.I.: Latent dirichlet allocation. J. Mach. Learn. Res. **3**, 993–1022 (2003)
2. Hofmann, T.: Unsupervised learning by probabilistic latent semantic analysis. Mach. Learn. **42**(1), 177–196 (2001)
3. Zuo, Y., Wu, J., Zhang, H., Lin, H., Wang, F., Xu, K., Xiong, H.: Topic modeling of short texts: a pseudo-document view. In: Proceedings of the 22nd ACM SIGKDD International Conference on Knowledge Discovery and Data Mining, pp. 2105–2114. ACM (2016)
4. Yin, J., Wang, J.: A dirichlet multinomial mixture model-based approach for short text clustering. In: Proceedings of the 20th ACM SIGKDD international conference on Knowledge discovery and data mining, pp. 233–242. ACM (2014)
5. Yan, X., Guo, J., Lan, Y., Cheng, X.: A biterm topic model for short texts. In: Proceedings of the 22nd International Conference on World Wide Web, pp. 1445–1456. ACM (2013)
6. Lin, T., Tian, W., Mei, Q., Cheng, H.: The dual-sparse topic model: mining focused topics and focused terms in short text. In: Proceedings of the 23rd International Conference on World Wide Web, pp. 539–550. ACM (2014)
7. Mikolov, T., Chen, K., Corrado, G., Dean, J.: Efficient estimation of word representations in vector space. arXiv preprint arXiv:1301.3781 (2013)
8. Mikolov, T., Sutskever, I., Chen, K., Corrado, G.S., Dean, J.: Distributed representations of words and phrases and their compositionality. In: Advances in Neural Information Processing Systems, pp. 3111–3119 (2013)
9. Li, C., Wang, H., Zhang, Z., Sun, A., Ma, Z.: Topic modeling for short texts with auxiliary word embeddings. In: Proceedings of the 39th International ACM SIGIR Conference on Research and Development in Information Retrieval, pp. 165–174. ACM (2016)
10. Zhao, W.X., Jiang, J., Weng, J., He, J., Lim, E.P., Yan, H., Li, X.: Comparing twitter and traditional media using topic models. In: European Conference on Information Retrieval, pp. 338–349. Springer (2011)
11. Pennington, J., Socher, R., Manning, C.: Glove: global vectors for word representation. In: Proceedings of the 2014 Conference on Empirical Methods in Natural Language Processing (EMNLP), pp. 1532–1543 (2014)
12. Mimno, D., Wallach, H.M., Talley, E., Leenders, M., Mccallum, A.: Optimizing semantic coherence in topic models. In: Conference on Empirical Methods in Natural Language Processing, pp. 262–272 (2010)
13. Phan, X.H., Nguyen, L.M., Horiguchi, S.: Learning to classify short and sparse text & web with hidden topics from large-scale data collections. In: Proceedings of the 17th International Conference on World Wide Web, pp. 91–100. ACM (2008)
14. McAuley, J., Leskovec, J.: Hidden factors and hidden topics: understanding rating dimensions with review text. In: Proceedings of the 7th ACM Conference on Recommender Systems, pp. 165–172. ACM (2013)
15. Mimno, D., Wallach, H.M., Talley, E., Leenders, M., McCallum, A.: Optimizing semantic coherence in topic models. In: Proceedings of the Conference on Empirical Methods in Natural Language Processing, pp. 262–272. Association for Computational Linguistics (2011)

Temporal Analysis of Twitter Response and Performance Evaluation of Twitter Channels Using Capacitor Charging Model

Sirisup Laohakiat(✉), Photchanan Ratanajaipan, Krissada Chalermsook, Leenhapat Navaravong, Rachanee Ungrangsi, Aekavute Sujarae, and Krissada Maleewong

School of Science and Technology, Shinawatra University, Pathumthani, Thailand
{sirisup.l, photchanan, krissada.c, leenhapat, rachanee, aekavute, krissada}@siu.ac.th

Abstract. As twitter is one of the highly popular social networks, analyzing the responses from users can allow us to study the behavior of users as well as evaluate the popularity of the twitter channels. In this study, we present a novel framework for analyzing twitter temporal responses using capacitor charging model. The proposed model, inspired from electrical circuit analysis, can reveal the temporal characteristic of the responses of each twitter post which can be a better option for measuring the channel popularity than the number of followers. Representing each post as a data point in the feature space, data clustering is used to determine the modal performance of each twitter channel that can reflect the channel's popularity. The study illustrates the use of the proposed framework in comparison five news twitter channels.

Keywords: Twitter · Temporal analysis · Capacitor charging model
Clustering · Algorithm · Modal performance · Active audience

1 Introduction

Twitter is one of the most popular social networks. Several twitter channels have garnered millions of followers from around the world. Due to its popularity and availability of data to the public, several scholar articles have been devoted to study various aspects of twitter data including popularity, network, influence user, etc.

Many studies have been extensively researched on measuring or predicting the popularity of tweets. Various types of measures are introduced including the content-based and temporal information of tweets, metadata of tweets and users, as well as structural properties of the users in social network [1]. Some are based on simple metrics that can be obtained by the Twitter API, while others are based on complex mathematical models [2].

Suh et al. [3] investigated the content-based features (i.e., numbers of URLs and hashtags), and the author-based features (i.e., number of followers and followees, and age of the account) in order to calculate the retweetability or the popularity of tweets.

H. Unger et al. (Eds.): IC2IT 2018, AISC 769, pp. 65–74, 2019.
https://doi.org/10.1007/978-3-319-93692-5_7

Zaman et al. [4] predicted the evolution of retweets based on the size of the retweeter network and the depth from the source of tweets using a probabilistic model. Maleewong [5] proposed a number of retweeters-based features (i.e., Activity Rate and Popularity Score of retweeters) and applied the features to the regression model for predicting the popularity of tweets.

Recent works studied on temporal aspects of Twitter information for various purposes. Gorrab et al. [6] investigated strategies to build Twitter user profiles over time, temporal user profiles, to cluster users according to their profiles. The clustering results can be used to suggest new hashtags and users to follows.

To evaluate the performance of each channel, one can consider the popularity and reach of the channels from the entry-level metrics such as number of followers, retweets, likes, and replies. However, these metrics are not very effective since they reveal so small amount of information regarding the channels. Moreover, the number of followers can be unnaturally manipulated by Twitter follower markets that can result in a large number of blank accounts of followers that are not real people [7].

This study presents a novel framework for evaluating Twitter channel based on the temporal response analysis utilizing capacitor charging model inspired from RC (resistor-capacitor) circuit model [8]. Analogous to the voltage across a capacitor in RC circuit, the cumulative response from twitter users is modeled using differential equation with respect to time. The parameters obtained from solving the differential equation can reveal several aspects of each tweet and the overall performance of the twitter channel that have not been addressed in other studies. The real-time stream feed of tweeters from five twitter news channels including CNN (@CNN), BBC (@BBC), Al Jazeera (@AJEnglish), XinHua news (@XHNews), and Fox news (@FoxNews) are used to analyze the temporal response of each channel under the proposed model.

This paper consists of five sections. Section 2 introduces the idea of the proposed model. Section 3 describes the parameters obtained from the model and the framework for analyzing the performance of the channel. Experimental results are shown in Sect. 4 and finally, conclusions and future works are presented in Sect. 5.

2 Capacitor Charging Model

When we connect a capacitor and a resistor with a voltage source as shown in Fig. 1, from circuit analysis, the voltage across the capacitor can be described in differential equation as

$$RC\frac{dv(t)}{dt} + v(t) - Vs = 0 \tag{1}$$

By solving Eq. (1), we get

$$v(t) = Vs\left(1 - e^{-\frac{t}{T}}\right) \tag{2}$$

Equation (2) indicates the growth of voltage due to the current from voltage source charging the capacitor, where T is time constant calculated by T = RC, *Vs* is the value

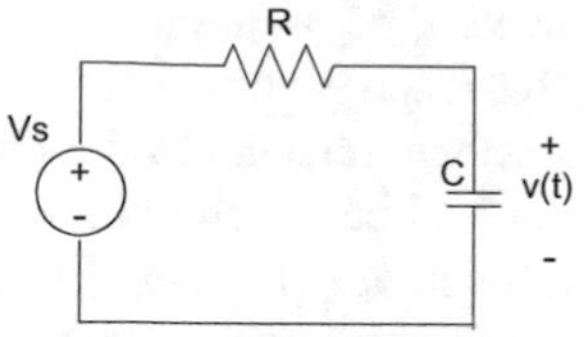

Fig. 1. RC circuit.

of the constant voltage source, *v(t)* is the voltage across the capacitor. The value of time constant indicates how fast the voltage across the capacitor becomes stable. Large time constant means the value of voltage would go up slowly and it takes long time to reach its final value *Vs*. The time response of the voltage across the capacitor in Eq. (2) is shown in Fig. 2.

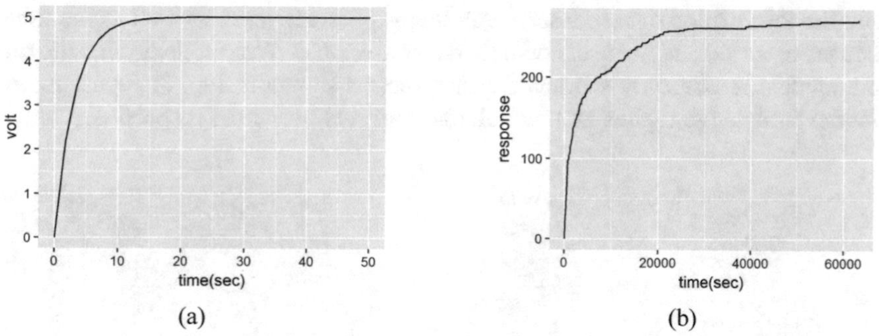

Fig. 2. Time response of the voltage and the twitter responses.

Since voltage across a capacitor corresponds with the charge in a capacitor as $v(t) = q(t)/C$, Eq. (2) becomes

$$q(t) = CVs\left(1 - e^{-\frac{t}{T}}\right) \tag{3}$$

where q(t) is the value of charge in the capacitor at time t, and C is the value of the capacitor. When considering the temporal response of a twitter post, we can see the similarity of the temporal pattern between the time responses in Fig. 2a and the cumulative temporal responses of twitter post in Fig. 2b.

Charge in the capacitor can be analogous to responses in tweeter post. Based on this idea, we propose the use of capacitor charging model to study temporal response of a tweet as follow.

$$r_i(t) = M_i\left(1 - e^{-\frac{t}{T_i}}\right) \tag{4}$$

where T_i is time constant of tweet i, M_i is the total number of people responding to tweet i, and $r_i(t)$ is the cumulative response of tweet i at time t. Similarly, time constant here indicates how fast the responders respond to the post. Let $r_i(t)$ be the cumulative response of the twitter post i and $r_j(t)$ be the cumulative response of the tweet j. When $T_i > T_j$, $r_i(t)$ would reach M_i faster than $r_j(t)$ reaches M_j, regardless of the values of M_i and M_j. Fitting the cumulative response profile with the model in Eq. (4), we get two parameters M_i and T_i which show the number of total responders and the response time respectively.

3 Framework for Analyzing Twitter Channel Using Capacitor Charging Model

Tweets from each channel can be collected in real-time fashion using streaming API connection available in several platforms such as R and Python. For each tweet we obtain the information of responses, including replies, retweets, as well as the time the responses occurred and the users who made the response. We can determine the profile of cumulative response over time of each tweet as shown in Fig. 3. Figure 3a and b show the cumulative replies and cumulative retweets over time respectively.

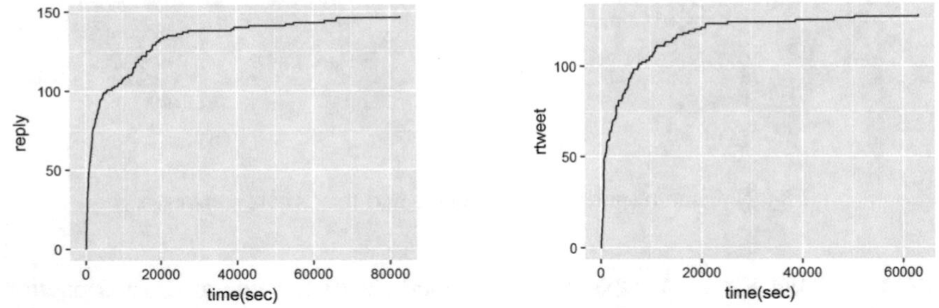

(a) Reply response (b) Retweet response

Fig. 3. Cumulative temporal responses of replies and retweets.

Fitting these data using the proposed model in Eq. (4) for reply and retweet responses separately for each tweet i, we obtain four parameters of each tweet i from the model namely: Mrpl_i, Trpl_i, Mrtw_i, and Trtw_i where Mrpl_i is the total number of people replying to tweet i, Trpl_i is the time constant of the reply of tweet i, Mrtw_i is the total number of people retweeting to tweet i and Trtw_i is the time constant of the retweet of tweet i.

The obtained parameters can be used to evaluate the performance of each tweet as well as the performance of the channel in several ways.

3.1 Analysis of Tweets in Mrpl, Trpl, Mrtw, and Trtw-Spaced

The four parameters can be used as features of four-dimensional feature-space for analyzing the characteristic of the tweets. For ease of computation, all four parameters are normalized to have value between 0 and 1. Figure 4a and b show the plots of 417 points representing tweets from CNN channel in Mrpl-Trpl axes and Mrtw-Trtw axes respectively.

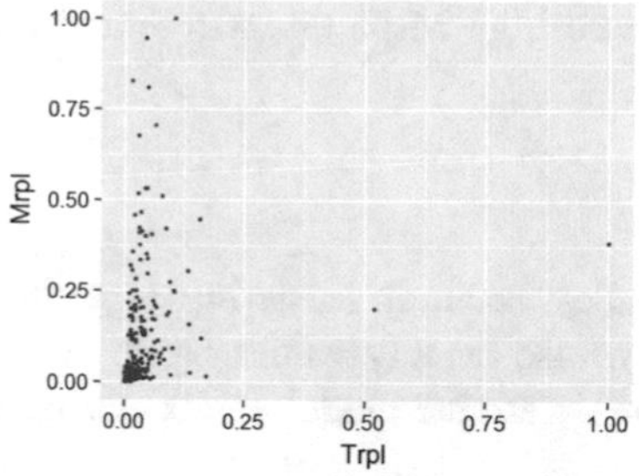

(a) Data points representing reply responses to twitter posts on Mrpl and Trpl.

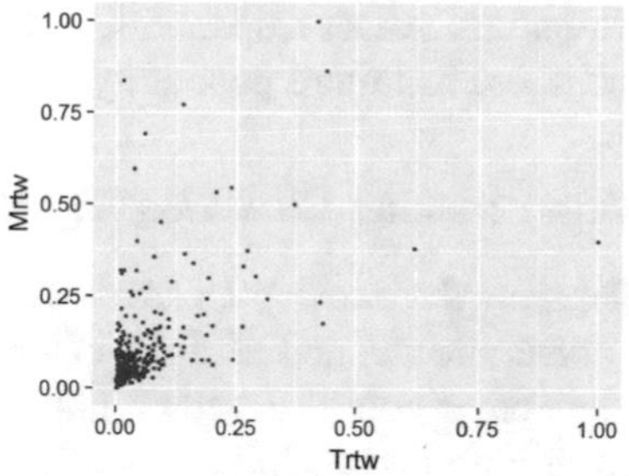

(b) Data points representing retweet responses to twitter posts on Mrtw and Trtw.

Fig. 4. Data points representing cumulative responses to each twitter post.

We can see that most data points flock near the origin where the values of M and T are relatively low, while twitter posts become sparser with larger M and T. We can roughly divide the feature space into three sections as shown in Fig. 5. Section A is the area where most points occur. Section B is the area with large M, but small T, and finally Section C where T is large but M is small. Points that occur in these three sections represent twitter posts with different characteristic. Points in Section A indicate twitter posts with modal performance, i.e., how many responders most of the twitter posts can garner and how fast those responders respond to most of the tweets. Those in Section B indicate highly popular tweets, i.e., tweets that can attract a lot of responders in short period of time. On the contrary, those in Section C represent tweets with low popularity that can draw small amount of responders using longer period of time.

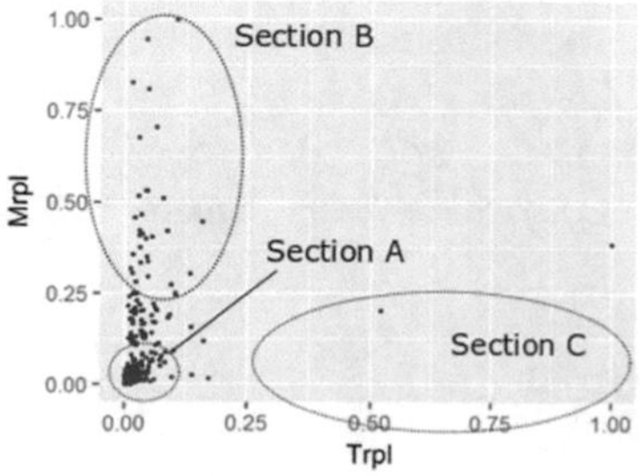

Fig. 5. Dividing the feature space into three sections based on the characteristic of the temporal responses.

Based on this idea, we propose the evaluation of a twitter channel by analyzing twitter posts on the feature space. We determine the most populated area denoted as Section A by using a clustering algorithm which will be described in the next section. From twitter points in Section A, we can determine the modal performance of the twitter channel, i.e., the average values of $Mrpl_i$, $Trpl_i$, $Mrtw_i$, and $Trtw_i$ of the points in Section A where $Mrpl_i$ and $Mrtw_i$ reflect how many responders most of the tweets of the channel can garner while $Trpl_i$ and $Trtw_i$ can indicate how fast the responders respond to most of the tweets as well as the life time of the tweets, i.e., how long most of the tweet can attract responders. The channel with high $Mrpl_i$ and $Mrtw_i$ and low $Trpl_i$ and $Trtw_i$ has more popularity.

3.2 Valley Seeking Clustering Algorithm

We determine the most populated area using valley seeking clustering algorithm. Modal points are defined as a set of data points whose density values are relatively higher than other points. Based on this idea, we can determine modal points clustering points with high density values as follow.

First, pair-wise distances are calculated. Then, density of each point is calculated using the following equation:

$$D(x_i) = \sum_j e^{\left(-\frac{d_{ij}}{\sigma}\right)^2} \tag{5}$$

where σ is determined from the average of the fourth nearest neighbor distance, d_{ij} is the distance between point i and point j. Then, valley seeking clustering is performed as follow.

Let $NN_{10}(x)$ be the set of ten nearest neighbors of x, and p is the peak with highest density obtained from $p = \max_i D(x_i)$. Let *seed* be a FIFO queue and *clust* be a set of the modal points.

AlgorithmValley seeking clustering algorithm for determining modal points
Input: $\{x_i\}$: a set of all points
Output:*clust*: a set of modal points

```
seed<- p
whileseed is not empty do
 pop sk from seed head
 for each xi in NN10(sk)
      If D(xi)>0.13*D(p)
           append xi to seed tail
      end if
 end for
 clust<- clust U {sk}
end while
```

Algorithm 1. Valley seeking clustering algorithm for determining modal points.

Starting from the peak p, we grow the cluster by including its ten nearest neighbors whose density is greater than the threshold value 0.13 * D(p). Modeling the cluster using normal distribution, we set the threshold at 13% of the peak in Line 6, which is equivalent to including points within $\pm 2\sigma$ or 95% of the point in the cluster. Final clustered points are stored in set *clust*.

4 Experimental Results

We use the proposed framework to examine the performance of five international news tweeter channels, namely, CNN (@CNN), BBC (@BBC), Al Jazeera (@AJEnglish), XinHua news (@XHNews), and Fox news (@FoxNews). The number of followers of each channel is shown in Table 1. The Data set is collected from streaming API connection. For this preliminary study, we collect the stream of each channel for 2 days.

Table 1. Number of followers of each twitter channel

News channel	Number of followers (Millions)
CNN	38.8
BBC	22.1
Fox news	16.7
XinHua news	11.6
Al Jazeera	4.67

4.1 Average Performance

Collecting tweets in the same time frame from five twitter channels, we identify the four parameters of the capacitor charging model, i.e., Mrtw, Trtw, Mrpl, and Trpl. The results are shown in Table 2.

Table 2. The average performances of the five twitter channels in term of temporal responses

News channel	Mrtw	Trtw (sec)	Mrpl	Trpl (sec)
Al Jazeera	125	4945	18	3360
BBC	175	3252	52	3202
CNN	243	7570	118	2724
Fox news	243	7600	231	3228
XinHua news	47	2501	7	2661

Collecting tweets in the same time frame from five twitter channels, we identify the four parameters of the capacitor charging model, i.e., Mrtw, Trtw, Mrpl, and Trpl. The results are shown in Table 2.

From Table 2, we find that CNN and Fox news have similar amount of active audience, users who either make a reply or retweet, despite the fact that CNN has much more followers than Fox news does. This result confirms our concern that the number of followers might not be an accurate indicator that can reflect the number of active audience. Similarly, despite less number of followers, Al Jazeera has more responses than XinHua news does in terms of both retweets and replies.

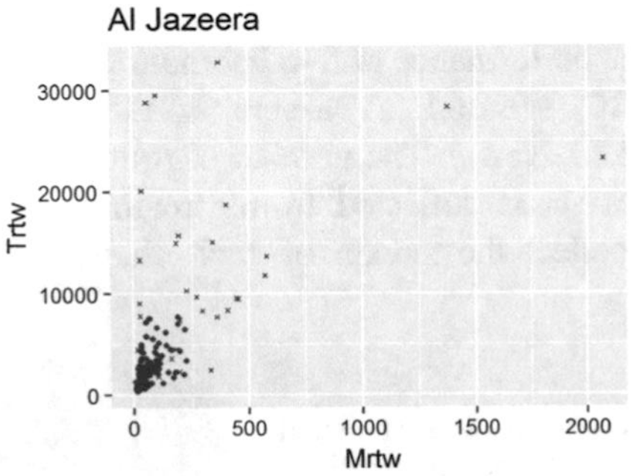

(a) Modal points of Al Jazeera channel.

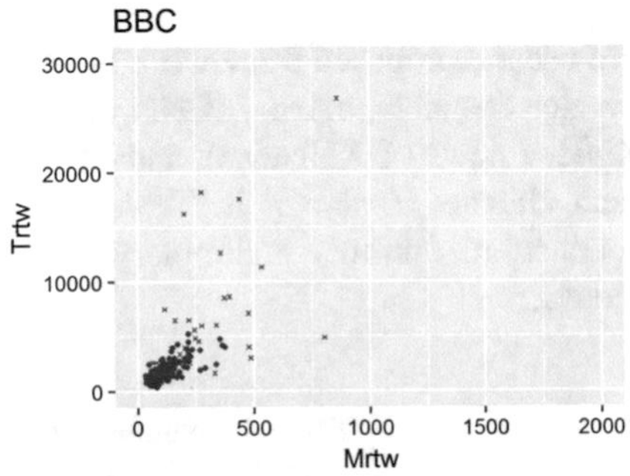

(b) Modal points of BBC channel.

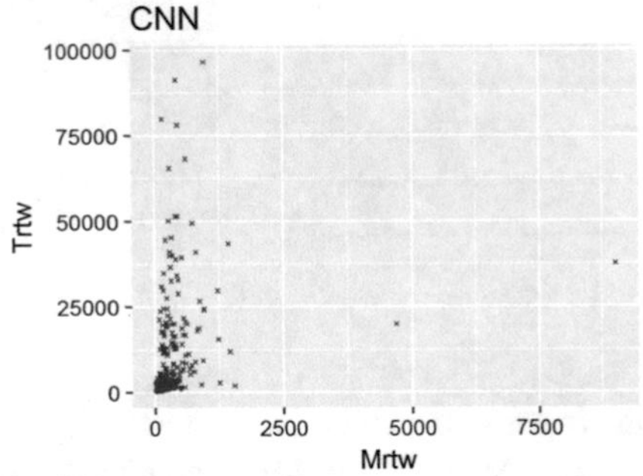

(c) Modal points of CNN channel.

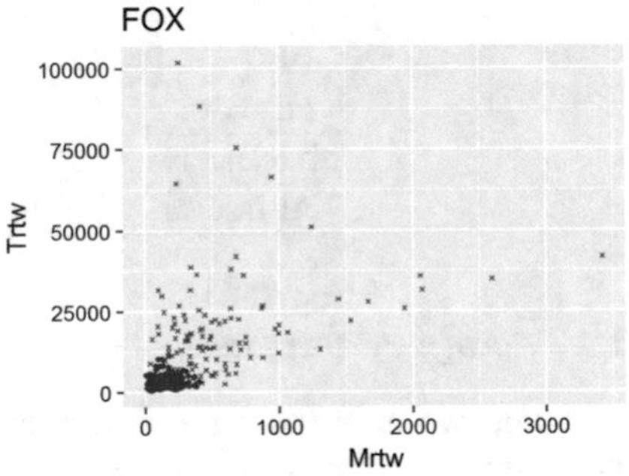

(d) Modal points of FOX channel.

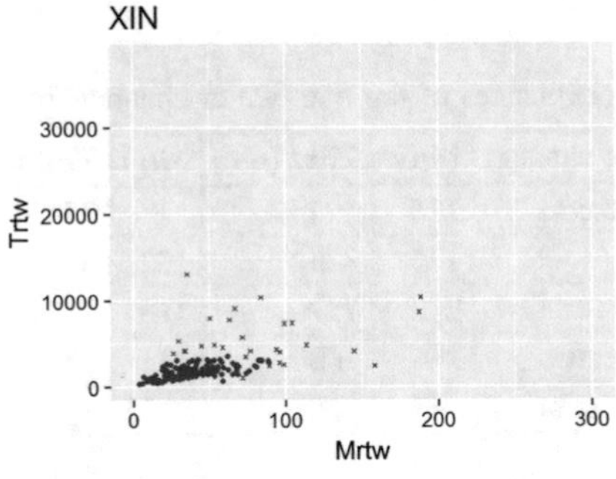

(e) Modal points of XinHua news channel.

Fig. 6. Clustering results from five twitter channels. (a) Al jazeera, (b) BBC, (c) CNN, (d) FOX news, (e) XinHua news. Blue points indicate modal points.

4.2 Modal Performance

In analyzing modal performance of each twitter channel, all four parameters, namely Trtw, Mrtw, Trpl, and Mrpl are normalized to have values between 0 and 1. Then, the clustering algorithm described in Algorithm 1 is applied to cumulative responses from each twitter channel. The clustering results are shown in Fig. 6a–e which are projected on two features, namely Trtw and Mrtw. The blue points indicate the clustered points which have high density that will be used to determine modal performance, while red points denote data points with less density.

Table 3 shows the average modal performances of the five channels. We can see that the modal performance of BBC and XinHua news are quite closed to those of the average performance in Table 2, indicating that most tweets from these two channels have similar performances. Also, we can see from Fig. 6b and e, most twitter posts from these two channels are labeled in blue (155 out of 204 total points in case of BBC and 150 out of 179 points in case of XinHua news), indicating that for these two channels most twitter posts can attract similar amount of audience.

Table 3. Average modal performance of the twitter channels

News channel	Mrtw	Trtw (sec)	Mrpl	Trpl (sec)
Al Jazeera	67	2857	10	1711
BBC	118	1826	23	1945
CNN	117	1842	59	1616
Fox news	117	2792	101	1607
XinHua news	39	1694	6	2214

Figure 7a and b show the comparative temporal responses of the five channels using parameters obtained from average modal performance. In term of retweets, we can see that both BBC and CNN channels have the fastest responses from engaging users who retweet the posts. In case of Fox news, despite the similar amount of retweets users, the users of Fox news response much slower.

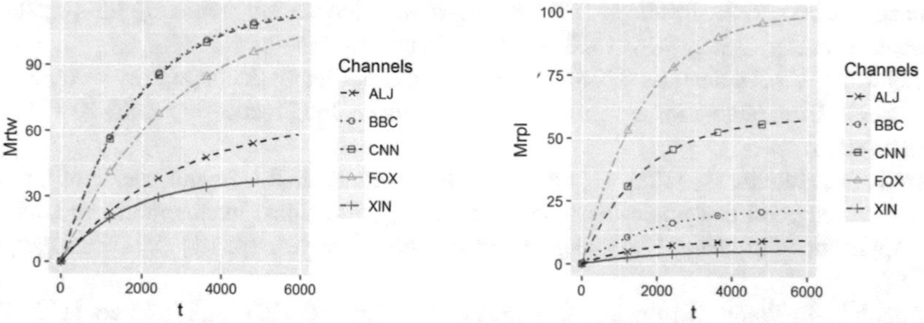

(a) Retweet responses (b) Reply responses

Fig. 7. The comparative responses of the five channels using the temporal parameters obtained from average modal performance

However, when considering reply responses, Fox news followers generate fastest and highest responses. Contrary to other four channels whose retweet responses are much greater than reply responses, the number of reply responses is almost as equal to that of retweet responses. These results can show the unique characteristic of Fox users that can be used for future analysis.

5 Conclusions

In this study, we propose the use of capacitor charging model to analyze temporal pattern of tweeter response as well as evaluate tweeter channels in term of the number of active audience who either retweet or reply to a given tweet post. The proposed model reveals several temporal characteristics that have not been addressed in previous study, including the modal performance of the twitter channel in terms of active audience and the temporal characteristic of the responses. The proposed framework presents alternative way to evaluate the popularity of the channel by considering the temporal responses of the active audience rather than simply use the number of followers.

In future study, we will use this framework to analyze the characteristic of twitter posts in the feature space of the temporal response which can allow us to determine what kind of twitter posts can draw more attention from users. With this information, we can get more understanding of the characteristic of different twitter channels and the preference of their users.

References

1. Hong, L., Dan, O., Davison, B.D.: Predicting popular messages in twitter. In: Proceedings of the 20th International Conference Companion on World Wide Web, pp. 57–58. ACM (2011)
2. Riquelme, F., González-Cantergiani, P.: Measuring user influence on Twitter: a survey. Inf. Process. Manag. **52**(5), 949–975 (2016)
3. Suh, B., Hong, L., Pirolli, P., Chi, E.H.: Want to be retweeted? Large scale analytics on factors impacting retweet in twitter network. In: 2010 IEEE Second International Conference on Social Computing (SocialCom), pp. 177–184. IEEE (2010)
4. Zaman, T., Fox, E.B., Bradlow, E.T.: A Bayesian approach for predicting the popularity of tweets. Ann. Appl. Stat. **8**(3), 1583–1611 (2014)
5. Maleewong, K.: An analysis of influential users for predicting the popularity of news tweets. In: Pacific Rim International Conference on Artificial Intelligence, pp. 306–318. Springer, Cham (2016)
6. Gorrab, A., Kboubi, F., Jaffal, A., Le Grand, B., Ghezala, H.B.: Twitter user profiling model based on temporal analysis of hashtags and social interactions. In: International Conference on Applications of Natural Language to Information Systems, pp. 124–130. Springer, Cham (2017)
7. Stringhini, G., Wang, G., Egele, M., Kruegel, C., Vigna, G., Zheng, H., Zhao, B.Y.: Follow the green: growth and dynamics in twitter follower markets. In: Proceedings of the 2013 Conference on Internet Measurement Conference, pp. 163–176. ACM (2013)
8. Hayt, W., Kemmerly, J., Durbin, S.: Engineering Circuit Analysis, 8th edn. McGraw-Hill Education, New York (2011)

On Evolving Text Centroids

Herwig Unger and Mario Kubek(✉)

Chair of Communication Networks, FernUniversität in Hagen, Hagen, Germany
{herwig.unger,mario.kubek}@fernuni-hagen.de
https://www.fernuni-hagen.de/kn/

Abstract. Centroid terms are single words that semantically and topically characterise text documents and thus can act as their very compact representation in automatic text processing tasks. In this paper, a novel brain-inspired approach is presented to first simplify the determination of centroid terms and second to generalise the underlying concept at the same time. As the precision of the centroid-based text representation is improved by this means as well, new applications for centroids are derived, too. Experimental results obtained confirm the validity of this new approach.

Keywords: Text centroid · Co-occurrence graph · Centroid trace
Text categorisation

1 Motivation and Definitions

From the literature, a plentitude of approaches to compare and categorise text documents are known [1,2]. Regarding their exactness and computational costs they differ significantly. The major disadvantage of most of them is not to utilise the general knowledge a human reader may have. Furthermore, these methods are usually not able to process and consider documents as sequences of words, whereby different sequences may significantly determine a text's contents and meaning — as well as quality. To solve these problems, the concept of text centroids as defined in [3] will be extended to refine the capabilities of text categorisation. To this end, some fundamentals are needed and shall be summarised shortly.

Let X be the set of pairwise different terms as found in a text corpus, and E the set of tuples (x_i, x_j) whose two terms x_i, x_j appear together in a sentence or paragraph of the text corpus. The connected subgraph G of the graph (X, E) is then called *co-occurrence graph* of the corpus. The number of a tuple's (x_i, x_j) co-occurrences in the corpus is denoted by $sig(x_i, x_j)$. It indicates a weight of the corresponding edge in G, whereas its reciprocal $d(x_i, x_j) = 1/sig(x_i, x_j)$ is used to define a distance between the terms x_i, x_j [3].

Given a text document D in form of an ordered set of not necessarily different words $w_1, w_2 ..., w_n$ with all $w_i \in X$. In [3], for D a centroid χ was defined as its node with the lowest average distance to any other node reachable in the

H. Unger et al. (Eds.): IC2IT 2018, AISC 769, pp. 75–82, 2019.
https://doi.org/10.1007/978-3-319-93692-5_8

subgraph of G induced by the elements of D. Properties and applications of centroids were discussed in [4–6]. It was found that, due to the discrete character of G, a 'centre of words' may not coincide with an element of D, just as the 'centre of mass' of a multi-body mechanical system does not necessarily coincide with one of the bodies. In order to improve the precision of text representation by centroids, in the sequel their position relative to the co-occurrence graph will be generalised.

This is motivated by the observation that for a simple document with just two co-occurring words the centroid's position would be expected to lie exactly at the centre of the edge connecting them. The original definition from [3] does not allow for this position, but assigns one of the two words to be the centroid. This effect causes imprecisions and misinterpretations in bigger settings.

Furthermore, the approach of [3] (and other methods known from literature) requires a text's complete processing within a single step. This contradicts the way in which humans read a text word by word from its very beginning and incrementally acquire the information contained. This is the reason why text documents D are considered here as ordered sets, and their centroids are constructed by a gradual process along the given orderings.

2 Centroids of Text Documents

2.1 Centroid Positions

As indicated above, generalised positions π of centroids in a graph shall be introduced, which either coincide with a node or lie somewhere on an edge connecting two nodes.

A **position** π in a graph $G = (X, E)$ is defined as a tripel

$$\pi = (x, y, \lambda)$$

with $x, y \in X$, $(x, y) \in E$ and the displacement λ, $0 \leq \lambda < d(x, y)$, of π from x into the direction of y (see also Fig. 1).

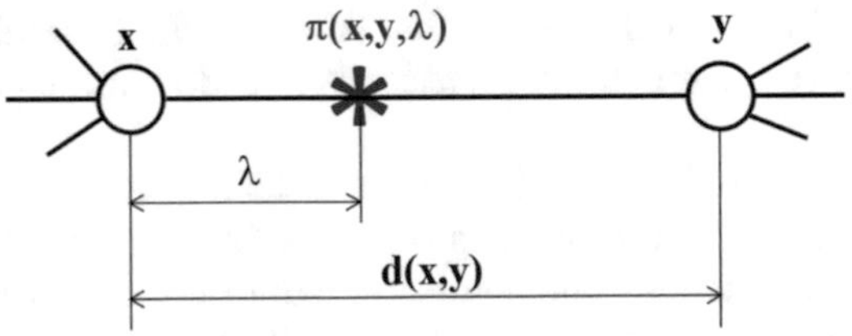

Fig. 1. Position π in a graph

For the special case of $x = y$ and, thus, $\lambda = 0$ the position $\pi = (x, x, 0)$ may be denoted just by x. Note that a position may be described by two different expressions, viz. $\pi = (x, y, \lambda) = (y, x, (d(y, x) - \lambda))$.

Distances $d(\pi_1, \pi_2)$ between any two positions $\pi_1 = (x_1, y_1, \lambda_1)$ and $\pi_2 = (x_2, y_2, \lambda_2)$ may now be defined and calculated according to the following rules.

1. **Both positions coincide with nodes.**
   $d(\pi_1, \pi_2) = d(x_1, x_2)$ as $(x_1 = y_1) \wedge (x_2 = y_2) \wedge (d_1 = d_2 = 0)$
2. **Both positions are located on a common edge, with the displacements oriented in equal directions.**
   $d(\pi_1, \pi_2) = \mid \lambda_1 - \lambda_2 \mid$ as $(x_1 = x_2) \wedge (y_1 = y_2)$
3. **Both positions are located on a common edge, with the displacements oriented in opposite directions.**
   $d(\pi_1, \pi_2) = \mid d(x_1, x_2) - \lambda_1 - \lambda_2 \mid$ as $(x_1 = y_2) \wedge (y_1 = x_2)$
4. **The two positions are located on different edges.**
   Then, the distance is the length of the shortest path between π_1 and π_2 along the $s \geq 1$ intermediate nodes $p_1, p_2, .., p_s \in X$:

$$d(\pi_1, \pi_2) = d(\pi_1, p_1) + \sum_{k=1}^{s-1} d(p_k, p_{k+1}) + d(p_s, \pi_2)$$

With these basic definitions and assumptions, the concept of centroids can now be generalised.

2.2 Evolution of Centroids

A **generalised centroid** ξ_D of a text document D is introduced as the position $\xi_D = (x, y, \lambda)$ in G, which has the minimum average distance to all nodes representing the document's words $w_1, w_2 ... w_n \in D$. It is easy to see that the definition of ξ refines the one of χ, since ξ may not only be situated on a node of G.

Nevertheless, at first glance its seems to be difficult to determine generalised centroids by a fast algorithm as shown, for instance, in [7]. Things turn much easier, however, when an evolutionary approach is employed, i.e. when a document's centroid is calculated in an iterative manner by taking one word $w_i \in D$ after the other into account. Such an **evolution of centroids** $\xi_D^{(i)}$, $i = 1, 2, ..., |D|$, can be carried out by the following algorithm.

1. *Initialisation*
   Set $i := 1$ and $\xi_D^{(1)} := w_1$ (or respectively $(w_1, w_1, 0)$).
2. **REPEAT**
   Consider the next word w_{i+1} of D.
   Calculate the shortest path $\{\xi_D^{(i)}, p_1, p_2, ..., p_s\}$ with $p_1, ..., p_s \in X$ and $p_s = w_{i+1}$ and determine its length $d(\xi_D^{(i)}, w_{i+1})$.
3. Determine $\delta = f(i) \cdot d(\xi_D^{(i)}, w_{i+1})$ with a damping factor f depending on i in order to weigh the influence of the $(i+1)$-st word in relation to the text processed so far.
4. Let $\xi_D^{(i)} = (x_b, x_e, \lambda_i)$.

Search along the shortest path $\{\xi_D^{(i)}, p_1, p_2, ..., p_s\}$ with $s \geq 1$, $p_1, ..., p_s \in X$ and $p_s = w_{i+1}$ for the smallest k where for the first time it holds $d(\xi_D^{(i)}, p_k) \geq \delta$. Execute the alternative applicable:
(a) If $(s = 1) \wedge (x_a = x_b)$, then set $\xi_D^{(i+1)} := (x_a, p_1, \delta)$.
(b) If $(s = 1) \wedge (p_1 = x_b)$, then set $\xi_D^{(i+1)} := (x_b, x_e, (\lambda_i - \delta))$.
(c) If $(s = 1) \wedge (p_1 = x_e)$, then set $\xi_D^{(i+1)} := (x_b, x_e, (\lambda_i + \delta))$.
(d) If $s > 1$, then set $\xi_D^{(i+1)} := (p_{k-1}, p_k, (\delta - d(\xi_D^{(i)}, p_{k-1})))$.

5. Increment $i := i + 1$.
6. **UNTIL** $i > n$

Figure 2 shows the positions of the centroids evolving when reading and processing one word of a text document D after the other in the order given.

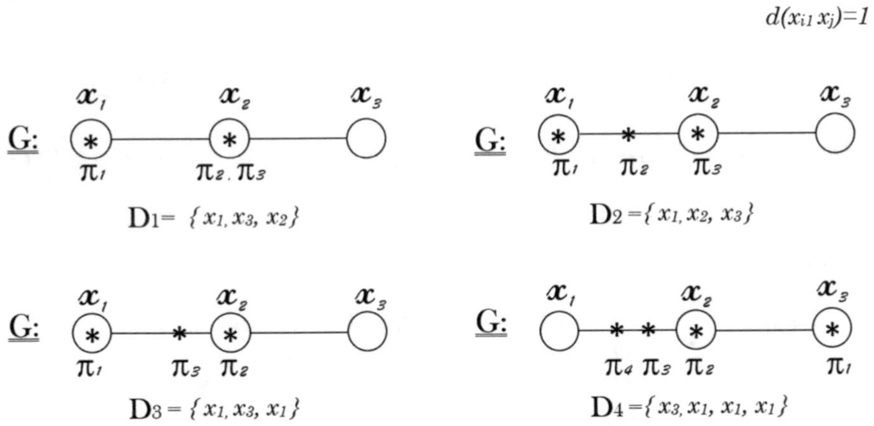

Fig. 2. Positions of centroids π_s resulting from word-by-word processing of documents D_s containing several permutations of three words x_1, x_2, x_3

Therefore, these generalised centroids can be called *evolving centroids* as well. Use and properties of the *evolving centroids* shall now be considered in detail.

2.3 Interpretation of Centroid Traces

Applying the above described algorithm results in a consecutive calculation of a text-representing centroid. While in former publications like [3] only the centroid itself was important, now the consecutive, intermediate states can and shall be considered, too. It is obvious to call

$$\{\xi_D^{(1)}, \xi_D^{(2)}, ..., \xi_D^{(n-1)}\}$$

the **centroid trace of document** D. Each $\xi_D^{(i)}$ on this trace might be interpreted as the centroid of a partially read document, i.e. the set of words $\{w_1, ..., w_i\} \in D$.

It is expected that the length and direction of the intermediate steps from $\xi_D^{(1)}$ to ξ_D are quite randomly distributed, and a selection of this information for a certain number of steps might (possibly) be employed as a fingerprint of the document. Further use cases of centroid traces are discussed in the next section.

3 Experimental Results and Discussion

For all of the exemplary experiments discussed herein, linguistic preprocessing has been applied on the documents to be analysed whereby stop words have been removed and only nouns (in their base form), proper nouns and names have been extracted. In order to build the undirected co-occurrence graph G (as the reference), co-occurrences on sentence level have been extracted. Their significance values have been determined using the Dice coefficient [8]. For the document analysis, 30 English Wikipedia articles[1] from an offline English Wikipedia corpus downloaded from http://www.kiwix.org have been used. The corpus used to create the reference co-occurrence graph G consisted of 100 randomly selected Wikipedia articles (including the mentioned 30 ones) from the same source.

3.1 Centroids of Wikipedia Articles

In order to get a first impression of the generalised centroids' quality in terms of whether they are actual useful representatives of documents, Table 1 presents those centroid terms of the mentioned 30 English Wikipedia articles. The damping factor f depending on the i-th read word of these articles has been set to $\frac{1}{\sqrt{i}}$. Furthermore, this table also lists the previously determined centroid terms [4] of the same documents using the initially described method of computation [7] inspired by the 'centre of mass'. They are referred to as classic centroids. This way, a direct comparison between these types of centroids is possible. In both cases, the articles' 25 most frequent terms have been taken into account for the centroid computation.

It can be seen that almost all centroids properly represent their respective articles. In case of the herein presented generalised centroids, it can be noted that they are in fact related to the classic centroids. Yet, the generalised centroids are more specific (e.g. in case of the article 'Malaria') and meaningful due to the way they have been computed. Unless such a centroid resides directly on a node, there are in each case two distinct terms (connected in G) to characterise it along with the displacement λ. In future contributions, the influence of different damping factors f on the resulting centroids will be investigated. Here, logarithmic or other non-linear functions could be factored in.

[1] Interested researchers can download these articles (1.5 MB) from: http://www.docanalyser.de/ec-articles.zip.

Table 1. Centroids of 30 Wikipedia articles

Title of Wikipedia article	Generalised centroid	Classic centroid term
Art competitions at the Olympic Games	[medal, artist, 3.92]	sculpture
Tay-Sachs disease	[carrier, gene, 6.08]	mutation
Pythagoras	[influence, Pythagoras, 3.16]	Pythagoras
Canberra	[Canberra, capital, 3.17]	Canberra
Eye (cyclone)	[cycle, eyewall, 6.87]	storm
Blade Runner	[Ridley Scott, film, 0.40]	Ridley Scott
CPU cache	[memory, cache, 0.08]	cache miss
Rembrandt	[Amsterdam, Rembrandt, 0.72]	Louvre
Common Unix Printing System	[print, system, 1.73]	filter
Psychology	[psychology, behavior, 0.59]	psychology
Religion	[religion, belief, 1.15]	religion
Universe	[universe, form, 2.03]	shape
Mass media	[media, term, 2.48]	database
Rio de Janeiro	[Rio, zone, 0.50]	sport
Stroke	[stroke, factor, 2.51]	blood
Mark Twain	[fiction, Twain, 1.16]	tale
Ludwig van Beethoven	[Beethoven, life, 2.37]	violin
Oxyrhynchus	[republic, Egypt, 2.47]	papyrus
Fermi paradox	[paradox, civilization, 0.57]	civilization
Milk	[milk, cow, 0.52]	dairy
Corinthian War	[Greece, Sparta, 3.48]	Sparta
Health	[WHO, health, 4.48]	fitness
Tourette syndrome	[tic, child, 0.71]	tic
Agriculture	[crop, production, 5.51]	crop
Finland	[island, Finland, 2.21]	tourism
Malaria	[malaria, parasite, 1.32]	disease
Fiberglass	[fiber, process, 5.04]	fiber
Continent	[continent, Europe, 1.85]	continent
United States Congress	[Senate, member, 1.07]	Senate
Turquoise	[crystal, property, 3.56]	turquoise

3.2 Centroid Traces of Wikipedia Articles

In Table 2, the centroid trace of the article 'Milk' is presented. The trace clearly shows that while reading the text word by word (those words must be among the 25 most frequent terms of the article) the centroid dynamically changes.

Table 2. Centroid trace of article 'Milk'

[[milk, milk, 0.0] → [milk, milk, 0.0] → [milk, glass, 4.48] → [milk, glass, 0.06] → [milk, glass, 0.03] → [milk, glass, 0.02] → [milk, glass, 0.01] → [milk, source, 2.75] → [milk, food, 1.74] → [milk, food, 1.19] → [milk, food, 0.83] → [milk, protein, 0.97] → [milk, protein, 4.00] → [protein, parasite, 1.71] → [protein, milk, 0.13] → [protein, milk, 1.45] → [protein, milk, 0.34] → [protein, milk, 1.54] → [protein, milk, 2.44] → [protein, milk, 3.11] → [protein, milk, 4.62] → [milk, product, 0.69] → [milk, product, 0.55] → [milk, product, 0.43] → [milk, cow, 0.52]]

It is also noteworthy that the centroid is almost always related to the article's main topic but touches several relevant thematic aspects as it evolves. These subtopics are of interest when it comes to suggesting related search words for the expansion of queries in information retrieval. A similar outcome is to be seen in Table 3 for the article 'Fermi paradox'.

Table 3. Centroid trace of article 'Fermi paradox'

[[Fermi, Fermi, 0.0] → [Fermi, paradox, 0.77] → [Fermi, paradox, 0.32] → [Fermi, paradox, 0.71] → [paradox, civilization, 2.36] → [paradox, civilization, 0.95] → [paradox, civilization, 0.59] → [paradox, civilization, 2.42] → [paradox, evidence, 0.18] → [paradox, civilization, 1.69] → [paradox, equation, 1.80] → [equation, planet, 2.54] → [equation, paradox, 1.04] → [equation, paradox, 1.83] → [equation, paradox, 2.40] → [paradox, civilization, 1.05] → [paradox, evidence, 0.50] → [paradox, probe, 0.96] → [paradox, emission, 1.27] → [paradox, emission, 0.74] → [paradox, life, 1.05] → [paradox, life, 0.59] → [paradox, life, 0.47] → [paradox, evidence, 0.72] → [paradox, civilization, 0.57]]

As depicted in these exemplary results, the centroid traces inherently reflect the topical organisation of the respectively analysed texts. This makes it possible to compare two texts regarding their structural and semantic similarity in a new way: The distance of two intermediate centroids indicates the degree of their semantic closeness. If the position of these centroids in the centroid traces is taken into account as well when doing so, a structural similarity of those texts can be detected. For this approach to work, the overlap of their respective sets of words can be low or even empty.

3.3 Discussion

Based on these findings and as suggested before, a document's centroid trace might be used as its (individual) fingerprint in further text processing tasks e.g. as a means for plagiarism detection. Also, a recommender system like DocAnalyser [9] could make use of these traces to find and suggest similar and related web documents. Furthermore, the visualisation of these traces in form of curves along with the used reference co-occurrence graph G will give interested users another possibility to explore and examine the structural and semantic closeness of text documents. Even though two traces might not even share one intermediate centroid, (at least partly) side-by-side running curves and similar curve shapes could suggest an existing text similarity, too. These application scenarios will be investigated in more detail in future contributions.

4 Conclusion

A novel, brain-inspired method for the computation of text-representing centroids has been presented. At the same time, the classic concept of centroid terms has been generalised such that a much more precise determination of centroids is possible. The derived concept of centroid traces allows for their usage in manifold interesting applications in the field of automatic text processing. Foremost, these traces can be applied to determine the degree of semantic similarity of text documents in a new way while considering their structure, too. Future research will investigate and evaluate this approach in greater detail.

References

1. Paaß, G.: Document classification, information retrieval, text and web mining. In: Handbook of Technical Communication, pp. 141–188. Walter de Gruyter (2012)
2. Mihalcea, R., Radev, D.: Graph-based Natural Language Processing and Information Retrieval. Cambridge University Press (2011)
3. Kubek, M., Unger, H.: Centroid terms as text representatives. In: Proceedings of the 2016 ACM Symposium on Document Engineering DocEng 2016, pp. 99–102. ACM, New York (2016)
4. Kubek, M., Unger, H.: Centroid terms and their use in natural language processing. In: Autonomous Systems 2016. Fortschritt-Berichte VDI, Reihe 10 Nr. 848, pp. 167–185. VDI-Verlag, Düsseldorf (2016)
5. Kubek, M., Unger, H.: Towards a librarian of the web. In: Proceedings of the 2nd International Conference on Communication and Information Processing (ICCIP 2016), pp. 70–78. ACM, New York (2016)
6. Kubek, M., Böhme, T., Unger, H.: Empiric experiments with text representing centroids. Lect. Notes Inf Theory **5**(1), 23–28 (2017)
7. Kubek, M., Böhme, T., Unger, H.: Spreading activation: a fast calculation method for text centroids. In: Proceedings of the 3rd International Conference on Communication and Information Processing (ICCIP 2017). ACM, New York (2017)
8. Dice, L.R.: Measures of the amount of ecologic association between species. Ecology **26**(3), 297–302 (1945)
9. Kubek, M.: DocAnalyser - Searching with web documents. In: Autonomous Systems 2014. Fortschritt-Berichte VDI, Reihe 10 Nr. 835, pp. 221–234. VDI-Verlag, Düsseldorf (2014)

Thai Word Safe Segmentation with Bounding Extension for Data Indexing in Search Engine

Achariya Klahan[1(✉)], Sukunya Pannoi[1(✉)], Prapeepat Uewichitrapochana[2(✉)], and Rungrat Wiangsripanawan[1(✉)]

[1] Department of Computer Science, Faculty of Science, King Mongkut's Institute of Technology Ladkrabang, Bangkok, Thailand
achariyakla@gmail.com, pnsukunya@gmail.com, kwrungrat@gmail.com

[2] THINKNET CO., Ltd., Bangkok, Thailand
prapeepat@thinknet.co.th

Abstract. Word segmentation ambiguity in Thai language affects data indexing process by creating the inverted index relatively to the segmentation results. This phenomenon leads to unreasonable search result. This article proposes Thai word Safe segmentation algorithm using dictionary to solve this problem so that all different terms in an ambiguous part of the sentence are queryable. Next, it shows the bounding extension to improve Safe segmentation performance. It also compares several off-the-shelf implementations of the trie data structure which -we believe- is the best data structure for dictionary-based Thai word segmentation and compares the efficiency of serializable libraries for de-serializing trie in the analyzer's initial state. Finally, it evaluates the Safe segmentation with several implementations called Safe Analyzer. The experimental results also show that the linked-list Trie and Protostuff library give the outstanding results. The Safe segmentation can definitely solve the ambiguity problem but still it could not solve the misspell within text accurately.

Keywords: Dictionary-based segmentation · Thai word segmentation
Trie structure · Text inverted indexing · Thai analyzer

1 Introduction

Inverted indexing a document in any languages has their specific way to analyze sentences differently. For Thai language, there are two categories of word segmentation, using dictionary based [1, 2] and non-dictionary based [3, 4]. This article is specifically interested in dictionary-based word segmentation.

The accuracy of the word segmentation module directly affects the outcome of the search engine. Sornlertlamvanich team has shown the ambiguity [3]. For example, the content of document A is abcbcb. By using a word segmentation module, the content is segmented into a|bc|bc|b (assuming that the correct segmentation is a|b|cb|cb). In this case, if the query is cb, it cannot be found in document A or if the query is bc, the document A will be incorrectly returned. Word segmentation ambiguity in Thai

H. Unger et al. (Eds.): IC2IT 2018, AISC 769, pp. 83–92, 2019.
https://doi.org/10.1007/978-3-319-93692-5_9

language, such as ตากลม, which can be read in two forms: ตาก-ลม (tak-lom) and ตา-กลม (ta-klom). So, if the result of word segmentation module is ตาก-ลม, ตา-กลม cannot be found in document. It is noted that unlike other application of word segmentation which aim at the most suitable words, word segmentation in invert indexing aims at all possible words in the ambiguity content.

This article proposes Safe segmentation algorithm to solve the ambiguity problem so that all different terms in an ambiguous part of the sentence are queryable. It also reviews various trie structure using in word segmentation process. Lastly, this article proposes Safe segmentation improvement using bounding extension to reduce the search space in the process.

2 Theories and Data Structures

2.1 Trie Structure

Trie structure [10] or prefix tree is an order tree that stores associative array where the keys are usually strings. Each node stores only one character and descend to the node that has a common prefix of string key as shown in Fig. 1.

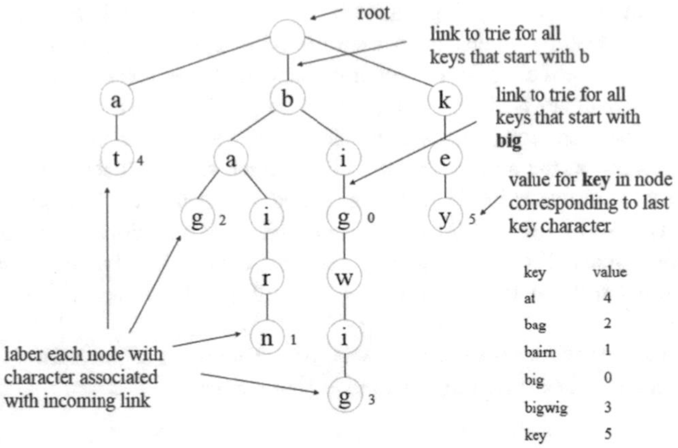

Fig. 1. Trie structure

By using Thai words as a key, trie structure could be used to find a prefix word in the input sentences efficiently.

2.2 Trie Implementation

There are 4 trie implementations in Java. This article names them as TrieA [7], TrieB [8], TrieC [9] and TrieD [10] and categorizes them into 3 groups based on fundamental data structure.

TrieA uses linked-list structure to sort the list in a sequential order. Each character is associated with the next character in a sequence. The referencing between each trie node is shown in Fig. 2.

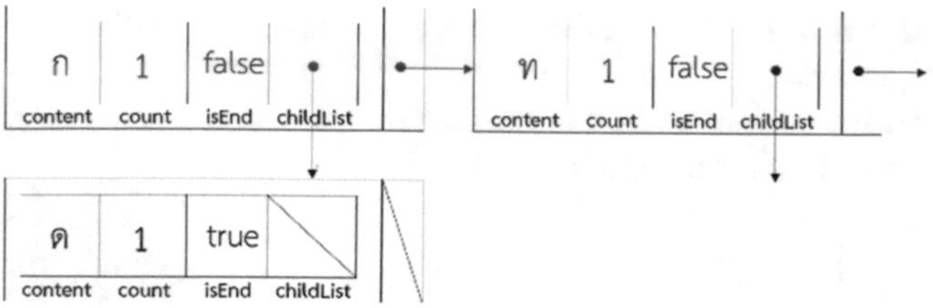

Fig. 2. Trie node using Linked list.

TrieB and TrieC using Hash Map structure stores variables in the form of key and value. This hash table in each trie node provides character look up when finding key. The referencing between each trie node is shown in Fig. 3.

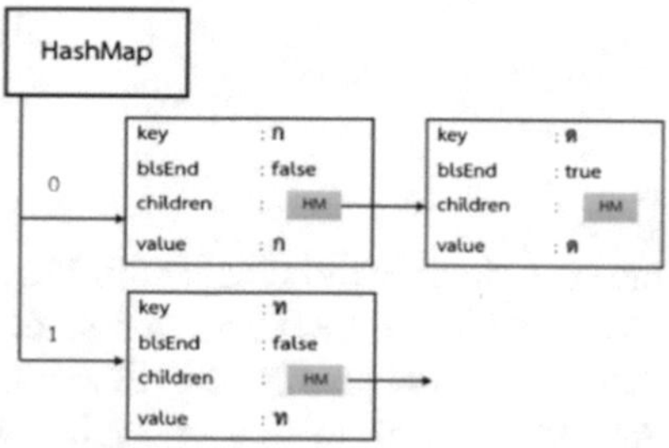

Fig. 3. Trie node using Hash map.

TrieD uses Tree structure to store data associated with the last letter of each word within the node. Its organizes the referencing to each trie node based on the appearance of character while constructing trie into left, mid or right node. Its trie node is shown in Fig. 4.

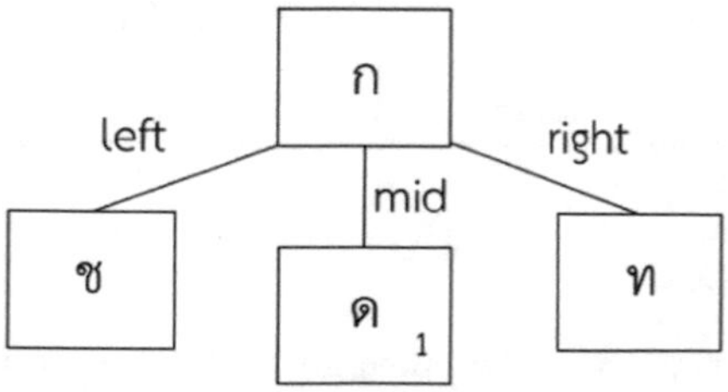

Fig. 4. Trie node using tree

*1 is a value associated with a word to indicate the end of which character.

2.3 Java Serializable Library

Using serializable library given a capable of transforming trie structures that contain dictionary into strings (Serialization). Therefore, word segmentation could use the prepared trie without constructing one when initialization by transforming serialized string into trie structure (De-serialization) for the speed of initializing trie in word segmentation process.

To optimize the de-serialization of the prepared trie, this article chose to measure 8 well-known Java serializable libraries [11].

2.4 Branch and Bound

The branch and bound [12] is an algorithm design paradigm. Starting from large problems and subdividing them, then consider the scope of the answers to the sub-problems, and consider the problem that cannot provide the best answer at the moment. And repeat with every minor problem. Until you find the sub-problem that provides the best answer.

The bounding technique is used to reduce the safe segmentation search space in the redundant segmented phrase in the sentence by using memorized result of the previous traversing branch.

3 Methodology

In Thai language, a sentence is written without separator between each word, which is distinct from English with a clear space between words. In Table 1, the result from longest matching segmentation [12] contains "ป่าดงดิบ" term which means rainforest. But using the longest-matching segmentation in text analyzing will not found this document with the query term "ป่า" which means forest.

Table 1. Sample sentence and wordlist

English sentence	I went to rainforest
Thai sentence	ฉันเดินป่าดงดิบ
By Safe segmentation	ฉัน, เดิน, ป่า, ป่าดงดิบ, ดง, ดงดิบ, ดิบ
By Longest-matching segmentation	ฉัน, เดิน, ป่าดงดิบ

3.1 Thai Word Corpus

Lexitron is an electronic dictionary developed by researchers from the National Electronics and Computer Technology Center (NECTEC) [5].

SWATH dictionary is a popular dictionary used in SWATH program development [6].

3.2 findAllPrefix Function

The off-the-shelf trie structures do not provide a function to find prefixes of a sentence, so findAllPrefix function is implemented as shown in Fig. 5.

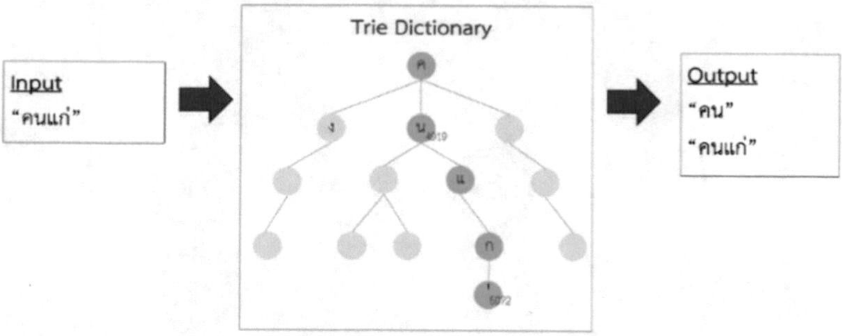

Fig. 5. findAllPrefix function execution with input sentence "คนแก"

3.3 Word Segmentation

Thai word segmentation is complex because the sentence in Thai language has no spaces between each word. The longest-matching segmentation does not find all different terms in ambiguous part of the sentence. One way to solve this ambiguity is to find all possible terms. The word segmentation has to traverse through search space of prefix terms.

Safe segmentation is designed to find the results of all the possibilities in a sentence preventing the missing term of ambiguous words in indexing.

The recurrence relation of the Safe segmentation term is shown in Eq. 1.

$$Segment_{trie}(\Sigma, \Omega) = \bigcup_{P \in prefix_{trie,\Omega}} Segment_{trie}(\Sigma_{append(P)}, \Omega_{substring(P)}) \tag{1}$$

Initial condition $Segment_{trie}(\Sigma, \varnothing) = \Sigma$

And the symbolic description is as shown in Table 2.

Table 2. Equation symbols

Symbol	Explanation
Σ	Sequence of words
Ω	Sentence or article
Prefix	The beginning of all the sentences in the article
Ø	Empty set

The recurrence relation is used to develop Safe segmentation program. By doing so, the implementation of traversing in recurrence relations could be chosen in several ways. For Example, the depth first search as shown at Fig. 6.

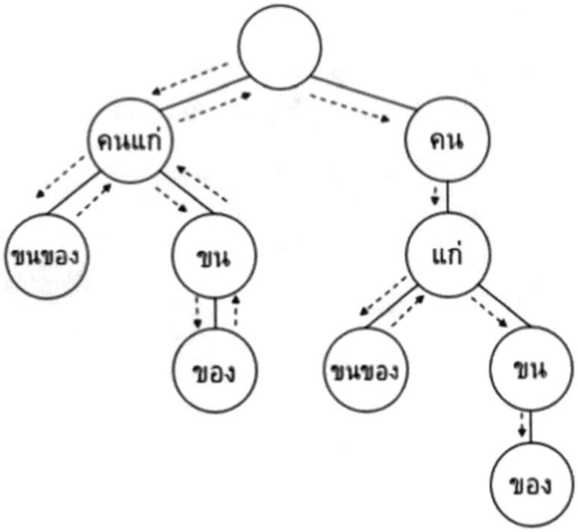

Fig. 6. The depth first search of Safe segmentation with input sentence "คนแก่ขนของ"

In the context of indexing document, the misspell text occurs within the document which could not be solved by Safe segmentation. The Unsafe segmentation had been proposed [13] to solve the case with roughly one segmentation result.

3.4 Bounding

To bound the search space of Safe segmentation, backtracking index is being used to optimize. Backtracking index stores the positions of last character in word when segmenting is possible. In addition, it is used to bound other possibility with the exact terms in the search space as shown at Fig. 7.

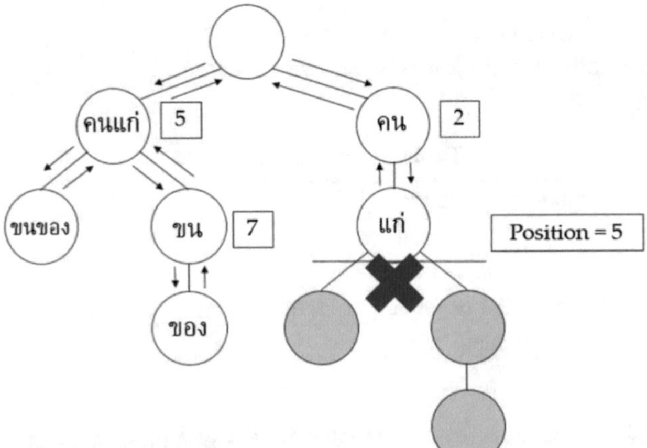

Fig. 7. Safe segmentation with bounding technique with input sentence "คนแก่ขนของ"

3.5 Program Code

The algorithm of safe segmentation is optimized with bounding technique. Its pseudocode is shown below.

```
SafeSegment_Bounding(curID, bt_list, segmTerm, sentence):
  prefixList = findAllPrefix(trie, sentence)
  if curID in bt_list:
    if segmentedSolutionList is not empty:
      add another new solution
    return
  if sentence is segmented completely:
    add the solution to segmentedSolutionList
  else:
    for each prefixTerm in prefixList:
      curID += prefixTerm.length
      segmTerm.add(prefixTerm)
      remSentence = sentence.substring(prefix.length)
      SafeSegment_Bounding(curID, segmTerm, remSentence)
      bt_list.add  (curID)
      segmTerm.remove(prefixTerm)
      curID -= prefixTerm.length
```

3.6 Analyzer

In Object Oriented programming, the module that specifically segments text based on local linguistic logic so, called -Analyzer-. We introduce 3 Safe analyzers which internally use based-data structure trieA with different word segmentation algorithm. The analyzer composition is shown in Table 3.

Table 3. Analyzer component

Analyzer	Word segmentation
ProtostuffSafeDFSAnalyzer	Depth first search
ProtostuffSafeBFSAnalyzer	Breadth first search
ProtostuffSafeBnBAnalyzer	Branch and bound

4 Performance

Time performance measurement uses an artificial form of sentence to show the difference between word segmentation using words in the dictionary.

4.1 Clean-Sentence

Clean-sentence follows the constraints of Safe segmentation tFhat a word must be spelled correctly by randomly generating sentences using words in the dictionary as shown in Fig. 8.

"บัตรห้องสมุดติดิงพิซซ่าฮัทซัลเฟตเพ่งดูร้องรำทำเพลงบ่าวสาว"

"หวันยิหวาผู้จ้างสำเนาสับสนปฏิกิริยาเคมีกุลาตีไม้อินทรชิตมหาชัย"

Fig. 8. Examples of clean-sentence

4.2 Results

The result from different segmentation algorithms can be shown as the example of input sentence "คนแก่ยืนอยู่ใต้ต้นไม้" in Table 4.

Table 4. Word segmentation

<table>
<tr><td>Safe Segment</td><td>คนแก่ | ยืน | อยู่ | ใต้ | ต้นไม้
คนแก่ | ยืน | อยู่ | ใต้ | ต้น | ไม้
คน | แก่ | ยืน | อยู่ | ใต้ | ต้นไม้
คน | แก่ | ยืน | อยู่ | ใต้ | ต้น | ไม้</td></tr>
<tr><td>Unsafe Segment
Branch and bound</td><td>คนแก่ | ยืน | อยู่ | ใต้ | ต้นไม้
คนแก่ [5] | ยืน [8] | อยู่ [12] | ใต้ [15] | ต้นไม้ [21]
คนแก่ [5] | ยืน [8] | อยู่ [12] | ใต้ [15] | ต้น [18] | ไม้ [21]
คน [2] | แก่ [5]</td></tr>
</table>

*[index] is stored backtracking index of position word.

4.3 Performance Measurement

Platform
Performance measurement on the Dell Vostro-3450 Notebook Series
Processor: Intel® Core ™ i3-2310 M CPU @ 2.10 GHz × 4
Memory: DDR3 4 GB 1333 MHz
Operating System: Ubuntu 14.04 LTS 64-bit
Disk: 312.8 GB

Performance Measurement of Trie Structure. Clean sentences have been generated and measured the average execution time of findAllPrefix against each trie. The measurement result is shown in Fig. 9.

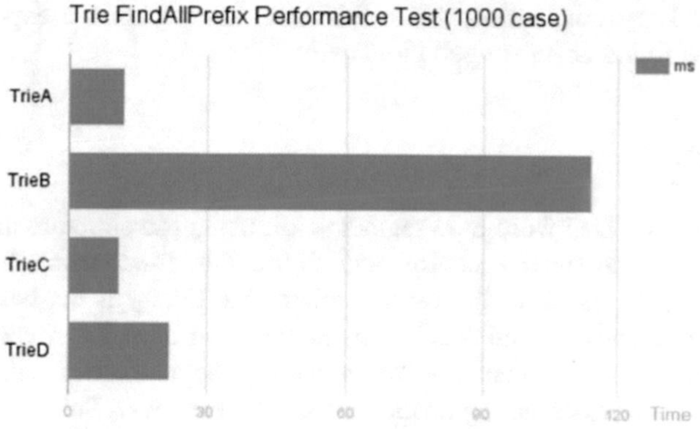

Fig. 9. Performance measurement of findAllPrefix

Performance Measurement of Java Serializable Library. 4 different tries have been constructed with the same word list and serialized them in order to measure the deserializing time of those serialized tries. The measurement graph is shown in Fig. 10.

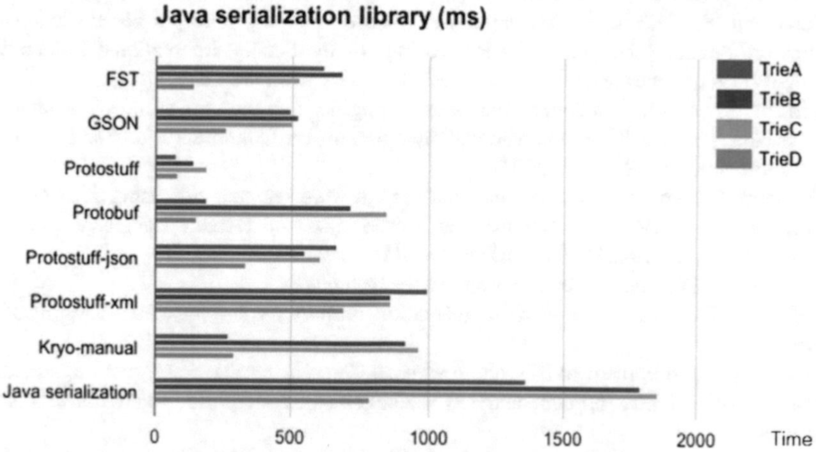

Fig. 10. Performance measurement of Java serializable library

Performance Measurement of Safe Segmentation. By using clean generated sentences, the analyzers have been measured. Table 5 shows the measurement results.

Table 5. Analyzer performance

	ProtostuffSafe DFSAnalyzer	ProtostuffSafe BFSAnalyzer	ProtostuffSafe BnBAnalyzer
Clean-sentence	2196 ms	4533 ms	22 ms

Source Code. The measurement that has been conducted in the experimental environment can be found at https://goo.gl/nKqveP.

5 Summary

The article proposes Thai word Safe segmentation using trie structure to support prefix finding. We have measured each trie and found that trieA using linked list have outperformed the others. And, Protostruff serializable library is the best fit for deserializing trie in the analyzer initialization. Moreover, we have improved Safe segmentation by using bounding technique for reducing the redundant sub-search branch resulting in the speedup of the segmentation process. However, Safe analyzer could not analyze the misspell text in document. In this case, Unsafe analyzer could roughly give one possibility.

References

1. Surapant, M., Paisarn, C., Boonserm, K.: Feature-based Thai word segmentation. In: Proceedings of the Natural Language Processing Pacific Rim Symposium, pp. 41–46 (1997)
2. Tanapong, P., Virach, S., Thatsanee, C.: Towards building a corpus-based dictionary for nonword-boundary languages. In: Proceedings of the Second International Conference on Language Resources and Evaluation (LREC2000), pp. 82–86 (2000)
3. Virach, S., et al.: Dictionary-less search engine for the collaborative database. In: Proceedings of the Third International Symposium on Communications and Information Technologies, pp. 177–182 (2003)
4. Thanaruk, T., Sasiporn, U.: Non-dictionary based word segmentation using decision tree. In: Proceedings of the First International Conference on Human Language Technology Research, pp. 1–5. ACM, Stroudsburg (2001)
5. LEXiTRON Dictionary. http://lexitron.nectec.or.th/2009_1
6. Software: SWATH – Thai word segmentation. http://www.cs.cmu.edu/~paisarn/software.html
7. Java Program to implement trie. http://www.sanfoundry.com/java-program-implement-trie/
8. Trie(treeform) in Java. http://codereview.stackexchange.com/questions/41630/trie-tree-form-in-java
9. Longest prefix matching – A trie based solution in Java, http://www.geeksforgeeks.org/longest-prefix-matching-a-trie-based-solution-in-java/
10. Robert, S., Kevin, W.: Algorithms, 4th edn., Massachusetts (2011)
11. JVM-serializers. https://goo.gl/47dcag
12. A branch and bound algorithm for the Knapsack problem. https://www0.gsb.columbia.edu/mygsb/faculty/research/pubfiles/4407/kolesar_branch_bound.pdf
13. Thai word safe segmentation for data indexing in dictionary-based search engine. http://sub.aucc2017.nu.ac.th/proceeding/pdfFile/DSA/207-615-camera-ready.pdf

Robust Kurtosis Projection Approach for Mangrove Classification

Dyah E. Herwindiati[1(✉)], Janson Hendryli[1], and Sidik Mulyono[2]

[1] Faculty of Information Technology, Tarumanagara University, Jakarta, Indonesia
{dyahh, jansonh}@fti.untar.ac.id

[2] The Indonesian Agency for the Assessment and Application of Technology (BPPT), Central Jakarta, Indonesia
sidik.mulyono@bppt.go.id

Abstract. Mangroves are coastal vegetations that grow at the interface between land and sea. It can be found in tropical and subtropical tidal areas. Mangrove ecosystems have many ecological roles spans from forestry, fisheries, environmental conservation. The Indonesian archipelago is home to a large mangrove population which has enormous ecological value. This paper discusses mangrove land detection in the North Jakarta from Landsat 8 satellite imagery. One of the special characteristics of mangroves that are distinguishing them from another vegetation is their growing location. This characteristic makes mangrove classification using satellite imagery non trivial task. We need an advanced method that can confidently detect the mangrove ecosystem from the satellite images. The objective of this paper is to propose the robust algorithm using projection kurtosis and minimizing vector variance for mangrove land classification. The evaluation classification provides that the proposed algorithm has a good performance.

Keywords: Kurtosis · Mangrove · Multivariate · Projection pursuit
Robust · Satellite imagery · Vector variance

1 Introduction

In this paper, we discuss a robust kurtosis projection minimizing vector variance for mangrove detection. Robust kurtosis projection minimizing vector variance is a procedure which combines two advantages of both kurtosis projection pursuit and robust estimation minimizing vector variance. Projection pursuit is a technique that explores a higher to lower dimensional space by examining the marginal distributions of low dimensional linear projections.

Robust statistics has been developed since almost sixty years ago. Huber [16] introduced the robust estimator because the assumptions of normality, linearity, and independence stucked on the classic estimation are frequently not satisfied. The major goal of robust statistics is to develop the statistics measures that are robust against one or more outliers hidden in the dataset [9]. There are several outlier definitions; the word "outlier" is closed to the word "inconsistent". Hawkins [6] defined an outlier as an

H. Unger et al. (Eds.): IC2IT 2018, AISC 769, pp. 93–103, 2019.
https://doi.org/10.1007/978-3-319-93692-5_10

observation that deviates so much from other observations as to arouse suspicion that it was generated by a different mechanism. Since the 20th century, the robust method has been used by scientists to eliminate the influence of anomalous observations in numerous applications.

The robust kurtosis projection minimizing vector variance method was proposed by Herwindiati et al. [4]. The algorithm works in two stages. In the first stage, it has been proposed the projection approach finding the orthonormal set of all vectors that maximize the kurtosis of the projected standardized data. This approach improves on the slow convergence rate proposed by Friedman [11]. In the second stage, we estimate robust covariance matrix minimizing vector variance to identify the data contamination which is the labeled outlier.

Robust minimum vector variance (MVV) is a measure minimizing vector variance to obtain the robust estimator. Robust MVV was proposed by [3]. The multivariate dispersion vector variance (VV) is a measure of dispersion. Geometrically, VV is a square of the length of the diagonal of a parallelotope generated by all principal components [13]. Robust kurtosis projection has a good performance and robust to detect data contamination in a low, moderate, high, even very high percentage.

Mangroves are coastal vegetations that grow at the interface between land and sea [12] and can be found in tropical and subtropical tidal areas. Mangrove ecosystems have many ecological roles; spans from forestry, fisheries, and environmental conservation [1]. The Indonesian archipelago is home to a large mangrove population which has enormous ecological values.

The input data of this research is derived from Landsat 8 satellite imagery. One of the special characteristics of mangroves that are distinguishing them from another vegetation is their growing location. Detecting mangrove might not be enough by focusing only on NIR, red, and green spectral band from satellite imagery, but also the existence of the water, particularly the sea, around the ecosystem. Because of that fact, existing approaches for detection using vegetation indices cannot effectively detect the mangrove ecosystem. We need an advanced method that is able to confidently detect the mangrove ecosystem from satellite images.

The objective of this paper is to propose a robust algorithm using kurtosis projection and minimizing vector variance for mangrove detection. The case study is the area of coastal and waters of North Jakarta, Indonesia.

2 Case Study - Mangrove

Mangroves are coastal vegetations that grow at the interface between land and sea [12] and can be found in tropical and subtropical tidal areas. Mangrove ecosystems have many ecological roles, spans from forestry, fisheries, and environmental conservation [1]. Nutrient cycling and fish spawning grounds are other services provided by the mangrove ecosystem [5]. The erosion of coastal areas can also be prevented by mangrove forests. Lately, mangrove ecosystems have also been studied as the object of conservation for reducing the effects of the tsunami in Asia [8].

The Indonesian archipelago is home to a large mangrove population which has enormous ecological value. The total area of mangrove forests in Indonesia accounts for 49% of mangroves in Asia, followed by Malaysia (10%) and Myanmar (9%) [2]. In spite of that, it is estimated that the area of mangrove forests in Indonesia has been reduced by about 120,000 hectares from 1980 to 2005, mainly due to the changes in land use for agriculture. In 2007, the ministry of forestry released a report stated that 70% of the total area of the mangrove ecosystem in Indonesia, which is 7,758,410.595 hectares, are damaged. Conservation of mangroves in Indonesia should be prioritized for the benefits of the environment and ecosystems [5].

Some of the mangrove conservation areas in Indonesia are in the northwest of Java, particularly in the North Jakarta, Tangerang, and Bekasi region. Figure 1 shows Google Earth image at North Jakarta, Indonesia. According to the Department of Forestry West Java province, 1,000 hectars of the mangrove forest at Jakarta bay in 1977 have been reduced to only 200 hectares. The mangrove forest at the area faces serious threats from urban development and waste, in addition to the lack of conservation plans from the government.

Fig. 1. The area of mangrove forest at North Jakarta, Indonesia

The near-infrared (NIR), red, and green spectral bands are often used for vegetation detection from satellite images. The normalized difference vegetation index (NDVI), which is the ratio of the difference of the red and NIR spectral band divided by their sum, is widely used in remote sensing studies of vegetation [14]. However, the characteristics of mangrove vegetation might be different, so that some vegetation indices may not be able to detect it well. One of the special characteristics of mangroves is their growing location, that is the coastal areas. Detecting mangrove might not be enough by focusing on NIR, red, and green spectral band, but also the existence of the sea around the ecosystem. Furthermore, the location of mangrove can coincide with another green vegetation, as in Fig. 2, and make it harder for remote sensing methods to correctly detect it.

Fig. 2. The challenge of mangrove detection from satellite imagery

3 The Kurtosis Projection Approach Using Minimum Vector Variance

The kurtosis coefficient is a measure of how peaked or flat the distribution is. The dataset with high kurtosis tends to have heavy tails and sharp density peak near the center point. Kurtosis can be formally defined as the standardized fourth population moment about the mean.

$$K = \frac{E(X-\mu)^4}{\left(E(X-\mu)^2\right)^2} = \frac{\mu_4}{\sigma^4} \tag{1}$$

Projection pursuit [11] is a technique aiming at identifying low dimensional projections of data that reveal interesting structures. The framework of projection pursuit is formulated as an optimization problem with the goal of finding projection axes that minimize or maximize a measure of interest called projection index. Projection pursuit is a technique that explores a high dimensional data by examining the marginal distributions of low dimensional linear projections. The two basic components of projection pursuit are its index and its algorithm [15].

The algorithm of robust kurtosis projection minimizing vector variance (MVV) works in two stages. In the first stage, the projection approach finds the orthonormal set of all vectors that maximize the kurtosis of the projected standardized data. The algorithm of projection is inspired by Pena and Prieto [7]. In the second stage, we estimate robust covariance matrix MVV to estimate spectral characteristic of mangrove

Let $X = (x_1, x_2, \ldots, x_n)'$ be a data matrix of size $n \times p$ as the observation result of p variables to n individual objects. The orthogonal projection of each observation results on to one dimensional space spanned by d is $y_i = d'x_i$ where d is a unit vector in $\mathbb{R}^p$.

As the projected data is written as $Y = (y_1, y_2, \ldots, y_n)$, the kurtosis of the projected data is formulated as in Eq. 2.

$$K = \frac{\frac{1}{n}\sum_{i\in\mathbb{N}_n}(y_i - t)^4}{s^4} \tag{2}$$

where $t = \frac{1}{n}\sum_{i\in\mathbb{N}_n} y_i$ and $s^2 = \frac{1}{n}\sum_{i\in\mathbb{N}_n}(y_i - t)^2$ are the sample mean and the sample of the data Y respectively. It should be noted that s^4 is the square of s^2 and $\mathbb{N}_n$ is the set of all natural numbers less than or equal to n.

Centering and scaling transformation of Y gives a new data $Z = (z_1, z_2, \ldots, z_n)$ where for all $i \in \mathbb{N}_n$, $z_i = \frac{y_i - t}{s}$. Since the sample mean of Z is 0 and the sample varaiance of Z is 1, the kurtosis of Z can be formulated as in Eq. 3.

$$K = \frac{1}{n}\sum_{i\in\mathbb{N}_n} z_i^4 \tag{3}$$

The sample mean t and the covariance matrix S of the data matrix X can be defined as in Eqs. 4 and 5. Then, the kurtosis K can also be written as in Eq. 6.

$$t = \frac{1}{n}\sum_{i\in\mathbb{N}_n} x_i \tag{4}$$

$$S = \frac{1}{n}\sum_{i\in\mathbb{N}_n}(x_i - t)(x_i - t)' \tag{5}$$

$$K = \frac{1}{n}\sum_{i\in\mathbb{N}_n}\left(\frac{d'(x_i - t)}{\sqrt{d'Sd}}\right)^4 \tag{6}$$

Consider the objection function f defined as in Eq. 7. We assume that κ_1 is the eigen value of $\frac{1}{n}\sum_{i\in\mathbb{N}_n}(d'y_i)^2 y_i y_i'$ and d is the corresponding eigenvector. Multiplying Eq. 7 by d' from the left gives us the following result in Eq. 8.

$$f(d) = \frac{1}{n}\sum_{i\in\mathbb{N}_n}(d'y_i)^4 - \lambda(d'd - 1) \tag{7}$$

$$\kappa_1 = d'\left(\frac{1}{n}\sum_{i\in\mathbb{N}_n}(d'y_i)^2 y_i y_i'\right)d = \frac{1}{n}\sum_{i\in\mathbb{N}_n}(d'y_i)^4 = K \tag{8}$$

So the unit vector d that maximizes K is the eigenvector of $\frac{1}{n}\sum_{i\in\mathbb{N}_n}(d'y_i)^2 y_i y_i'$ corresponding to the maximum eigenvalue of that matrix. We will call such eigen vector by d_1. The Fig. 3 illustrates the good performance of kurtosis projection from $X = (x_1, x_2, \ldots, x_n) \in \mathbb{R}^3$ to d_1.

The measures of kurtosis relate to the fourth moment of the data and emphasize the tails of the distribution. This paper use the algorithm of robust projection of maximizing kurtosis which was discussed by [4] for classification of mangrove. The good

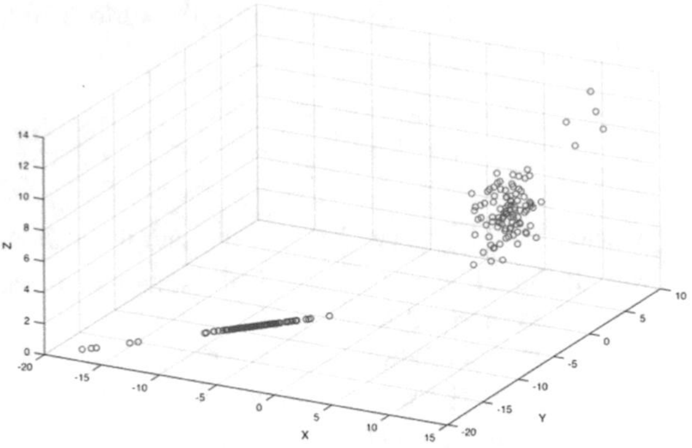

Fig. 3. Illustration of kurtosis projection based on simulation

performance of robust approach will be applied for detection and mapping of mangrove land in Jakarta.

The kurtosis projection approach is used in the initial step of robust kurtosis projection minimizing vector variance. Suppose a matrix data $X = (x_1, x_2, \ldots, x_n)'$ of size $n \times p$, t is the sample mean, and S is the sample covariance matrix of X, the kurtosis projection algorithm can be derived as follows:

1. Standardize the matrix X such that the projected data has mean 0 and variance 1.

$$y_i = S^{-\frac{1}{2}}(x_i - t) \tag{9}$$

2. Find the first unit vector d_1 as the eigenvector of $\frac{1}{n}\sum_{i \in \mathbb{N}_n} (d'y_i)^2 y_i y_i'$ corresponding to the maximum eigenvalue of that matrix.
3. For $k = 2, 3, \ldots, p$, find the k-th unit vector d_k as the eigenvector of $\left(I - \sum_{j=1}^{k-1} d_j d_j'\right) \frac{1}{n}\sum_{i \in \mathbb{N}_n} (d'y_i)^2 y_i y_i'$ corresponding to the maximum eigen value of that matrix. The output is an orthonormal set of all vectors that maximize the kurtosis $d_1, d_2, \ldots, d_p$.
4. Find the projection $z_i = d'y$.

Kurtosis projection provides excellent results for detection of multivariate labeling outlier with very small and small contamination in the dataset, but it fails on a moderate percentage of data contamination. To improve the performance, the algorithm of robust kurtosis projection minimizing vector variance was proposed by [4].

Robust minimum vector variance was proposed by Herwindiati [3] for identifying multivariate outlier labeling. MVV estimator is the robust high breakdown point, i.e. $\frac{n-2(p-1)}{2n}$. Geometrically, vector variance (VV) is a square of length of parallelotope diagonal generated by all principal components of X [14].

Assume a random sample $X_1, X_2, \ldots, X_n$ of a p-variate, the estimator MVV is the pair (T_{MVV}, C_{MVV}) having minimum square of diagonal length of the parallelogram. The robust MVV algorithm was described by [3].

$$T_{MVV} = \frac{1}{n}\sum\nolimits_{i \in H} X_i \tag{10}$$

$$C_{MVV} = \frac{1}{n}\sum\nolimits_{i \in H} (X_i - T_{MVV})(X_i - T_{MVV})^T \tag{11}$$

Robust kurtosis projection minimizing vector variance is an effective method to identify the multivariate outliers in the high dimension. In the simulation experience, robust kurtosis projection using MVV has a breakdown point nearing 0.5. The measure indicates that the robust kurtosis projection method provides results that can be trusted until the half of the data is contaminated [5].

The algorithm of robust kurtosis method can be briefly explained as follows:

1. Find the orthonormal set of all vectors that maximizing kurtosis $\left\{\vec{d}_1, \vec{d}_2, \ldots, \vec{d}_p\right\}$.
2. Compute MVV robust estimators: the location and scale estimator.

The detailed computation for MVV robust estimator can be found in [3–5].

4 The Mangrove Classification

The supervised mangrove classification will be done in two processes: training and testing process. To conduct the training process, the algorithm of robust kurtosis projection is used for the estimation of mangrove and non-mangrove's spectral.

The mangrove and non-mangrove area for the training and testing purposes can be located visually from Google Earth; and by using the Global Mapper and ENVI software, the Landsat images can be cropped according to the coordinates. The data retrieved from the satellite images are the spectral values of each pxiel.

The mangrove vegetation is detected from the Landsat 8 satellite imagery which can be freely downloaded from http://glovis.usgs.gov. For the model training process, we use North Jakarta area which is cropped using the Global Mapper software. Subsequently, 300 points are selected by visual inspection of the area from Google Earth which proportionally represents mangrove, water area, and soil. Meanwhile, for testing and evaluating the trained model, we use the Greater Jakarta area. Therefore, the advantage of our model is that it can learn from a small training dataset size of 300 locations to detect much larger area (more than a million pixels in the testing data). The illustration of the training data can be seen in Fig. 4.

The classification step is conducted by using the spectral reference in the training step. For each satellite image pixels from the training data, our model learns the characteristics of mangrove vegetation, water, and soil.

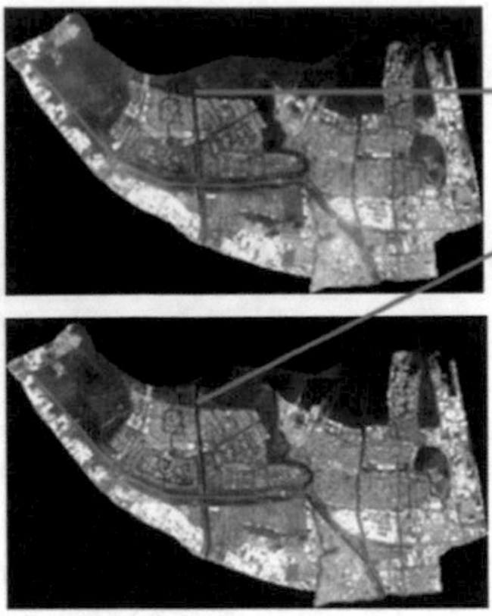

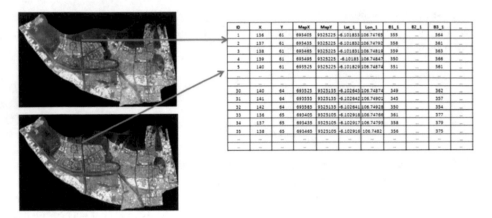

ID	X	Y	MapX	MapY	Lat_1	Lon_1	B1_1	B2_1	B3_1	…
1	136	61	693405	9325225	-6.101833	106.74765	355	…	364	…
2	137	61	693435	9325225	-6.101832	106.74792	358	…	361	…
3	138	61	693465	9325225	-6.101831	106.74819	359	…	363	…
4	139	61	693495	9325225	-6.10183	106.74847	350	…	366	…
5	140	61	693525	9325225	-6.101829	106.74874	351	…	361	…
…	…	…	…	…	…	…	…	…	…	…
…	…	…	…	…	…	…	…	…	…	…
30	140	64	693525	9325135	-6.102643	106.74874	349	…	362	…
31	141	64	693555	9325135	-6.102642	106.74901	345	…	357	…
32	142	64	693583	9325135	-6.102641	106.74928	350	…	354	…
33	136	65	693405	9325105	-6.102918	106.74766	361	…	377	…
34	137	65	693435	9325105	-6.102917	106.74793	358	…	379	…
35	138	65	693465	9325105	-6.102916	106.7482	356	…	375	…
…	…	…	…	…	…	…	…	…	…	…
…	…	…	…	…	…	…	…	…	…	…

Fig. 4. Illustration of cropping data for training process from the Landsat 8 satellite imagery (band 1 and band 3)

The spectral references are very useful for mangrove classification process. Let $W_1, W_2, \ldots, W_n$ be the pixels of the North Jakarta area having p-variates. The mapping of mangrove is conducted by checking the similarity between each pixel W_k, for $k = 1, 2, \ldots, n$, and the spectral references. Each pixel is classified into one of the three classes based on the value of the smallest distance. The pixel of W_k is classified as mangrove if its distance with the mangrove spectral reference is the smallest one. Figure 5 shows the mapping result of mangrove detection in North Jakarta.

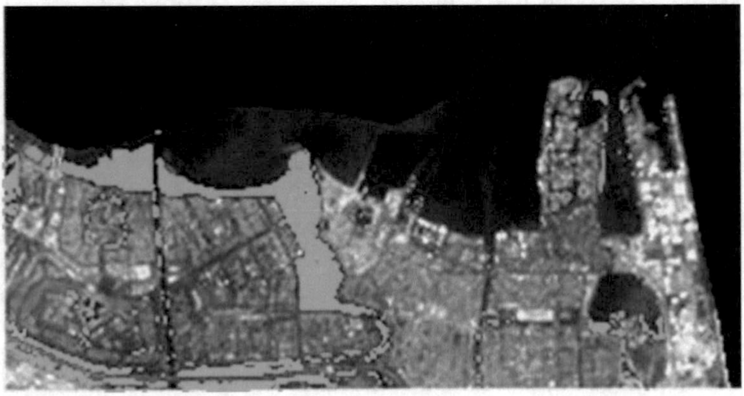

Fig. 5. The mapping of mangrove in the north of Jakarta on 2015

The classification result is evaluated in two experiments. Google Earth imagery and our ground truth are used as the evaluation data. In the first experiment, we choose two areas that are classified as mangrove and non-mangrove. We compare the results of our classification model to the actual land cover. The visualization of this experiments can be seen in Fig. 6.

Note:

Information	Coordinat	The Existing Condition
[1]: Mangrove Land	Lat: -6.102668 Lon: 106.749994	
[2]: Non Mangrove Land	Lat: -6.121816 Lon: 106.790116	

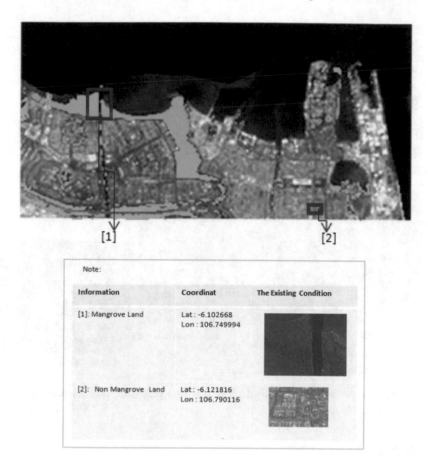

Fig. 6. Evaluation of the first experiments on the performance of mangrove classification using robust kurtosis projection

In the second experiment, we compare two results of classification, the classification based on the robust kurtosis projection and the classification using the classical method. Figure 7 illustrates the comparison. The failure of detection is found in the classical method; the existing mangrove land is detected as non mangrove land at the coordinates (Lat: 6.111449; Long: 106.768597). This tells us that robust approach has better performance than the classical method, see Fig. 7.

The arithmetic mean $\overrightarrow{\overline{X}}$ is often considered as a location estimator in the classical method. The arithmetic mean is the sum of a collection of numbers divided by the number of numbers in the collection. In this research we build the classical distance for classical detection of mangrove land. The distance measures the similarity of an every pixel to the arithmetic mean $\overrightarrow{\overline{X}}$. The classical distance is very sensitive to an anomolous observation or an outlier. The occurrence of one or more outliers shifts the mean vector toward outliers and the covariance matrix becomes to be inflated.

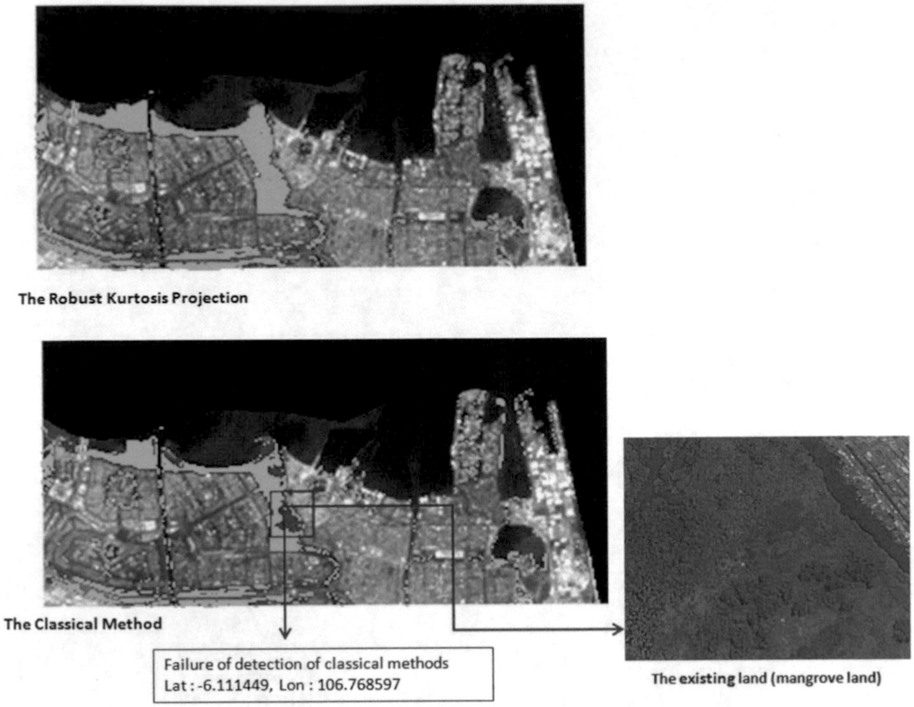

Fig. 7. The performance comparison of robust kurtosis projection performance and classical method for mangrove classification

5 Remark

The experiments and the evaluations prove that the robust kurtosis projection is an effective method that can be considered for mangrove classification. The process of mangrove land detection is not trivial. The mangrove land detection is different from the detection of green land. From the research experience, we need an advanced method that can confidently detect the mangrove ecosystem from satellite imagery. The special characteristics of mangrove is its growing location that can coincide with other green vegetation. This characteristic causes failure in mangrove detection. The empirical results provide strong evidence that the robust kurtosis projection has good performance for mangrove classification.

This study is our preliminary research. Regarding to the study results, we will use the robust kurtosis projection method for our further research; that is the mapping of mangrove land in the Central Java Province Indonesia.

References

1. Vaiphasa, C.: Remote sensing techniques for mangrove mapping. Unpublished thesis, Wageningen University (2006)
2. Center for International Forestry Research. Mangroves: a global treasure under threat. http://blog.cifor.org/31178/indonesian-mangroves-special-fact-file-a-global-treasure-under-threat
3. Herwindiati, D.E., Djauhari, M.A., Mashuri, M.: Robust multivariate outlier labeling. Commun. Stat. Simul. Comput. **36**(6), 1287–1294 (2007)
4. Herwindiati, D.E., Sagara, R., Hendryli, J.: Robust kurtosis projection for multivariate outlier labeling. In: International Conference on Advanced Computer Science and Information Systems (ICACSIS), pp. 97–102. IEEE, Depok (2015)
5. Murdiyarso, D., Purbopuspito, J., Kauffman, J.B., Warren, M.W., Sasmito, S.D., Donato, D. C., Manuri, S., Krisnawati, H., Taberima, S., Kurnianto, S.: The potential of Indonesian mangrove forests for global climate change mitigation. Nat. Clim. Change **5**(12), 1089–1092 (2015)
6. Hawkins, D.M.: The feasible solution algorithm for the minimum covariance determinant estimator in multivariate data. Comput. Stat. Data Anal. **17**(2), 197–210 (1994)
7. Peña, D., Prieto, F.J.: Multivarate outlier detection and robust covariance matrix estimation. Technometrics **43**(3), 286–310 (2001)
8. Danielsen, F., Sørensen, M.K., Olwig, M.F., Selvam, V., Parish, F., Burgess, N.D., Hiraishi, T., Karunagaran, V.M., Rasmussen, M.S., Hansen, L.B.: The Asian Tsunami: a protective role for coastal vegetation. Science **310**(5748), 643 (2005)
9. Hampel, F.R., Ronchetti, E.M., Rousseeuw, P.J., Stahel, W.A.: Linear models: robust estimation. In: Robust Statistics: The Approach Based on Influence Functions, pp. 307–341. Wiley (1986)
10. Colwell, J.E.: Vegetation canopy reflectance. Remote Sens. Environ. **3**(3), 175–183 (1974)
11. Friedman, J.H.: Exploratory projection pursuit. J. Am. Stat. Assoc. **82**(397), 249–266 (1987)
12. Kathiresan, K., Bingham, B.L.: Biology of mangroves and mangrove ecosystems. Adv. Mar. Biol. **40**, 81–251 (2001)
13. Djauhari, M.A.: Improved monitoring of multivariate process variability. J. Qual. Technol. **37**(1), 32–39 (2005)
14. Brown, M.E., Pinzón, J.E., Didan, K., Morisette, J.T., Tucker, C.J.: Evaluation of the consistency of long-term NDVI time series derived from AVHRR, SPOT-vegetation, SEAWIFS, MODIS, and Landsat ETM+ sensors. IEEE Trans. Geosci. Remote Sens. **44**(7), 1787–1793 (2006)
15. Huber, P.J.: Robust Statistics: Wiley Series in Mathematical Statistics. Wiley, New York (1980)
16. Huber, P.J., et al.: Robust estimation of a location parameter. Ann. Math. Stat. **35**(1), 73–101 (1964)
17. U.S.G.S.: Landsat 8 data user handbook. https://landsat.usgs.gov/sites/default/files/documents/Landsat8DataUsersHandbook.pdf

Machine Learning

Diabetes Classification with Fuzzy Genetic Algorithm

Wissanu Thungrut and Naruemon Wattanapongsakorn(✉)

Department of Computer Engineering, King Mongkut's University of Technology Thonburi, 126 Pracha-utid Rd., Bangmod, Thoongkru, Bangkok, Thailand
thungrut@gmail.com, naruemon.wat@kmutt.ac.th

Abstract. In this research work, we consider the diabetes classification on PIMA Indian dataset with Fuzzy Genetic Algorithm. We applied two algorithms consisting of Fuzzy Algorithm and Genetic Algorithm to combine the process to enhance the classification performance. In addition, we used Synthetic Minority Over-sampling Technique (SMOTE) to handle the imbalance data set. We conducted the experiments and found out that 5-fold cross-validation is the suitable approach, providing very good results compared with those obtained from other techniques.

Keywords: Fuzzy Genetic Algorithm · Diabetic · SMOTE · BaggingLOF Imputation technique

1 Introduction

Nowadays, diabetes is known as a difficult disease to completely cure. The symptom can be varied in many stages of severity. The human body is a mysterious thing although medical profession cannot explain everything in our body. Many factors of human body affect the internal system to the owner such as blood sugar, fat, insulin, blood pressure, pedigree and other, that cause diabetes. However, the prime factor comes from insulin. When considering deep into the human body, the pancreas produces insulin to assimilate sugar through the body. However, a problem occurs when the insulin cannot be produced at a regular level and cause digestion to deteriorate. Other factors that may affect to diabetes are exercise activities, age, consumption as a sub-factor.

However, several types of researches have proposed various methods such as rule-based classification systems that widely used for diabetes diagnosis. A challenge is to producing a comprehensive optimal rule-set while balancing accuracy, sensitivity and specificity values. Most researches attempt to improve or get the actual useful model with accuracy. However, a problem is not easy to solve from different dataset because of the differences of nationality or patient case. Diabetes has three different levels and corresponding factors. The level of type 1 consists of a ratio of waist/hip, fat accumulation, overweight, high blood pressure. Type 2 or pre-diabetes consists of the

H. Unger et al. (Eds.): IC2IT 2018, AISC 769, pp. 107–114, 2019.
https://doi.org/10.1007/978-3-319-93692-5_11

problem of consuming sugar and abnormal insulin produce, and type 3 is the last stage of diabetes consisting of complications in several diseases. We need to classify a useful rule-set if those datasets are reliable.

In 2002, Chawla et al. [1] studied the performance of Naive Bayes, C4.5 Decision Tree and Ripper Rule using Synthetic Minority Over-sampling Technique (SMOTE) to handle imbalance data set. Their valuation methods are Receiver Operating Characteristic (ROC) and Area Under the Curve (AUC). The results gave better performance, compared to those with over-sampling and under-sampling.

Handling missing value is a strategy to solve misleading classification. Christobel and Prakasam [2] did research on the negative impact of missing value imputation. They found that the mean substitution, data scaling and K-nearest neighbor (KNN) algorithm can enhance the accuracy of their works.

In 2013, Wattanapongsakorn et al. [3] studied the performance of online intrusion using Fuzzy Genetic Algorithm to create a rule-set with a trapezoidal membership function. They tested on various experiments including offline testing with the KDD99 dataset and online testing in their university environment. In 2015, Guan et al. [4] studied the performance of their experiments in big data process with vector similarity based local outlier factor (SLOF) and bagging feature method to improve the precision and recall rates. Their report showed the comparison of outlier detection methods. The winner is SLOF that is suitable for big data process.

Then in 2016, Behera and Rani [5] studied about the most effective outlier detection method in breast cancer data by comparing several methods such as Density-based spatial clustering of applications with noise (DBSCAN). Ordering points to identify the clustering structure (OPTICS) and outlier detection method. The result showed that LOF is the best for detecting noise with high accuracy.

In addition, Palwisut [6] studied the improvement of the decision tree using SMOTE for Internet Addiction Disorder Data. SMOTE was considered to test on classification, with Decision Tree, Random Forest. The evaluation method were applied with confusion matrix to compare all models. The results give improvement +5.24% accuracy, +13.82% sensitivity, and +8.47% specificity.

Barale and Shirke [7] considered using deleting missing value together with KNN technique to classify data.

A hybrid model of Fuzzy ARTMAP and Genetic Algorithm for data classification and Rule Extraction [8], Pourpanah et al. studied the Fuzzy ARTMAP with Genetic Algorithm (QFAM-GA) to compare the performance using different dataset with noise and noise free.

Next, Brodinová et al. [9] studied the imbalanced clustering (IClust) in high dimension media data by merging multi-group based on outlier detection method using local outlier factor (LOF).

Interpretable and accurate medical data classification – a multi-objective genetic-fuzzy optimization approach [10], Gorzałczany and Rudzinski showed the Multi-Objective Genetic Algorithm (MOEA) with Fuzzy Algorithm to produce a rule and test on several datasets. The result got satisfied.

SM-RuleMiner: Spider monkey based rule miner using novel fitness function for diabetes classification [11], Cheruku, Edla and Kuppili studied the performance of SM-RuleMiner against PIMA Indian dataset. The accuracy and sensitivity tested with 10 folds cross-validation provided good performance but with low specificity. The result is satisfied in term of sensitivity analysis.

In this paper, we classify the diabetes into 2 classes: "yes" (with the diabetes), and "no" (without the diabetes). A Fuzzy Genetic Algorithm (Fuzzy GA) has recently been proposed to solve many problems with satisfying results [4, 5, 10]. The technique can help generating rules to classify disease when the medical data is available.

The remaining parts in this paper is presented as follows. In Sect. 2, we discuss research methodology, dataset, data preprocessing and fuzzy genetic algorithm. The algorithm is applied to classify the PIMA Indian dataset. Section 3 presents the classification result. Finally, in Sect. 4, we give a conclusion of this research work.

2 Introduction

2.1 Dataset

We consider the Pima Indian dataset from UCI Machine Learning Repository [12] as our input dataset. This dataset has 9 attributes and 768 instances. The description of all attributes is shown in Table 1.

Table 1. Pima Indian dataset.

No	Attribute name	Mean	Standard deviation	Min…Max
1	Number of time pregnant	3.8	3.4	0…17
2	Plasma glucose concentration 2 h	120.9	32.0	0…199
3	Diastolic blood pressure (mm Hg)	69.1	19.4	0…122
4	Triceps skin fold (mm)	20.5	16.0	0…99
5	2-h serum insulin (mu U/ml)	79.8	115.2	0…846
6	Body mass index (height in m) ^ 2)	32.0	7.9	0…67.1
7	Diabetes pedigree function	0.5	0.3	0.078…2.42
8	Age (years)	33.2	11.8	21…81
9	Diastolic blood pressure (mm Hg)	–	–	–

2.2 Data Preprocessing

The major problem with this dataset is the missing value. We studied several methods on missing value imputation [8, 10, 12], and found out that the best approach for this problem solving is by using mean imputation in each class. The original dataset contains two output classes: yes (with diabetes) and no (without diabetes). The ratio of the two classes is 0.53:1 as shown below (Fig. 1).

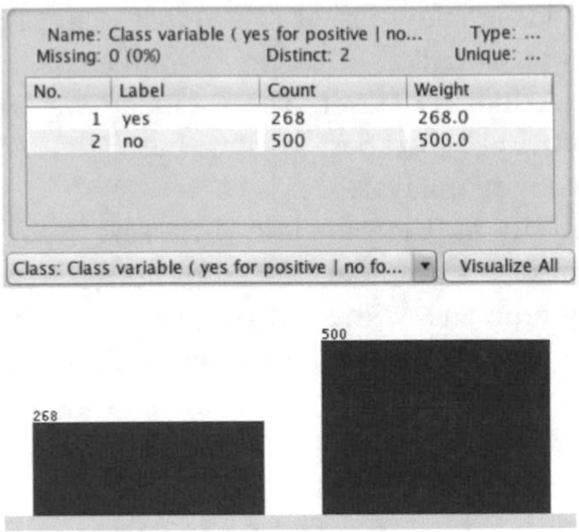

Fig. 1. The original dataset.

This original dataset is imbalance. We used SMOTE to deal with this problem of imbalance datasets giving satisfied outcomes. SMOTE tuning is necessary to synthesis density-base population of the instance group. However, the synthesis data may contain some outliers. So, we used BaggingLOF to identify the outliers which is directly improved accuracy of the classification. Our approach is explained in detail as follows.

Mean Imputation

We used the mean imputation to deal with missing value. For the implementation, we split data into 2 sections, for class "yes" and class "no". Then calculate mean value of each attribute of those two classes and replace the missing value with its attribute mean value. This method can reduce the error from data replacement better than when using other methods such as global mean imputation, K-mean imputation, EM Imputation and median imputation.

SMOTE

We used SMOTE [7] in Weka tool to generate sampling. The parameter setting of percentage create data = 86% and nearest neighbors = 15% of all instances. This method performs a synthesis sampling from the neighborhood of dataset. The new sampling is then generated.

BaggingLOF

We used Weka tool to generate the probability of BaggingLOF [2, 3, 6]. The threshold of defined outlier is a probability of mean value from the overall dataset. In the original version of this algorithm, there is no explanation of the threshold of noise detection. We need to test it on our own for the best result. The outlier can be occurred when performing SMOTE. Thus, the outlier detection method is needed to clean the data.

2.3 Fuzzy Genetic Algorithm

We considered using a Fuzzy Genetic algorithm (FuzzyGA) with a trapezoidal membership function based on Wattanapongsakorn [3]. In this work, the Fuzzy GA can classify the diabetic output class effectively.

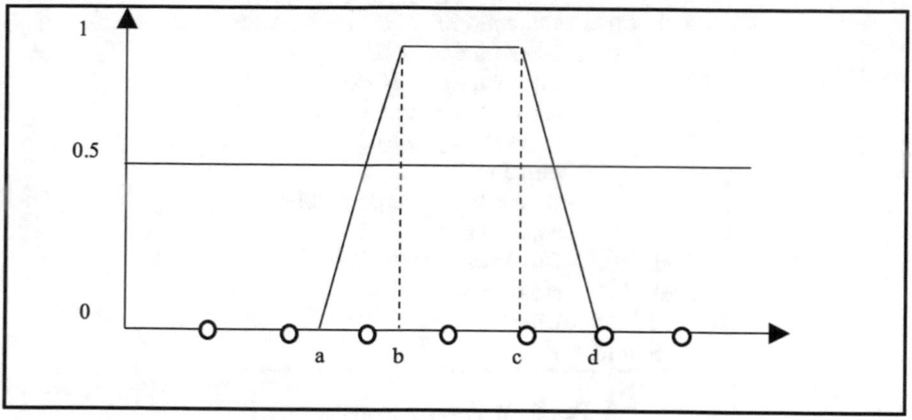

Fig. 2. A trapezoidal membership function.

From Fig. 2, the goal of this work is to find accuracy along with specificity and sensitivity. We used maximum fitness function to increase the correct output as "yes" and reduce incorrect output as *n*. The fitness function is described as follows.

$$fitness = \frac{y}{Y} - \frac{n}{N} \tag{1}$$

where
Y is the total number of "yes" records in the dataset.
N is the total number of "no" records in the dataset.
"y" is the total number of "yes" records correctly classified as "yes".
"n" is the total number of "yes" records incorrectly classified as "no".

2.4 Fuzzy Algorithm

This paper considered the trapezoidal membership function working with the genetic algorithm. The trapezoidal membership function offers a flexible shape/rule depending on the value of adjusting parameters to identify/predict the output class. According to Fig. 2, the probability to determine "yes" and "no" outputs are described with a trapezoidal membership function. The trapezoidal scale is assigned as the probability in each attribute considered for a rule. The parameters consist of a, b, c and d. The interval of each individual is defined as 1.0–7.0 for the range in this work. We briefly explain the concept of the methodology as shown in Fig. 3.

The shape of trapezoidal membership can be changed depending on the values of a, b, c, d. The graph shown in Fig. 2 has x-axis presenting value of an attribute and y-axis for the probability of having "yes" for the output class. The forms of a graph should be correctly sorting as a $\leq$ b $\leq$ c $\leq$ d. Then the probability of being "yes" is calculated according to the pseudocode shown in Fig. 4.

From Fig. 3, the fuzzy algorithm in the fifth line is a trapezoidal membership function that is related as a function in Fig. 4. In addition, Fig. 4, the gene is a subtle of the chromosome in a genetic algorithm that we will discuss next.

```
threshold = manualsetting
for each record i in A do
    for each rule j in B do
        for each attribute k in C do
            prob = fuzzy()
        end for
        If prob >= threshold Then
        class = yes
        Else prob < threshold
        class = no
    end for
end for
```

Fig. 3. A Fuzzy Algorithm

```
INPUT : number_attribute = 8
for each gene i in A do
    IF data value between b and c THEN
    prob = 1.0
    Else IF data value between a and b THEN
    prob = (value − a)/(b − a)
    Else IF data value between c and d THEN
    prob = (d − value)/(d − c)
    Else
    prob = 0.0
end for
for each gene i in A do
    totalprob + = prob[i]
end for
totalprob + = totalprob/number_attribute
```

Fig. 4. A pseudocode—trapezoidal membership function.

2.5 Genetic Algorithm (GA)

A chromosome contains N genes and each gene has its corresponding a, b, c, d for determining a rule for an attribute value (Fig. 5).

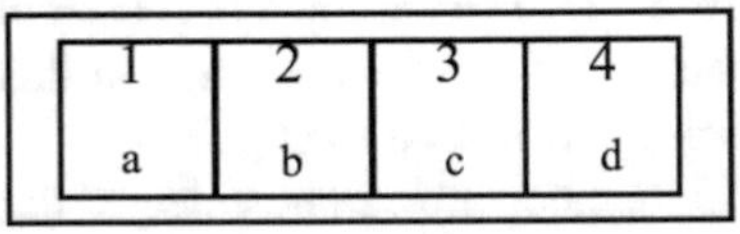

Fig. 5. One gene with 4 blocks refers to one attribute with a range of 1–7.

The length of the chromosome is equal to the total number of attributes/genes. Total number of blocks in the chromosome is equal the number of attributes multiply by four. We considered an integer encoding for a simple implement. The GA is used to find appropriate values of a, b, c and d for each attribute, forming a rule for classification, which maximizes the accuracy of prediction. We used GA to generate various solutions as rules for finding the best classification rule. The GA parameter setting is as follows.

- Crossover rate = 90%
- Mutation rate = 80%
- Population size (number of solutions considered at the same time) = 100
- Number of Generations/iterations = 300

GA has the main process consisting of reproduction and mutation. We used single-point crossover method. For mutation, we randomly change a bit value in the range of 1 to 7. For parent selection method, the tournament selection is chosen to perform a good ratio of choice to select the candidate to create their offspring.

3 Classification Result

The test is on Pima Indian Dataset. We designed the evaluation criteria using 5-fold cross-validation. We tested on a personal computer with CPU 2.9 GHz Intel i5 and 8 GB RAM. The results are reported in Tables 2, 3 and 4, with 87.40% of accuracy, 86.82% of sensitivity and 88% of specificity, respectively. The accuracy describes the overall correctness of classification as disease and no disease.

Table 2. Considering SMOTE and 5-fold cross-validation test

Dataset	Measure		
	Accuracy (%)	Sensitivity (%)	Specificity (%)
Original dataset	75.91%	62.31%	83.03%
Dataset with SMOTE	**83.26%**	**84.33%**	**82.19%**

Table 3. Considering SMOTE and BaggingLOF with 5-fold cross-validation test

	Accuracy (%)	Sensitivity (%)	Specificity (%)
SMOTE	83.26%	84.33%	82.19%
SMOTE+BaggingLOF	**85.82%**	**84.71%**	**86.82%**

Table 4. Detailed results of the 5-fold cross-validation

Fold	TP rate (%)	TN rate (%)	FP rate (%)	FN rate (%)	Accuracy (%)	Sensitivity (%)	Specificity (%)
1	91.23%	88.57%	11.42%	8.77%	89.76%	91.23%	88.57%
2	89.55%	86.67%	13.33%	10.44%	88.19%	89.55%	86.67%
3	78.13%	92.06%	7.93%	21.87%	85.04%	78.13%	92.06%
4	89.47%	88.57%	11.42%	10.52%	88.98%	89.47%	88.57%
5	85.71%	84.51%	15.49%	14.28%	85.04%	85.71%	84.51%
Average	86.82%	88.08%	11.91%	13.17%	87.40%	86.82%	88.08%

We compared our results using FuzzyGA with those from QFAM-GA [9], MOEA-GA [11] and SM-RuleMiner [12]. The best accuracy obtained from the QFAM-GA, MOEA-GA and SM-RuleMiner were 90.35%, 81.5%, and 89.87% respectively. However, when testing with noise in the dataset, the QFAM-GA with noise gave 85.61% accuracy which is lower than from our FuzzyGA combing with noise. In addition, the results from our research work are superior and balance both in terms of sensitivity and specificity.

4 Conclusion and Future Work

This paper applied FuzzyGA to classify diabetes. Data preprocessing is required to take care of missing value and imbalance dataset. Classifying or predicting patients with critical disease requires high accuracy method with high sensitivity and specificity. Our Fuzzy GA gave superior performance for this problem solving. We plan to apply this algorithm to other research problems in the near future.

References

1. Chawla, N.V., Bowyer, K.W., Hall, L.O., Kegelmeyer, W.P.: SMOTE: Synthetic Minority Over-sampling Technique. J. Artif. Intell. Res. **16**, 321–357 (2002)
2. Christobel, A., Prakasam, S.: The negative impact of missing value imputation in classification of diabetes dataset and solution for improvement. IOSR J. Comput. Eng. **7**(4), 16–23 (2012)
3. Wattanapongsakorn, N., Jongsuebsuk, P., Charnsripinyo, C.: Real-time intrusion detection with fuzzy genetic algorithm. In: 10th International Conference on Electrical Engineering/Electronics, Computer, Telecommunications and Information Technology, art no. 6559603 (2013)
4. Guan, H., Li, Q., Yan, Z., Wei, W.: SLOF: identify density-based local outliers in big data. In: 12th International Conference on Web Information System and Application, art no. 7396608, pp. 61–66 (2015)
5. Behera, S., Rani, R.: Comparative analysis of density based outlier detection techniques on breast cancer data using hadoop and map reduce. In: International Conference on Inventive Computation Technologies, art no. 7824883 (2016)
6. Palwisut, P.: Improving decision tree technique in imbalanced data sets using SMOTE for internet addiction disorder data. Inf. Technol. J. **12**, 54–63 (2016)
7. Barale, M.S., Shirke, D.T.: Cascaded modeling for PIMA Indian diabetes data. Int. J. Comput. Appl. **139**(11), 1–4 (2016)
8. Pourpanah, F., Peng Lim, C., Saleh, J.M.: A hybrid model of Fuzzy ARTMAP and Genetic Algorithm for data classification and Rule Extraction. Expert Syst. Appl. **49**, 74–85 (2016)
9. Brodinová, S., Zaharieva, M., Filzmoser, P., Ortner, T., Breiteneder, C.: Clustering of imbalanced high-dimensional media data. Adv. Data Anal. Classif. 1–24 (2017)
10. Gorzałczany, M.B., Rudzinski, F.: Interpretable and accurate medical data classification – a multi-objective genetic-fuzzy optimization approach. Expert Syst. Appl. **71**, 26–39 (2017)
11. Cheruku, R., Edla, D.R., Kuppili, V.: SM-Rule Miner: spider monkey based rule miner using novel fitness function for diabetes classification. Comput. Biol. Med. **81**, 79–92 (2017)
12. Sigillito, V.: Machine learning repository (UCI). https://archive.ics.uci.edu/ml/datasets/pima+indians+diabetes

Decision Fusion Using Fuzzy Dempster-Shafer Theory

Somnuek Surathong[1,2], Sansanee Auephanwiriyakul[1,4](✉), and Nipon Theera-Umpon[3,4]

[1] Department of Computer Engineering, Faculty of Engineering, Chiang Mai University, Chiang Mai, Thailand
Somnuek_s@cmu.ac.th, sansanee@ieee.org
[2] Graduate School, Chiang Mai University, Chiang Mai, Thailand
[3] Department of Electrical Engineering, Faculty of Engineering, Chiang Mai University, Chiang Mai, Thailand
nipon@ieee.org
[4] Biomedical Engineering Institute, Chiang Mai University, Chiang Mai, Thailand

Abstract. One of the popular tools in decision making is a decision fusion since there might be several sources that provide decisions for one task. The Dempster's rule of combination is one of the decision fusion methods used frequently in many research areas. However, there are so many uncertainties in classifier output. Hence, we propose a fuzzy Dempster's rule of combination (FDST) where we fuzzify the discounted basic probability assignment and compute the fuzzy combination. We also have a rejection criterion for any sample with higher belief in both classes, not only one of the classes. We run the experiment with 2 classifiers, i.e., support vector machine (SVM) and radial basis function (RBF). We test our algorithm on 5 data sets from the UCI machine learning repository and SAR images on three military vehicle types. We compare our fusion result with that from the regular Dempster's rule of combination (DST). All of our results are comparable or better than those from the DST.

Keywords: Fuzzy number · Dempster-Shafer Theory
Dempster's rule of combination · Fuzzy Dempster's rule of combination
Fuzzy belief · Fuzzy basic probability assignment

1 Introduction

Results from several classifiers can be combined using a decision fusion tool in order to improve the performance of the decision system because of the coherence or contradiction among them. The Dempster's rule of combination in the Dempster-Shafer Evidence theory has been used in several decision fusion methods [1–12]. This method considers each classification output as a degree of evidence or basic probability assignment (bpa), then the method combines the bpas from 2 sources. There are some other research works believing that the source might not be certain. Hence, these works focus on the fuzzy focal element in the Dempster's rule of combination [13–16].

H. Unger et al. (Eds.): IC2IT 2018, AISC 769, pp. 115–125, 2019.
https://doi.org/10.1007/978-3-319-93692-5_12

However, sometimes the uncertainty does not occur at the source but on the decision resulting from the feature collection process or a classifier performance itself. Also, it is normal that classifiers provide incorrect classification results, hence, each classifier will have an associated credibility. Hence in this paper, we propose a fuzzy Dempster's rule of combination (FDST) with credibility. In particular we calculate the bpa from each classifier's output and adding the uncertainty into the number. We also assign a credibility for each classifier based on its performance. Then the FDST will combine all the fuzzy decisions to produce the final decision.

2 Background

In Dempster-Shafer Theory or Evidence theory [17–19], the belief measure can be interpreted as the degree of belief that the element x belongs to subset A. It is described as

$$\mathrm{Bel}: \mathrm{P}(\mathrm{X}) \rightarrow [0, 1] \tag{1}$$

such that $\mathrm{Bel}(\varnothing) = 0$ and $\mathrm{Bel}(\mathrm{X}) = 1$, and

$$\begin{aligned} \mathrm{Bel}(\mathrm{A}_1 \cup \mathrm{A}_2 \cup \ldots \cup \mathrm{A}_n) \geq & \sum_j \mathrm{Bel}(\mathrm{A}_j) - \sum_{j<k} \mathrm{Bel}(\mathrm{A}_j \cap \mathrm{A}_k) \\ & + \ldots + (-1)^{n+1} \mathrm{Bel}(\mathrm{A}_1 \cap \mathrm{A}_2 \cap \ldots \cap \mathrm{A}_n) \end{aligned} \tag{2}$$

where X is the universal set or frame of discernment. The belief measure can be characterized by a basic probability assignment (bpa). The function is defined as

$$m : 2^{\mathrm{X}} \rightarrow [0, 1] \tag{3}$$

such that

$$m(\varnothing) = 0, \tag{4}$$

$$\sum_{\mathrm{A} \subseteq \mathrm{X}} m(\mathrm{A}) = 1 \tag{5}$$

The value $m(\mathrm{A})$ normally represents the proportion of relevant evidence supporting that a particular element of X belonging to the set A not any subset of A. Any set $\mathrm{A} \in \mathrm{P}(\mathrm{X})$ with $m > 0$ is called a focal element of m. The belief measure can be expressed as

$$\mathrm{Bel}(\mathrm{A}) = \sum_{\mathrm{B}|\mathrm{B} \subset \mathrm{A}} m(\mathrm{B}) \tag{6}$$

This means that the degree of belief that a particular element is in set A depending on the bpas assigned to all the subsets of A and including A.

If there are several sources producing an evidence, the Dempster's rule of combination is used to produce a combined bpa as follows:

$$m_{1,2}(\mathrm{C}) = \frac{\sum\limits_{\mathrm{A}\cap\mathrm{B}=\mathbf{C}} m_1(\mathrm{A})m_2(\mathrm{B})}{1-K} \quad (7)$$

where

$$K = \sum_{\mathrm{A}\cap\mathrm{B}=\phi} m_1(\mathrm{A})m_2(\mathrm{B}) \quad (8)$$

m_1(A) and m_2(B) are the degrees of evidence (or bpa) from the first source focusing on set A and the second source focusing on set B, respectively. Both sets A and B are subset of X.

In our case, bpa is no longer a number but a fuzzy number. The definition of fuzzy number is as follows. A fuzzy number $\boldsymbol{A}$ is a normal convex fuzzy set defined on the real line, $\Re$ [19]. The support of $\boldsymbol{A}$ is bounded in $\Re$. Suppose that $\boldsymbol{A}$ and $\boldsymbol{B}$ are two fuzzy numbers. Let the symbol $\otimes$ represent any of the algebraic operations +, −, * or /. According to the extension principle [20], a fuzzy number $\boldsymbol{Z}$ is produced by the algebraic operation that maps fuzzy sets in $\Im(\Re)$ (the fuzzy power set of $\Re$) to a fuzzy subset in $\Im(\Re)$. The result membership function is

$$\mu_{\boldsymbol{Z}}(y) = \sup_{x_1\otimes x_2=y} \min(\mu_{\boldsymbol{A}}(x_1), \mu_{\boldsymbol{B}}(x_2)) \quad (9)$$

where x_1 and x_2 satisfy the mapping constraint. If we discretize the continuous support into a finite number of points and approximate the fuzzy set on the discrete domain, the resultant fuzzy set is often irregular and inaccurate compared with the exact result. Therefore, the operation is done based on interval arithmetic [21–24] and the decomposition theorem [19]. It has been proved in [19] that for a fuzzy set ($\boldsymbol{A} \otimes \boldsymbol{B}$), its α-cut ($[\boldsymbol{A} \otimes \boldsymbol{B}]_\alpha$) equals $[\boldsymbol{A}]_\alpha \otimes [\boldsymbol{B}]_\alpha$ for all $\alpha \in (0, 1]$. Then, by the decomposition theorem,

$$\boldsymbol{A} \otimes \boldsymbol{B} = \bigcup_{\alpha\in[0,1]} [\boldsymbol{A} \otimes \boldsymbol{B}]_\alpha \quad (10)$$

Since, for any of these operations, $[\boldsymbol{A} \otimes \boldsymbol{B}]_\alpha$ is a closed interval for each $\alpha \in (0, 1]$ and $\boldsymbol{A}$, $\boldsymbol{B}$ are fuzzy numbers, $\boldsymbol{A} \otimes \boldsymbol{B}$ is also a fuzzy number.

3 System Description

The summary of the fuzzy Dempster's rule of combination (FDST) used in decision fusion is shown in Fig. 1. C_i^A and C_i^B are the classifier outputs in class i from classifiers A and B, respectively. The classifiers used in the experiment are the support vector machine (SVM) [25, 26] and the radial basis function (RBF) [25, 26]. In our case, only

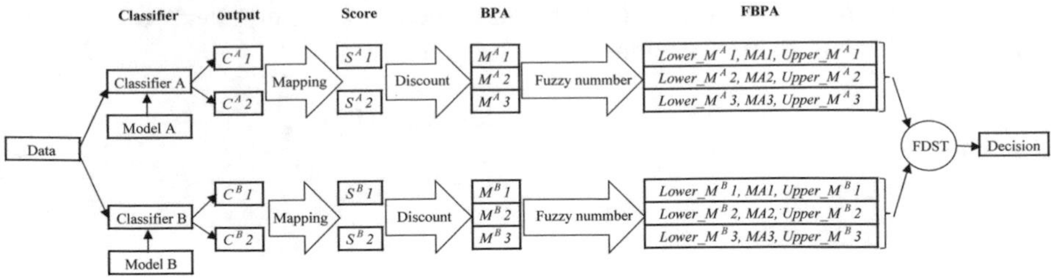

Fig. 1. Fuzzy Dempster's rule of combination system

2 classifiers are combined at a time. We also map each classifier output within the range of 0 and 1, the value $\left(C_i^A\left(C_i^B\right)\right)$ is mapped to 0 and 1 $\left(S_i^A\left(S_i^B\right)\right)$.

The value S of each classifier is as follows.

SVM: The classifier output of feature vector $\mathbf{x}$ in class i can be any real number, hence,

$$S^{SVM}i(\mathbf{x}) = \frac{1}{1 + \exp^{-C^{SVM}i(\mathbf{x})}} \tag{11}$$

where $C^{SVM}i(\mathbf{x}) = \sum_{i=1}^{N} \alpha_i K(\mathbf{x}, z_i) + w_0$, $K(\mathbf{x}, z_i)$ is a linear kernel function and α_i are the trained coefficients from SVM.

RBF: The mapped value in this case is

$$S^{RBF}i(\mathbf{x}) = \frac{1}{1 + \exp^{-C^{RBF}i(\mathbf{x})}} \tag{12}$$

Although, the mapped value is in the range of 0 and 1, the properties of bpa are not held. Hence, for each classifier A, the bpa value of feature vector $\mathbf{x}$ in class i is

$$bpa^A i(\mathbf{x}) = \frac{S^A i(\mathbf{x})}{\sum_{j=1}^{C} S^A j(\mathbf{x})} \tag{13}$$

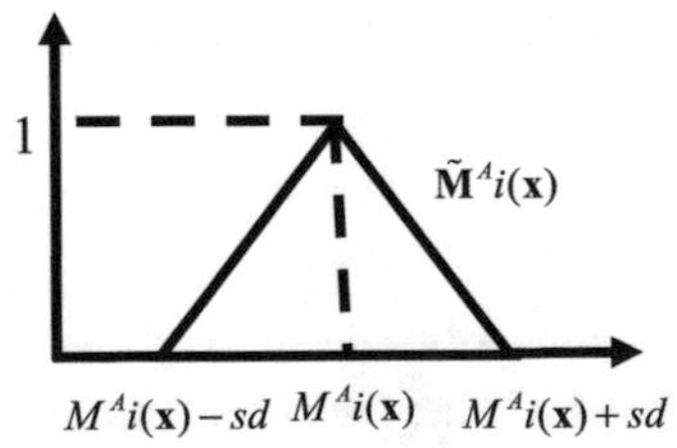

Fig. 2. Fuzzified version of $M^A i(\mathbf{x})$

1st classifier (model A) discounted with α_{cre}

2nd classifier (model B) discounted with α_{cre}

Focal element	Class 1 with $\tilde{\mathbf{M}}^A 1(\mathbf{x})$	Class 2 with $\tilde{\mathbf{M}}^A 2(\mathbf{x})$	Class 1 and class 2 with $\tilde{\mathbf{M}}^A \mathrm{X}(\mathbf{x})$
Class 1 with $\tilde{\mathbf{M}}^B 1(\mathbf{x})$	1	$\varnothing$	1
Class 2 with $\tilde{\mathbf{M}}^B 2(\mathbf{x})$	$\varnothing$	2	2
Class 1 and class 2 with $\tilde{\mathbf{M}}^B \mathrm{X}(\mathbf{x})$	1	2	1 and 2

Fig. 3. Combination of 2 classifiers with discounting strategy

Because of the incorrect classification result from classifiers, we decide to relate that to the credibility of the classifiers. Since we perform the fusion on the test data set, we utilize the accuracy of the classifier on the training data set as our credibility. Hence, the bpa of each classifier will be given the credibility of $(1-\alpha_{\text{cre}})$ accurate rate. This is actually similar to discount the classifier with α_{cre}. Therefore [27]

$$M^A i(\mathbf{x}) = (1 - \alpha_{\text{cre}}) bpa^A i(\mathbf{x}) \;\; \text{for} \;\; i \in \{1, \ldots, C\} \tag{14}$$

$$M^A \mathrm{X}(\mathbf{x}) = \alpha_{\text{cre}} + (1 - \alpha_{\text{cre}}) bpa^A \mathrm{X}(\mathbf{x}) \quad \text{for } \mathrm{X} = \{1, \ldots, C\} \tag{15}$$

where C is the number of classes, and X is the frame of discernment, in our case, the set of classes. In our case, $bpa^A \mathrm{X}(\mathbf{x})$ equals to 0.

Then, we fuzzify each M to be a triangular fuzzy number shown in Fig. 2. We generate sd manually in the experiment. Now, we are ready to describe our fuzzy Dempster's rule of combination (FDST). The example of combination of 2 classifiers with discounting strategy is shown in Fig. 3. The combined basic probability assignment is

$$\tilde{\mathbf{M}} i(\mathbf{x}) = \frac{\sum_{j \cap l = i} \tilde{\mathbf{M}}^A j(\mathbf{x}) \tilde{\mathbf{M}}^B l(\mathbf{x})}{1 - K} \quad \text{for all } i, j, l \in \{1, \ldots, C\} \tag{16}$$

where

$$K = \sum_{i \cap j = \varnothing} \tilde{\mathbf{M}}^A i(\mathbf{x}) \tilde{\mathbf{M}}^B j(\mathbf{x}) \tag{17}$$

Then from the decomposition theorem and interval analysis [19, 21–24], the combined basic probability assignment is

$$\tilde{\mathbf{M}}i(\mathbf{x}) = \bigcup_{\alpha\in[0,1]} \left[\tilde{\mathbf{M}}i(\mathbf{x})\right]_\alpha \text{ for all } i \in \{1, 2, \ldots, C\} \tag{18}$$

where

$$\left[\tilde{\mathbf{M}}i(\mathbf{x})\right]_\alpha = \frac{\sum_{j\cap l=i} \left[\tilde{\mathbf{M}}^A j(\mathbf{x})\right]_\alpha \left[\tilde{\mathbf{M}}^B l(\mathbf{x})\right]_\alpha}{1 - \left(\sum_{i\cap j=\phi} \left[\tilde{\mathbf{M}}^A i(\mathbf{x})\right]_\alpha \left[\tilde{\mathbf{M}}^B j(\mathbf{x})\right]_\alpha\right)} \quad \text{for } 0 \le \alpha \le 1. \tag{19}$$

The decision rule depends on the belief measure or the degree of belief of feature vector $\mathbf{x}$ in class i ($\widetilde{\mathbf{Bel}}i(\mathbf{x})$) and it is described as

$$\widetilde{\mathbf{Bel}}i(\mathbf{x}) = \bigcup_{\alpha\in[0,1]} \left[\widetilde{\mathbf{Bel}}i(\mathbf{x})\right]_\alpha \tag{20}$$

where

$$\left[\widetilde{\mathbf{Bel}}i(\mathbf{x})\right]_\alpha = \left[\sum_{j=i} \tilde{\mathbf{M}}j(\mathbf{x})\right]_\alpha \quad \text{for } 0 \le \alpha \le 1. \tag{21}$$

Hence, the decision rule will be

$$\mathbf{x} \text{ is assigned to class } i \text{ if } i = \operatorname*{agrmax}_{j=1,\ldots,C} \widetilde{\mathbf{Bel}}j(\mathbf{x}) \text{ and } i \ne \mathrm{X}(\mathbf{x}). \tag{22}$$

The rejection criterion is when $\widetilde{\mathbf{M}}\mathrm{X}(\mathbf{x})$ is maximum because the evidence points to universal set are more than any of the classes. If this is a case, we will reject the assignment of vector $\mathbf{x}$ into any classes.

We utilize the index by area [28] to find the maximum fuzzy number. An index of a strictly convex fuzzy set $\tilde{\mathbf{A}}$ is

$$R(\tilde{\mathbf{A}}) = \frac{1}{2}\left[h_{\tilde{\mathbf{A}}} - \frac{D\left(\mu_{\tilde{\mathbf{A}}}, \mu_{\tilde{\mathbf{U}}_\mathbf{A}}\right) - D\left(\mu_{\tilde{\mathbf{A}}}, \mu_{\tilde{\mathbf{L}}_\mathbf{A}}\right)}{d - a}\right] \tag{23}$$

where $h_{\tilde{\mathbf{A}}}$ is the height of $\tilde{\mathbf{A}}$, d is the maximum barrier (value greater than or equal to the largest level cut endpoint) and a is the minimum barrier (value less than or equal to the smallest level cut endpoint). Let x_m be a number such that $\mu(x_m) = h_{\tilde{A}}$ and let $\mu_{n+1} = \mu_0 = 0$. Then the area is computed by

$$D\left(\mu_{\tilde{\mathbf{A}}}, \mu_{\tilde{\mathbf{U}}_{\mathbf{A}}}\right) = \sum_{i=m}^{n} (\mu_i - \mu_{i+1})(d - x_i) \tag{24}$$

and

$$D\left(\mu_{\tilde{\mathbf{A}}}, \mu_{\tilde{\mathbf{L}}_{\mathbf{A}}}\right) = \sum_{i=1}^{m} (\mu_i - \mu_{i-1})(x_i - a). \tag{25}$$

4 Experiment Results

We run experiments on 5 public data sets from the UCI machine learning repository [29] shown in Table 1. We perform the experiment in 10-fold cross validation style with each classifier. We also run the experiment on the MSTAR public data set from the DARPA/WRIGHT laboratory Moving and Stationary Target Acquisition and Recognition (MSTAR) program. The details of this data set are shown in Table 2. Since there are 3 classes in the MSTAR data set, we utilize the one against all strategy for each classifier called MSVM.

Table 1. The details of data sets used in the experiment.

Data set	Features	Class	Instances
Pima: Pima Indians Diabetes	8	2	768
Haberman: Haberman's Survival	3	2	306
Australian: Statlog (Australian Credit Approval)	14	2	690
Bupa: Liver Disorders Data Set	7	2	345
Ionosphere	34	2	351

Table 2. Details of MSTAR data set.

Vehicle type	Training data set	Blind test data set	Total images
BMP2	233	196	429
BTR70	233	196	429
T72	228	191	419
Total images	694	583	1277

The feature set generation of this data set is described in [30]. For this data set, we fuse the result of 10-best models [30] from MSVM and RBF. Each classifier's class output is mapped to *bpa* and discounted with the classifier's miss rate as described in previous section. Then we fuzzify each discounted *bpa* to a triangular fuzzy number. The value of SD is varied from 0.01 to 0.4 with the increasing step of 0.01.

Table 3. Example of the FDST accuracy rate of Ionosphere 10th validation set.

SD	0.01	0.02	0.03	0.04	0.05
{SVM,RBF}	77.14	77.14	77.14	77.14	77.14

We combine 2 classifiers at a time. Example of the correct classification from fuzzy Dempster's rule of combination (FDST) of the Ionosphere 10th validation set is shown in Table 3. We also compare our result with the numeric discounted *bpa* using the regular Dempster's rule of combination (DST). The experiment is also described in Fig. 4. Table 4 shows the best validation results of 5 UCI data sets. Table 5 shows the fusion result of the 10-best model of the MSTAR data set. We can see that the result from the FDST is better than the DST for the Pima data set and equally efficient to the DST for the Haberman, Australian, Bupa and Ionosphere data sets. Again, the result from the MSTAR data set, the FDST is better than that from the DST. However, the fusion result from the DST of the 4th and 7th models is a little bit better than the FDST.

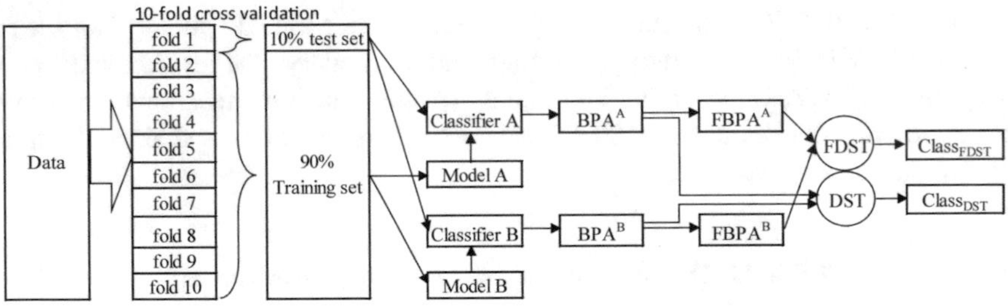

Fig. 4. The description of the experiment

Table 4. The best validation result of the 5 UCI data sets.

Algorithm	Pima data set	Haberman data set	Australian data set	Bupa data set	Ionosphere data set
SVM	81.82	74.19	**81.16**	77.14	91.67
RBFN	65.79	70.97	56.52	64.71	74.29
DST	80.52	**77.42**	**81.16**	**80.00**	**94.29**
FDST	**81.82** (SD = 0.29)	**77.42** (SD = 0.01)	**81.16** (SD = 0.01)	**80.00** (SD = 0.01)	**94.29** (SD = 0.01)

Table 5. The best blind test result of the MSTAR data set.

Model	MSVM	RBFN	DST	FDST
1	85.25	89.02	91.77	**92.10 (SD = 0.21)**
2	80.62	88.16	88.51	**88.67 (SD = 0.25)**
3	89.02	87.82	92.45	**92.62 (SD = 0.2)**

(*continued*)

Table 5. (*continued*)

Model	MSVM	RBFN	DST	FDST
4	87.48	89.88	**93.14**	93.13 (SD = 0.21)
5	88.16	83.53	**93.65**	**93.65 (SD = 0.23)**
6	84.91	87.99	88.85	**89.02 (SD = 0.21)**
7	84.56	84.05	**91.60**	91.59 (SD = 0.2)
8	87.65	86.11	90.91	**91.25 (SD = 0.2)**
9	83.70	83.70	**91.25**	**91.25 (SD = 0.18)**
10	81.82	**89.37**	88.68	89.19 (SD = 0.22)

5 Conclusion

The fuzzy fusion algorithm named fuzzy Dempster's rule of combination (FDST) is introduced in this paper. The method is an extension of the Dempster's rule of combination (DST). In the experiment, we also implement a discounting method to compute a discounted basic probability assignment. This can provide a rejection criterion for our algorithm as well. The result shows that the FDST is able to give a better correct classification than the DST.

References

1. Li, W., Leung, H., Kwan, C., Linnell, B.R.: E-nose vapor identification based on Dempster-Shafer fusion of multiple classifiers. IEEE Trans. Instrum. Measur. **57**(10), 2273–2282 (2008)
2. Mejdoubi, M., Aboutajdine, D., Kerroum, M.A., Hammouch, A.: Combining classifiers using Dempster-Shafer evidence theory to improve remote sensing images classification. In: International Conference on Multimedia Computing and Systems. IEEE, Ouarzazate (2011)
3. Verma, P., Yadava, R.D.S.: Evidence generation for Dempster-Shafer fusion using feature extraction multiplicity and radial basis network. In: International Conference on Emerging Trends in Electrical and Computer Technology, Chunkankadai, India, pp. 542–545 (2011)
4. Mohandes, M., Deriche, M.: Arabic sign language recognition by decisions fusion using Dempster-Shafer theory of evidence. In: Computing, Communications and IT Applications Conference (ComComAp), pp. 90–94. IEEE, Hong Kong (2013)
5. Bagheri, M.A., Hu, G., Gao, Q., Escalera, S.: A framework of multi-classifier fusion for human action recognition. In: International Conference on Pattern Recognition, pp. 1260–1265. IEEE (2014)
6. Perez, A., Tabia, H., Declercq, D., Zanotti, A.: Using the conflict in Dempster-Shafer evidence theory as a rejection criterion in classifier output combination for 3D human action recognition. Image Vis. Comput. **55**(2), 149–157 (2016)
7. Ladjal, M., Bouamar, M., Djerioui, M., Brik, Y.: Performance evaluation of ANN and SVM multiclass models for intelligent water quality classification using Dempster-Shafer theory. In: 2nd International Conference on Electrical and Information Technologies, Movenpick Hotel Tangier Tangier, Morocco, pp. 191–196 (2016)

8. Ming, Y., You, L., Xueshan, H.: Probabilistic wind generation forecast based on sparse Bayesian classification and Dempster-Shafer theory. IEEE Trans. Ind. Appl. **52**(3), 1998–2005 (2016)
9. Saha, S., Saha, S.: Combined committee machine for classifying dengue fever. In: International Conference on Microelectronics, Computing and Communication. IEEE, Durgapur (2016)
10. Liu, Z., Pan, Q., Dezert, J., Han, J., He, Y.: Classifier fusion with contextual reliability evaluation. IEEE Trans. Cybern. **99**, 1–14 (2017)
11. Li, L., Tang, J., Liu, Y.: Partial discharge recognition in gas insulated switchgear based on multi-information fusion. IEEE Trans. Dielectr. Electr. Insul. **22**(2), 1080–1087 (2015)
12. Jiao, L., Denoux, T., Pan, Q.: Fusion of pairwise nearest-neighbor classifiers based on pairwise-weighted distance metric and Dempster-Shafer theory. In: 17th International Conference on Information Fusion. IEEE, Salamanca (2014)
13. Yen, J.: Generalizing the Dempster-Shafer theory to fuzzy sets. IEEE Tran. Syst. Man Cybern. **20**(3), 559–570 (1990)
14. Binaghi, E., Madella, P.: Fuzzy Dempster-Shafer reasoning for rule-based classifiers. Int. J. Intell. Syst. **14**, 559–583 (1999)
15. Zhu, H., Basir, O.: A K-NN associated fuzzy evidential reasoning classifier with adaptive neighbor selection. In: Third IEEE International Conference on Data Mining, pp. 709–712 (2003)
16. Dutta, P., Ali, T.: Fuzzy focal elements in Dempster-Shafer theory of evidence: case study in risk analysis. Int. J. Comput. Appl. **34**(1), 46–53 (2011)
17. Dempster, A.P.: Upper and lower probabilities induced by a multivalued mapping. Ann. Math. Stat. **38**(2), 325–339 (1967)
18. Shafer, G.: A Mathematical Theory of Evidence. Princeton University Press, Princeton (1976)
19. Klir, G., Yuan, B.: Fuzzy Sets and Fuzzy Logic: Theory and Application. Prentice Hall, New Jersey (1995)
20. Zadeh, L.: Outline of new approach to the analysis of complex systems and decision processes. IEEE Trans. Syst. Man Cybern. **3**(1), 28–44 (1973)
21. Dong, W., Shah, H., Wong, F.: Fuzzy computations in risk and decision analysis. Civ. Eng Syst. **2**, 201–208 (1985)
22. Dong, W., Wong, F.: Fuzzy weighted averages and implementation of the extension principle. Fuzzy Sets Syst. **21**, 183–199 (1987)
23. Mizumoto, M., Tanaka, K.: Some properties of fuzzy numbers in advances in fuzzy sets theory and applications. In: Gupta, M.M. (ed.) Advances in Fuzzy Set Theory and Applications, pp. 153–164. North-Holland, Amsterdam (1979)
24. Moore, R.: Interval Analysis. Prentice-Hall, New Jersey (1966)
25. Ludmila, K.: Combining Pattern Classifiers Methods and Algorithms, 2nd edn. Wiley, Hoboken (2014)
26. Haykin, S.S.: Neural Networks: A Comprehensive Foundation. Prentice Hall, Upper Saddle River (1999)
27. Shafer, G.: Perspectives on the theory and practice of belief functions. Int. J. Approx. Reason. **4**(5–6), 323–362 (1990)
28. Choobineh, F., Li, H.: An index for ordering fuzzy numbers. Fuzzy Sets Syst. **54**(3), 287–294 (1993)

29. Blake, C., Keogh, E., Merz, C.J.: UCI repository of machine learning databases, Department of Information and Computer Science, University of California, Irvine, CA (1998). http://www.ics.uci.edu/~mlearn/MLRepository.html
30. Auephanwiriyakul, S., Munklang, Y., Theera-Umpon, N.: Synthetic Aperture Radar (SAR) Automatic Target Recognition (ATR) using fuzzy co-occurrence matrix texture features. In: Abielmona, R., Falcon, R., Zincir-Heywood, N., Abbass, H. (eds.) Recent Advances in Computational Intelligence in Defense and Security. Studies in Computational Intelligence, vol. 621. Springer, Cham (2016)

Enhanced Manhattan-Based Clustering Using Fuzzy C-Means Algorithm

Joven A. Tolentino(✉), Bobby D. Gerardo, and Ruji P. Medina

Technological Institute of the Philippines, Quezon City, Philippines
jatolentino@tau.edu.ph, Bobby.gerardo@gmail.com,
ruji.medina@tip.edu.ph

Abstract. Fuzzy C-Means is a clustering algorithm known to suffer from slow processing time. One factor affecting this algorithm is on the selection of appropriate distance measure. While this drawback was addressed with the use of the Manhattan distance measure, this sacrifice its accuracy over processing time. In this study, a new approach to distance measurement is explored to answer both the speed and accuracy issues of Fuzzy C-Means incorporating trigonometric functions to Manhattan distance calculation. Upon application of the new approach for clustering of the Iris dataset, processing time was reduced by three iterations over the use of Euclidean distance. Improvement in accuracy was also observed with 50% and 78% improvement over the use of Euclidean and Manhattan distances respectively. The results provide clear proof that the new distance measurement approach was able to address both the slow processing time and accuracy problems associated with Fuzzy C-Means clustering algorithm.

Keywords: Fuzzy C-Means · Enhanced Fuzzy C-Means
Enhanced Manhattan · Trigonometric approach

1 Introduction

Corporate databases and social media [1] generate vast amounts of data that provide meaningful information using data mining techniques. Learning from a given dataset such as association, correlation of certain groups of record was done using clustering. Clustering splits a large amount of data and performs grouping considering the similarities of the individual data supplied [2, 3]. Several Clustering algorithms have developed such as K-Means, K nearest neighbor etc.

A clustering algorithm that can be used is Fuzzy C-Means. It is a simple convenient method of clustering [4]. However, Fuzzy C-means also suffers with its accuracy and speed specially when the data is high dimensional or not [5, 6]. One way of solving this problem is to change the distance measure that the algorithm used.

The Euclidean distance measure is commonly used in clustering because of its higher accuracy compared to the other distance measure developed [7]. However, Euclidean distance measure sacrifices its speed when applied to clustering algorithm [8]. The use of Euclidean distance for the selection of the actual points near the centroid

H. Unger et al. (Eds.): IC2IT 2018, AISC 769, pp. 126–134, 2019.
https://doi.org/10.1007/978-3-319-93692-5_13

requires higher computational cost and when it is used in fuzzy C-means, Euclidean distance measures can unequally weight underlying factors [9].

On the other hand, Manhattan distance measure has the capability of clustering a dataset faster compared to another distance measure when supplied to a clustering algorithm [10]. However, it has a problem in terms of accuracy compared to other distance measures [10]. Selecting the appropriate distance measure will also increase the accuracy and/or processing speed of a certain algorithm.

The study aims to propose an Enhanced Fuzzy C-Means Clustering by implementing an enhanced Manhattan distance measure using trigonometric approach in order to increase the speed of the algorithm and increase the accuracy of the distance measure used.

2 Related Studies

2.1 Clustering

Several clustering algorithms developed such as K-Means, K-Nearest Neighbor, and Fuzzy C-means are categorized as a partitioning algorithm ideal for creating partitions for several points identified by a distance measure [7]. Selecting the right algorithm for clustering is important. Reliable results can be achieved when a correct algorithm is selected. The Fuzzy C-means algorithm can be used for clustering data. However, several algorithms outperform this algorithm [11]. Providing a result with accurate clusters and lesser computational cost can be addressed using different strategies. Observing the distance measure used by the algorithm and its performance can be a basis of identifying what distance measure is applicable for the high dimensional dataset. Selecting an appropriate distance measure plays a vital role in providing a good cluster [12]. Due to this problem of the algorithm, a proposed enhanced Fuzzy C-Means Algorithm was developed.

2.2 Manhattan Distance Measure

The use of Euclidean distance is accurate when a data supplied is not multi-dimensional. On the other hand, Manhattan distance measure can be used with a dataset is high dimensional. Studies were also done to see the behavior of different distance measures with high dimensional dataset and surprisingly Manhattan distance measure outperforms other distance measures [13]. Manhattan distance is defined as the distance between two points or strings and it is the sum of the absolute differences of their Cartesian coordinates [14].

2.3 Trigonometric Approach

A study of [15] shows how trigonometry is used to calculate the area of two parallel lines and it reflects that these procedures were effective. The same principle will be applied in order to calculate the actual distance of the centroid from the data points.

Employing trigonometric function will be used to address the problem of Manhattan Distance Measure and incorporate it to Fuzzy C-Means.

3 Design Architecture

3.1 Fuzzy C-Means Algorithm

Figure 1 shows the actual process of how Fuzzy C-Means Clustering works.

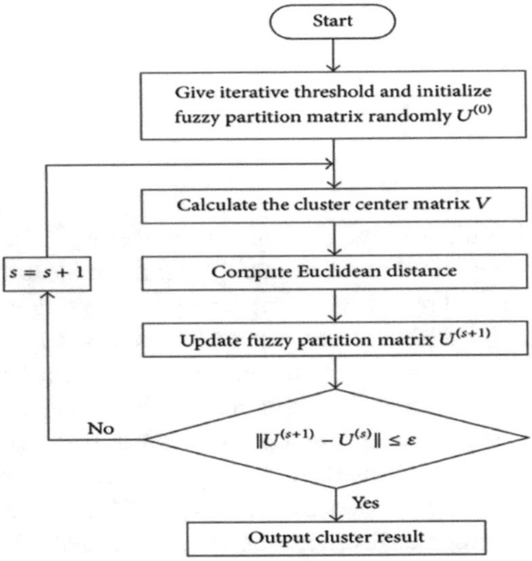

Fig. 1. Procedures in fuzzy f-means

First, Fuzzy C-Means selects the number of cluster and membership functions ranging from zero to one. The second step is to calculate the actual centroid with the corresponding parameter. The computation of the actual centroid plays a vital role in creating the clusters [16]. This will identify how many iterations will be done. The next step is to update the actual cluster with the specific distance measure and lastly validate the result then iteration take place until convergence is achieved [17]. With this given process of Fuzzy C-Means algorithm, changing the distance measure can improve the performance of the said algorithm.

3.2 Euclidean Distance Measure Calculation

On the other hand, Euclidean distance is used to compute the actual distance to the different point of the datasets.

$$d = \sqrt{\sum_{i-1}^{n} (qi - pi)^2} \quad (1)$$

3.3 Manhattan Distance Measure Calculation

The Manhattan distance measure is commonly used when the point that is generated was vertically or horizontally connected. Equation 2 shows the actual computation for Manhattan distance.

$$d = \sum_{i=1}^{n} |x_i - y_i| \quad (2)$$

where n is the number of variables, and Xi and Yi are the values of the ith variable, at points X and Y respectively [14].

3.4 Comparison of the Two Distance Measures

With the two distance measures presented, we can now investigate what is the weakness of Manhattan distance measure. With that, it can be addressed a side by side comparison on how the two measures differs from each other (Fig. 2).

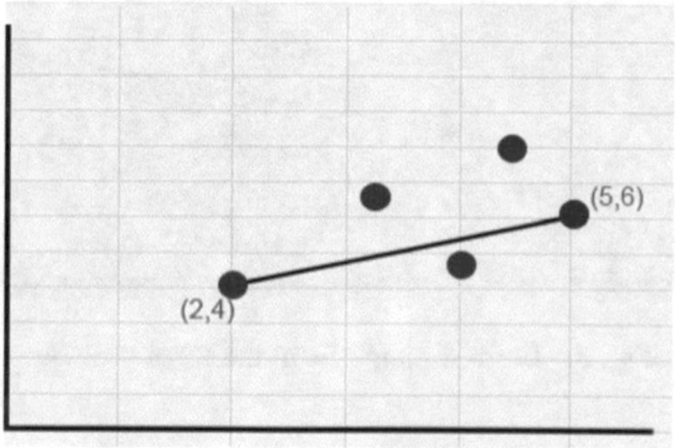

Fig. 2. Point plotted along with centroid

Given that this distance measure is less accurate from the other distance measure, the following computations can be observed to identify the behavior of the equation. Suppose we have two points given, (2,4) and (5,6) with the corresponding points applied to Manhattan $|x_1 - x_0| + |y_1 - y_0|$ by substitution $|5 - 2| + |6 - 4|$ the distance between the two points is 5. When applied to a standard distance formula $\sqrt{(5-2)^2 + (6-4)^2} = 3.61$. Manhattan distance measure is less accurate than the standard distance formula since points were connected diagonally. Other studies show that the calculated differences of the two equations were also doubled in value in some cases.

3.5 Proposed Modification

The use of trigonometric function can be done in order to solve the problem of Manhattan distance measure. Considering the different trigonometric function, we can solve the problem of this distance measure. Given two vertices and an imaginary line to form a triangle, SOHCAHTOA can now be used to calculate the actual distance of the original line. SOHCAHTOA stands for Sine equals Opposite over Hypotenuse; Cosine equals Adjacent over Hypotenuse; Tangent equals Opposite over adjacent [18]. In order to calculate the actual distance of the two points given, an equation was derived.

$$d = \sum \frac{|(x_3 - x_2) + (y_3 - y_3)|}{cosine(\emptyset)} \tag{3}$$

Where X_3 and Y_3 is the point of intersection from the created imaginary line and X_2 and Y_2 is the centroid. After solving the absolute differences of X_3, X_2 and Y_3, Y_2, it will be divided to $cosine(\emptyset)$. $Theta(\emptyset)$ is used since the actual angle is not yet solved. To calculate the actual distance the following steps were considered (Fig. 3).

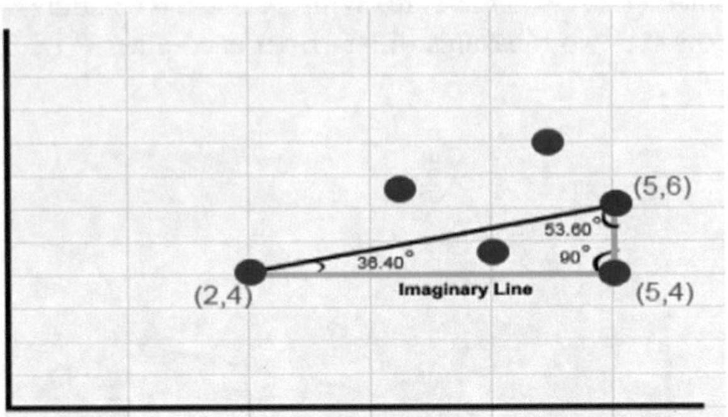

Fig. 3. Derived image from the steps provided

Step 1. *Create an imaginary Line to form a right triangle*
Step 2. *Identify the point of intersection*
Step 3. *Compute the Distance of the Imaginary line using Manhattan.*
(5,6) & (5,4) = 2
Step 4. *Compute for the distance*
2/Cosine(53.60) = 3.61

Using trigonometric function over Manhattan distance can solve the problem of this distance measure and might increase its accuracy.

3.6 Simulation

To further investigate the result of the modified Manhattan Distance measure, a java application was developed. It will be implemented by the Fuzzy C-Means. The Iris dataset will be used. It has 5 attributes and 150 rows of records. The last attribute will be used for label evaluation for clustering.

To simulate the actual records of Iris dataset, the records were stored in an array. The 6th columns hold the actual positioning of the record after clustering. The code was run using three distance measures for comparison. Euclidean distance, Manhattan distance measure and Modified Manhattan distance measure.

```
Start
Required Array of Points and Centroid
Declare Distance
   For counter=0; to LengthofPoints step 1
    //Euclidean Distance measure
    Get the difference of points and the centroid
    Get the square of the difference and update it to dis-
    tance
   EndFor
    Update Distance by getting the square root value
End
```

```
Start
Required Array of Points and Centroid
Declare Distance
   For counter=0; to LengthofPoints step 1
    //Manhattan Distance measure
      Get the absolute difference of points and the centroid
      Add the existing value of Distance to the difference
   EndFor
End
```

The actual dataset shows that the result of the algorithm varies as expected. Euclidean distance outperformed Manhattan distance measure while the proposed Manhattan distance measure increased its accuracy. On the other hand, the following pseudo-code was used to implement the Modified Manhattan distance measure for Fuzzy C-Means Algorithm.

```
Start
Required Array of Points and Centroid
Declare Distance

   For counter=0; to LengthofPoints step 2

   If Centroid is equal to Points
      Get the absolute difference of points and the centroid
   Else
      Get the absolute difference of centroid and Imaginary
line
      Divide the absolute difference to cosine(∅)
   EndIf
      Update distance by adding the difference
      Iterate to each column and pair it with (x,y) format
      and do calculation for the distance.
End
```

With the proposed modification for Manhattan distance Measure, it increases its accuracy.

4 Evaluation

After simulating the algorithm, Evaluation takes place to check the accuracy as labeled, the clustered dataset can be evaluated by checking the actual label compare to the original cluster [19, 20]. The last column represents the cluster name to which the particular record or data belong.

The correct cluster was divided to the actual number of population to calculate its results accurately.

For the speed of the algorithm and iteration, the actual cluster was monitored.

```
Start
Declare Threshold=1, iteration=0
   While Threshold is not equal to 0
           Update value of iteration +1
           Assign new center
    EndWhile
End
```

Each distance measure implemented on the algorithm was tested through this processes and the result is reflected on Table 1. The number of records is 150 and 5 attributes including the label having 3 clusters.

Table 1. Result of the simulation

	Euclidean	Manhattan	Modified Manhattan
Accuracy	50.66%	42.66%	76%
No. of correct matches	76	64	114
Iteration	15	10	12

The result of the algorithm shows that, it has improved the enhancement done in the distance measure.

5 Conclusion and Recommendation

Fuzzy C-Means Algorithm paired with Manhattan Distance Measure makes the processing of the actual algorithm faster. The said distance measure needed to be improved to increase its accuracy with the idea of incorporating trigonometric function to Manhattan distance algorithm. It shows that the algorithms improved its accuracy and iterations. Results show that accuracy was increased from 42% to 76% which is ideal compared to the Euclidean distance having a 51% accuracy when incorporated into the algorithm making the clustering algorithm more reliable.

Having the algorithm and distance measure modified, the following recommendations can also be considered for the improvement of the algorithm. Multiple dataset can be used and high dimensional dataset can be considered in order to test the accuracy and speed of the algorithm. Another measure can also be considered to check how fast and accurate the algorithm is.

Acknowledgement. This study would not be possible without the financial support of the Commission on Higher Education Kto12 Transition Program Unit Quezon City, Philippines and Tarlac Agricultural University Tarlac, Philippines.

References

1. Chen, H., Storey, V.C.: Business intelligence and analytics: from big data to big impact. MIS Q. **36**(4), 1165–1188 (2012)
2. Jung, Y.G., Kang, M.S., Heo, J.: Clustering performance comparison using K-Means and expectation maximization algorithms. Biotechnol. Biotechnol. Equip. **28**(sub1), 44–48 (2014)
3. Raval, U.R., Jani, C.: Implementing and improvisation of K-Means clustering. Int. J. Comput. Sci. Mob. Comput. **4**(11), 72–76 (2015)
4. Han, D., Ji, J., Dai, Y., Li, G., Fan, W., Chen, H.: Improved Fuzzy C-Means algorithm and its application to classification of remote sensing image in Chengdu City. In: 2016 International Conference on Progress in Informatics and Computing (PIC), pp. 437–443. IEEE, Shanghai (2016)
5. Cebeci, Z., Yildiz, F.: Comparison of K-Means and Fuzzy C-Means algorithms on different cluster structures. J. Agric. Inf. **6**(3), 13–23 (2015)
6. Winkler, R., Klawonn, F., Kruse, R.: Problems of Fuzzy C-Means clustering and similar algorithms with high dimensional data sets. In: Gaul, W., Geyer-Schulz, A., Schmidt-Thieme, L., Kunze, J. (eds.) Challenges at the Interface of Data Analysis, Computer Science, and Optimization. Studies in Classification, Data Analysis, and Knowledge Organization, pp. 1–8. Springer, Heidelberg (2012)
7. Fahad, A., et al.: A survey of clustering algorithms for big data : taxonomy & empirical analysis. IEEE Trans. Emerg. Topics Comput. **2**(3), 267–279 (2014)
8. Sharma, S.K., Kumar, S.: Comparative analysis of Manhattan and Euclidean distance metrics using A* algorithm. J. Res. Eng. Appl. Sci. **1**(4), 196–198 (2016)
9. Kalaivani, E.R., Suganya, G., Kiruba, J.: Review on K-Means and fuzzy C means clustering algorithm. Imp. J. Interdiscip. Res. **3**(2), 1822–1824 (2017)
10. Kouser, K.: A comparative study of K means algorithm by different distance measures. Int. J. Innov. Res. Comput. Commun. Eng. **1**(9), 2443–2447 (2013)
11. Kapoor, A.: A comparative study of K-Means, K-Means ++ and Fuzzy C-Means clustering algorithms. In: 3rd International Conference on Computational Intelligence & Communication Technology (CICT), pp. 1–6. IEEE (2017)
12. Son, L.H.: Generalized picture distance measure and applications to picture fuzzy clustering. Appl. Soft Comput. J. **46**(C), 284–295 (2016)
13. Aggarwal, C.C., Hinneburg, A., Keim, D.A.: On the surprising behavior of distance metrics in high dimensional space. In: Van den Bussche, J., Vianu, V. (eds.) ICDT 2001. LNCS, vol. 1973, pp. 420–434. Springer, Heidelberg (2001). https://doi.org/10.1007/3-540-44503-X_27

14. Rathod, A.B.: A comparative study on distance measuring approaches for permutation representations. In: 2016 IEEE International Conference on Advances in Electronics, Communication and Computer Technology (ICAECCT), pp. 251–255 (2016)
15. Wang, X., Yu, F., Pedrycz, W.: An area-based shape distance measure of time series. Appl. Soft Comput. J. **46**(C), 650–659 (2016)
16. Wang, Z., Zhao, N., Wang, W., Tang, R., Li, S.: A fault diagnosis approach for gas turbine exhaust gas temperature based on Fuzzy C-Means clustering and support vector machine. Math. Prob. Eng. **2015**, 1–11 (2015)
17. Grover, N.: A study of various fuzzy clustering algorithms. Int. J. Eng. Res. **5013**(3), 177–181 (2014)
18. Leddo, J.: Comparing the relative effects of homework on high aptitude VS average aptitude students **2**(9), 59–62 (2016)
19. Dom, B.E., Jose, S.: An information-theoretic external cluster-validity measure. In: Proceedings of the Eighteenth Conference on Uncertainty in Artificial Intelligence (UAI 2002), pp. 137–145, Alberta, Canada (2002)
20. Färber, I., et al.: On using class-labels in evaluation of clusterings. In: Proceedings of 1st International Workshop on Discovering, Summarizing and Using Multiple Clusterings (MultiClust 2010) Held in Conjunction with 16th ACM SIGKDD Conference on Knowledge Discovery and Data Mining (KDD 2010), p. 9. Washington, D.C., USA (2010)

The Cooperation of Candidate Solutions Vortex Search for Numerical Function Optimization

Wirote Apinantanakon(✉), Siriporn Pattanakitsiri, and Pochra Uttamaphant

Rajabhat Chaiyaphum University, Meuang, Chaiyaphum, Thailand
{wirote, Siriporn, Pochara}@cpru.ac.th

Abstract. This study presents the cooperation of candidate solutions vortex search called CVS that has been used for solving numerical function optimization. The main inspiration of CVS is that there have been some drawbacks of the Vortex Search (VS) algorithm. Although, the results from the proposal of VS are presented with a high ability but it could produce some drawbacks in updating the positions of vortex swarm. The VS used only single center generating the candidate solutions. The disadvantages happened when VS suffers from multi-modal problems that contain a number of local minima points. To overcome these drawbacks, the proposed CVS generated some cooperation of swarms which created from the diverse points. The experiments were conducted on 12 of benchmark functions. The capability of CVS was compared among the 5 algorithms: DE, GWO, MFO, VS and MVS. The results showed that CVS outperformed all of the comparisons of algorithms used.

Keywords: Vortex search · Differential evolution · Gray wolf optimization
Moth-flame optimization · Metaheuristic optimization algorithm

1 Introduction

In the past two decades, the metaheuristic optimization algorithms have been proposed and widely applied to exact the difficult of real-life problems [1–4]. In order to solve the complex targets of any problems, most of these algorithms have been used as an attempt to provide an approximately method in obtaining the optimal solution. In the most successful way, the metaheuristic algorithms mimic the behaviors of its nature called nature-inspired method. Some famous novel nature-inspired algorithms such as, genetic algorithm (GA) [5], simulated annealing (SA) [6], ant colony (ACO) [7], particle swarm optimization (PSO) [8], differential evolution (DE), artificial bee colony algorithm (ABC) [9] etc. The differences in nature is a primary factor why the algorithms possess different level of performance in producing results. Moreover, this factor may be the reason why some algorithms can best produce solution to certain problems, while others do not. Apparently, it is the limitation that one algorithm cannot be good enough to solve all kinds of problems.

H. Unger et al. (Eds.): IC2IT 2018, AISC 769, pp. 135–144, 2019.
https://doi.org/10.1007/978-3-319-93692-5_14

In regard to the properties of various natures, mentioned in the above paragraph, which have been in correspondence to the various solutions of real-life problems, metaheuristic algorithms and their methods are in being developed. The vortex search algorithm (VS) [10] is a recent proposal of metaheuristic optimization algorithm. The simple structure of VS is consisted of two important keys which are 'center' and 'radius'. First, center is a current position from which VS algorithm can be evaluated based on the problem search space where iterations pass. In regard to exploration for optimal solution which has been conducted so far, VS used this position to identify the 'center' in order to update a new position of the populations. Secondly, 'radius' is a method used to simplify the problems—making a large-radius problem to become a small-radius problem. In additional, a Gaussian distribution is a VS method which is used to balance the exploration and exploitation at each iteration pass. The details of VS algorithm will be mentioned in the next section.

Since the proposed VS structure has been simplified, the proposed VS—which has been derived from the proposal and was tested on all 50 test mathematical benchmark functions—was able to produce good results. Based on a Wilcoxon-Signed Rank Test values, the results showed that VS outperformed the compared algorithms—SA, PS and ABC. However, VS used only a single 'center' that is called the single solution to generate candidate solutions around the current best solution. However, the disadvantages of VS cannot be ignored from the local point when it suffers from the problems which have several local minimum values. In the same time, the radius used to update the best solution were able to decrease the iteration pass by using a Gaussian distribution, making it easier to trap the VS in local optima. This explains some drawbacks of VS algorithm.

To overcome the drawbacks of VS, this study proposed the cooperation of candidate solutions vortex search (CVS). The advantages of CVS is that it has the diverse points to generate the cooperative candidate solutions at iteration pass and each point is shifted from the current best solution. At the same time, the radius between exploration and exploitation can still adjusted using Gaussian distribution. The capability of CVS has been tested with 12 benchmark functions which was compared with the original VS, three novel algorithms—DE, GWO [11], MFO [12], and a modified MVS [13] respectively. The results showed that CVS capability outperforms every compared algorithms and also produced fast convergence to the optimal solution.

The remaining of this paper is presented in the next section. The details of VS algorithm are presented in the following section and the proposed CVS is described in Sect. 3 while in Sect. 4, the results are presented. Finally, the Sect. 5 contains the conclusion of this work.

2 Overview and Related Work

2.1 The Vortex Search Algorithm

The Vortex Search (VS) is a recently proposed metaheuristic optimization algorithm which was inspired by the vortical flow of the stirred fluids. The VS procedures consists of the simplified generation phases like other single-solution algorithms.

The generation of VS populations is changed to any generations by using values solely form current single solution. Moreover, the efficiency of each update and search of iteration pass on the search space is an important phase in rendering single-solution. In the proposed VS algorithm, this balance is achieved by using a vortex-like search pattern. The strategies of vortex pattern are simulated by a number of nested circles. The details of VS strategies can be briefly described in four steps as follow.

Algorithm 1.

```
Inputs: Initial center μ0 is calculated using Eq. (1)
Initial radius r0 (or the standard deviation, σ0) is computed using Eq. (2)
Fitness of the best solution found so far f(s_best) = inf
t = 0;
Repeat
/* Generate candidate solutions by using Gaussian distribution around the center μt
with a standard deviation (radius) rt */
Generate (Ct(s));
if f(ŝ) < f(s_best)
        s_best = ŝ
        f(s_best) = f(ŝ)
else
        keep the best solution so far s_best
end
/* Center is always shifted to the best solution found so far */
μ_{t+1} = s_best
/* Decrease the standard deviation (radius) for the next iteration */
r_{t+1} = Decrease (rt)
t = t+1;
Until the maximum number of iterations is reached
Output: Best solution found so far s_best
```

Fig. 1. The pseudocode of the proposed VS algorithm (Source: Doğan, 2015, p.128).

The first step: the initial process initials 'center' μ_0 and 'radius' r_0. In this phase, the initial 'center' μ_0 can be calculated using Eq. (1) as follow.

$$\mu_0 = \frac{upperlimit + lowerlimit}{2} \tag{1}$$

where *upperlimit* and *lowerlimit* is the bound constraints of problem which can be defined in vector of $d \times 1$ dimensional space. Whereas, σ_0 is considered as the initial radius r_0 generated with Eq. (2).

$$\sigma_0 = \frac{\max(upperlimit) - \min(lowerlimit)}{2} \tag{2}$$

The second step: generating the candidate solutions process is used to render the generation of populations $C_t(s)$ in any iterations where t is the *t-th* iteration. VS used was randomly generated around the initial center μ_0 by using a Gaussian distribution, where $C_0(s) = \{s_1, s_2, \ldots, s_m\}$ m = 1, 2, 3,…, n represents the solution and n is the total number of candidate solutions. The equation of multivariate Gaussian distribution is showed in Eq. (3) as follow.

$$p\left(x|\mu,\sum\right) = \frac{1}{\sqrt{(2\pi)^d|\sum|}} exp\left\{-\frac{1}{2}(x-\mu)^T\sum^{-1}(x-\mu)\right\} \quad (3)$$

where d is the dimension, x represents the $d \times 1$ vector of a random variable, μ is the $d \times 1$ vector of sample mean (center) and Σ is the covariance matrix respectively. Equation (4) presented that if the diagonal elements (variances) of the values of Σ are equal and if the off-diagonal elements (covariance) are zero (uncorrelated), then the resulted shape of the distribution will be spherical. Then, the value of Σ can be computed by using equal variances with zero covariance.

$$\sum = \sigma^2.[I]_{dxd} \quad (4)$$

where the representation in Eq. (4), σ^2 is the variance of the distribution, I represents the $d \times d$ identity matrix and σ_0 is the initial radius (r_O) as can see in Eq. (2).

The third step: replacement of the current solution, for the selection process. A solution (which is the best one) $\acute{s} \in C_O\,(s)$ is selected and memorized from $C_O\,(s)$ to replace the current circle center (μ_0). Prior to the selection process, the candidate solutions must be guaranteed to be inside the search spaces. The solutions that out of the boundaries are controlled within the boundaries, as in Eq. (5).

$$s_k^i = \begin{cases} rand\,x(upperlimit^i - lowerlimit^i) + lowerlimit^i, s_k^i < lowerlimit^i \\ lowerlimit^i \le s_k^i \le upperlimit^i \\ rand\,x(upperlimit^i - lowerlimit^i) + lowerlimit^i, s_k^i > upperlimit^i \end{cases} \quad (5)$$

where k = 1, 2, …, n and i = 1, 2, …, d and *rand* is a uniformly distributed random number. To select the next solutions, VS algorithm used $\acute{s}$ as a new center and reduce the size of vortex by using Eq. (3) generated the new set of solutions $C_1(s)$. If the selected solution is better than the best solution found so far, then this solution is determined to be the new best solution and it is memorized. This process repeats until the termination condition is reached. A description of the VS algorithm is provided in Fig. 1 (Algorithm 1.) (For further study: [10]).

3 The Proposed Method

Corresponding to the literature of VS algorithm, VS generates candidate solutions by using only a single 'center' (μ). The generation of 'center' is transformed to new center as each iteration pass through the limitation of upper and lower bound of problems. As mention above section, this mechanism produces some drawbacks. One of which is that VS algorithm tend to be trapped in local minima when it suffer from a local points of minimum problems. To overcome the drawback as mentioned above, the cooperation of candidate solution vortex search (CVS) is proposed. The proposed of this study focuses on the enhancement of creating some cooperation swarms which is added into the existing VS swarm. For the advantages in this study, the cooperation of candidate solutions are provided from diverse center (μ_t^g) while the VS swarm are generated from current center (μ_t). To provide the cooperation of candidate solutions, the steps of the CVS strategy can be presented as follow. Moreover, the structure of CVS algorithm is described in Fig. 2 (Algorithm 2.).

Algorithm 2.

```
Inputs: Initial center μ0 is calculated using Eq. (1)
Initial radius r0 (or the standard deviation, σ0) is computed using Eq. (2)
Fitness of the best solution found so far f(s_best) = inf
Initial number of CVS group (G)
t = 0;
Repeat
/* Generate candidate solutions by using Gaussian distribution around the center μt
with a standard deviation (radius) rt */
Generate (Ct(S));
if f(ś) < f(s_best)
        s_best = ś
        f(s_best) = f(ś)
else
        keep the best solution so far s_best
end
/* Center is always shifted to the best solution found so far */
μt+1 = s_best

/* Begin the cooperative solutions CVS */
If t > 1
        Ĉt(S_new) = Ct(S)-G;
/*Provide the member of CVS according to the proposed section
Generate C_sortedt(S);
/* Calculated each the CVS center by using Eq. (6)
Calculated (μt^g);
/*The cooperative solutions CVS are generated by using Eq. (7)
CVSt(S^g) = (μt − μt^G) * Gaussian distribution;
/*Combine CVS and VS populations
Ct(S) = { Ĉt(S_new), CVSt(Ŝ)};
end
/* End of the cooperative solutions CVS algorithm */

/* Decrease the standard deviation (radius) for the next iteration */
rt+1 = Decrease (rt)
t = t+1;
Until the maximum number of iterations is reached
Output: Best solution found so far s_best
```

Fig. 2. A description of the proposed CVS algorithm.

(1) After the initial iteration of VS algorithm: the best solution $\acute{s}$ from $C_t(S)$ is selected to replace the current center μ_t and, update the new positions of $C_t(S)$.
(2) For updating phases, CVS algorithm sorted the member of $C_t(S)$ which corresponds to the index of sorted fitness $f(S)$ and put them in the new variable $C\_sorted_t(S)$.
(3) Divided members of $C\_sorted_t(S)$ in to G group, where G is a number of group that user can define. The member in each group could be defined by the size of $C_t(S)/G$. After that, $\hat{C}_t(S_{\text{new}})$ is generated from size of $C_t(S) - G$;
(4) For each *g-th group*, the center (μ_t^g) of each group from (2) are calculated using Eq. (6).

$$\left(\mu_t^g\right) = \frac{\sum_i^n C\_sorted\left(S_i^g\right)}{n} \tag{6}$$

where $i = 1, 2, 3, \ldots n$ is a total member in *g-th* groups.
(5) The cooperation of candidate solutions CVS are generated with Eq. (7). In this step, CVS swarm is shifted from current VS position by using the diverse points μ_t^G.

$$CVS_t(S) = \left(\mu_t - \mu_t^G\right) * \text{Gaussian distribution}, \tag{7}$$

where Gaussian distribution is the vortex equation for updating phase.
(6) Use the cooperation of candidate solutions CVS combined with the VS swarm where

$$C_t(S) = \left\{\hat{C}_t(S_{\text{new}}), CVS_t(S)\right\}.$$

(7) Calculate for the best solution that is obtained so far.

4 Experimental and Results

In this study experiment, the performance of algorithms is measured on the same parameters such as 50 population size, 1,000 max-iteration and 100 dimension. For fairly presenting the efficiently of all algorithms, a new experiment of 12 benchmark functions [11, 12, 14] are established. The experiments were conducted on windows 7 with Intel Core i5 2.5 GHz 8 GB memory and Matlab R2017a 64 bit. Moreover, the capability of CVS is compared to the well-known optimization algorithms DE, GWO, MFO, original VS and the modified MVS algorithm.

4.1 Results and Discussion

The proposed CVS algorithm is compared the performance to the novel DE, GWO, VS and the modified MVS through the 12 benchmark functions on the 30 different runs. The averages of mean and standard deviation accuracy which obtained from the evaluation of each algorithms were recorded. The results in Table 1 showed the performance of CVS which outperform the comparison algorithms as can be seen in the bold-face values. Moreover, Fig. 3 are used to present the convergence curve of VS and comparison algorithms for finding the optimum solutions.

As in Table 1, function *f1* – *f7* are the uni-modal and *f8* – *f12* are the multi-modal function. Uni-modal test functions have only one optimum and there are no local optima. Multi-modal have more than one optimum which make them suitable for benchmarking the local optima avoidance and explorative behaviour of optimization algorithms. The results on uni-modal testing show that CVS outperforms other algorithms accept DE in the *f6* values. For example in Table 1, function *f1*, the optimum values (Mean/Std) of VS = 8.74e−01/3.16e−01, MVS = 3.4e−16/1.86e−15, GWO = 2.23e−34/3.39e−34, DE = 2.18e−05/6.29e+06 and the outperform values (Mean/Std) of CVS = 0/0. Whereas, on multi-modal, the proposed CVS still show the result values that outperform all comparison algorithms.

Table 1. Results of algorithms on the 100 dimensional benchmark functions.

Name	Criteria	CVS	VS	MVS	GWO	MFO	DE
F1	Mean	**0**	8.74e−01	3.4e−16	2.23e−34	3.14e+04	2.18e−05
	Std	**0**	3.16e−01	1.86e−15	3.39e−34	1.45e+04	6.29e+06
F2	Mean	**0**	2.14e+02	2.61e−09	7.8e−21	1.70e+02	6.35e−04
	Std	**0**	1.22e+02	1.41e−08	4.92e−21	4.91e+01	8.42e−05
F3	Mean	**0**	4.05e+04	9.62e−19	8.25e−01	1.78e+05	3.62e+05
	Std	**0**	6.70e+03	4.36e−18	2.93e+00	5.91e+04	2.85e+04
F4	Mean	**0**	4.65e+01	4.12e−09	2.15e−04	9.16e+01	3.91e+01
	Std	**0**	8.71e+00	2.19e−08	4.50e−04	2.47e+00	2.62e+00
F5	Mean	**9.84e+01**	8.62e+02	1.77e+03	9.74e+01	5.38e+07	2.68e+02
	Std	**7.93e−02**	7.15e+02	2.30e+03	8.41e−01	4.87e+07	1.10e+02
F6	Mean	5.01e+00	9.11e−01	2.62e+01	8.17e+00	2.55e+04	2.23e−05
	Std	7.09e−01	3.74e−01	2.53e+01	9.70e−01	1.33e+04	6.15e−06
F7	Mean	**2.50e−05**	4.64e−01	2.65e−03	1.55e−03	1.22e+02	1.18e−01
	Std	**2.34e−05**	7.88e−02	5.44e−03	5.05e−04	8.41e+01	1.30e−02
F8	Mean	**−4.09e+04**	−2.97e+04	−3.75e+04	−1.68e+04	−2.44e+04	−4.64e+05
	Std	**1.81e+03**	1.71e+03	2.41e+03	2.41e+03	2.70e+03	6.85e+05
F9	Mean	0	3.65e+02	0	7.13e−02	6.96e+02	5.26e+02
	Std	0	6.21e+01	0	3.90e−01	6.57e+01	1.71e+01
F10	Mean	**8.88e−16**	4.30e+00	6.35e−11	6.89e−14	1.98e+01	5.83e−04
	Std	**0**	3.93e+00	1.89e−10	4.73e−15	3.46e−01	7.00e−05
F11	Mean	0	4.57e−01	0	9.14e−04	2.28e+02	1.58e−05
	Std	0	8.27e−02	0	3.60e−03	1.18e+02	5.27e−06
F12	Mean	**1.57e−01**	1.04e+01	1.75e+00	1.86e−01	8.55e+07	3.21e−01
	Std	**3.47e−02**	2.45e+00	1.39e+00	5.18e−02	1.19e+08	3.86e−02

Figure 3 also shows clearly the different curves of the compared algorithms. The CVS, described with a red line, possess a convergence curve and is capable of finding a solution than the comparison algorithms.

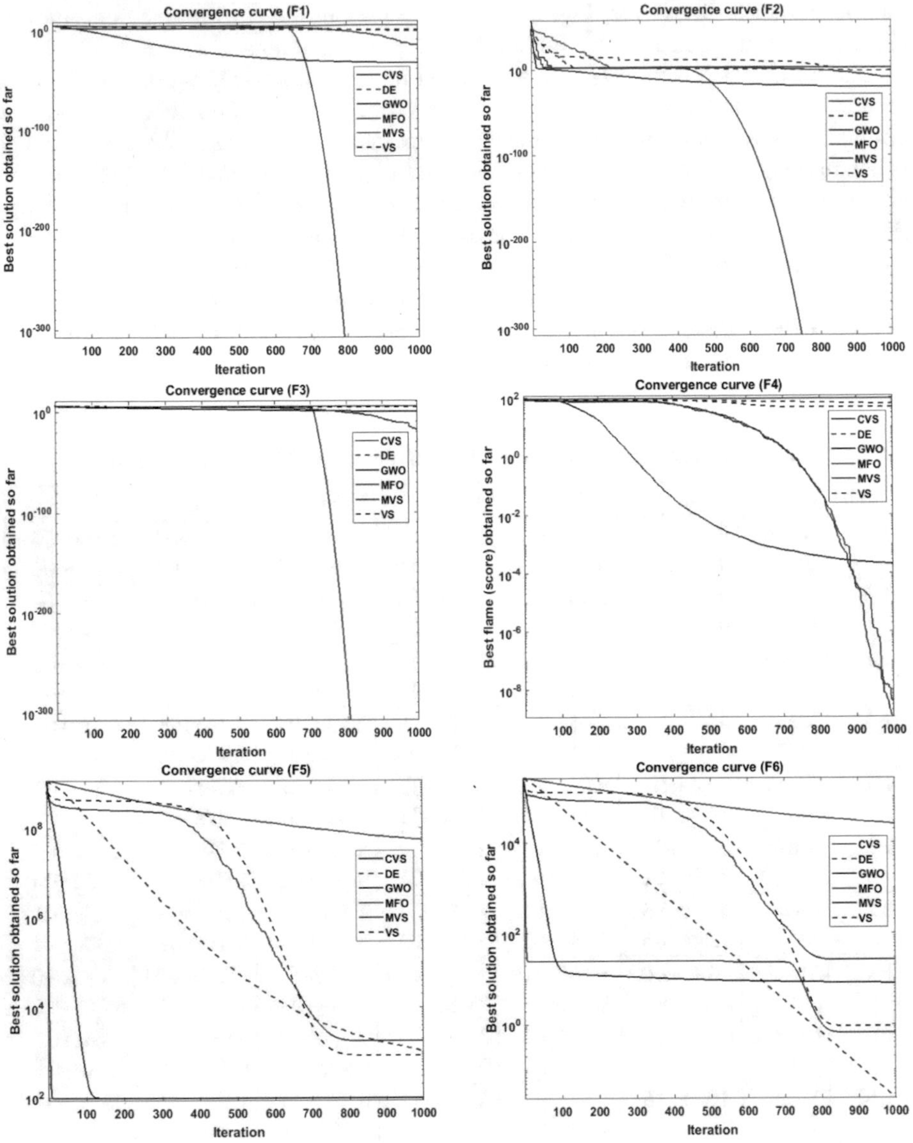

Fig. 3. The convergence curve of function *f*1 – *f*12 compared between CVS, DE, GWO, MFO, MVS and original VS algorithm on 12 benchmark function.

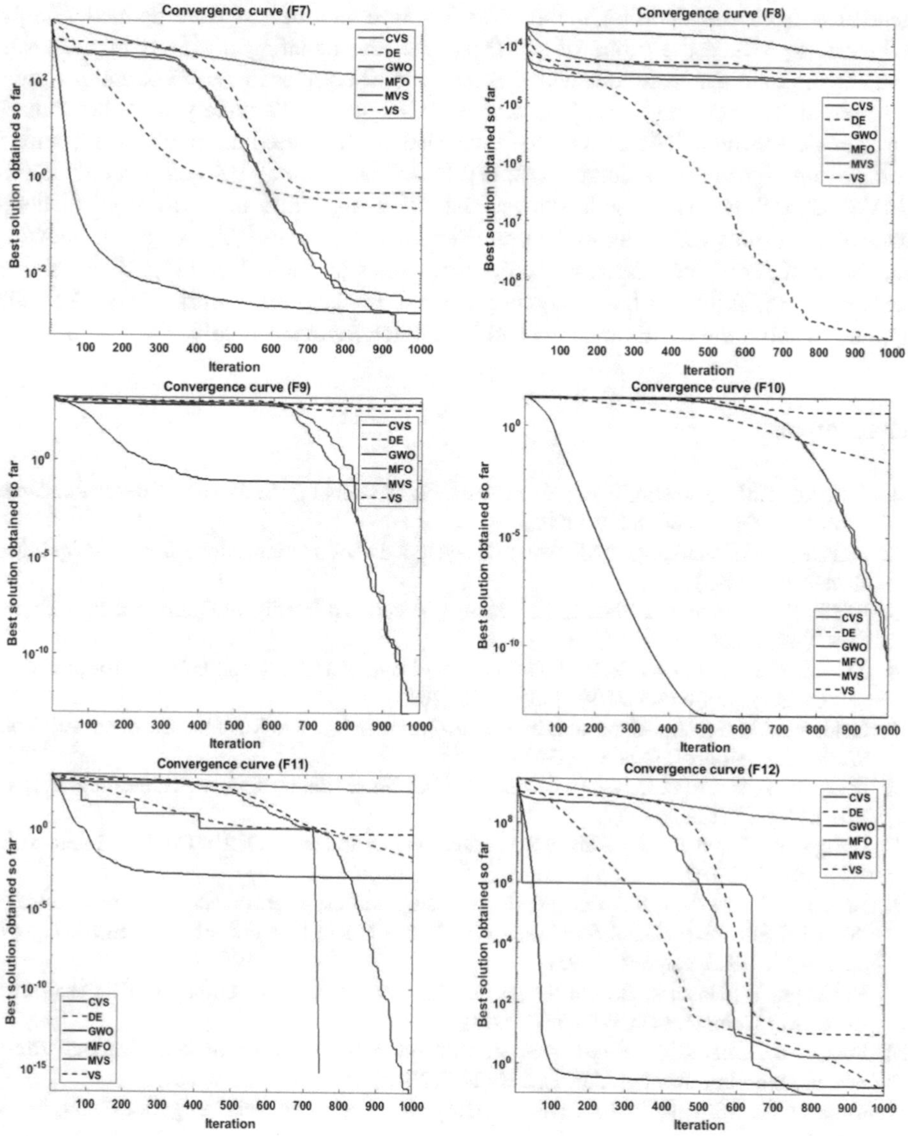

Fig. 3. (*continued*)

5 Conclusion

This study proposed the cooperation of candidate solutions vortex search (CVS) for numerical function optimization. The idea in this study is inspired from some drawbacks of VS algorithm. To overcome the drawback, CVS presented the achievement model by adding some candidate solutions added into the existing group of vortex candidate solutions. The advantages of CVS is that some candidate solutions are generated from the dividing member of VS into number of user defined groups. Whereas, to reduce the scope of search spaces, the updating positions of CVS still updated the candidate solutions with Gaussian distribution as in the VS technique used. To present the capability of CVS, new experiments on 12 widely used benchmark functions are provided. Moreover, the CVS also are measured the performance with 5 comparison algorithms including DE, GWO, MFO, original VS and a modified VS (MVS) algorithms. The results showed that CVS algorithm outperform all of comparison algorithms according to the promising of the proposed. The evidence showing capability of CVS can be seen in the optimal values in Table 1 and Fig. 3 also shows clearly the capabilities of convergence curves. In the future works, the CVS will applied to solve the multi-objective and its application respectively.

References

1. Pardalos, P.M., Romeijn, H.E. (eds.) Handbook of Global Optimization. Kluwer Academic Publishers, Dordrecht, vol. 2 (2002)
2. Floudas, C.A., Pardalos, P.M. (eds.): Encyclopedia of Optimization, 2nd edn. Springer, Heidelberg (2009)
3. Battiti, R., Brunato, M., Mascia, F.: Reactive Search and Intelligent Optimization. Springer, New York (2009)
4. Kvasov, D.E., Sergeyev, Y.D.: Lipschitz global optimization methods in control problems. Autom. Remote Control **74**(9), 1435–1448 (2013)
5. Goldberg, D.: Genetic Algorithms in Search, Optimization, and Machine Learning. Addison-Wesley Professional, Reading (1989)
6. Kirkpatrick, S., Gellet Jr., C.D., Vecchi, M.P.: Optimization by simulated annealing. Science **220**, 671–680 (1983)
7. Dorigo, M., Birattari, M., Stutzle, T.: Ant colony optimization. IEEE Comput. Intell. Mag. **1**(4), 28–39 (2006)
8. Kennedy, J., Eberhart, R.: A new optimizer using particle swarm theory. In: Proceedings of the IEEE 6th International Symposium on Micro Machine and Human Science, Nagoya, pp. 39–43. IEEE, Nagoya (1995)
9. Karaboga, D., Basturk, B.: On the performance of artificial bee colony (ABC) algorithm. Appl. Soft Comput. **8**(1), 687–697 (2008)
10. Doğan, B., Ölmez, T.: A new metaheuristic for numerical function optimization: Vortex Search algorithm. Inf. Sci. **239**(1), 125–145 (2015)
11. Mirjalili, S., Mirjalili, S.M., Lewis, A.: Grey wolf optimizer. Adv. Eng. Softw. **69**, 46–61 (2014)
12. Mirjalili, S.: Moth-flame optimization algorithm: A novel nature-inspired heuristic paradigm. Knowl.-Based Syst. **89**, 228–249 (2015)
13. Doğan, B.: A modified vortex search algorithm for numerical function optimization. Int. J. Artif. Intell. Appl. (IJAIA) **7**(3), 37–54 (2016)
14. Mirjalili, S.: The ant lion optimizer. Adv. Eng. Softw. **83**, 80–98 (2015)

Integrating Multi-agent System, Geographic Information System, and Reinforcement Learning to Simulate and Optimize Traffic Signal Control

Hoang Van Dong[1,2], Bach Quoc Khanh[1], Nguyen Tran Lich[1], and Nguyen Thi Ngoc Anh[1,2(✉)]

[1] Hanoi University of Science and Technology, Hanoi, Vietnam
donghv.mi@gmail.com, bachkhanh24@gmail.com, lichntbk@gmail.com, anh.nguyenthingoc@hust.edu.vn

[2] Sorbonne Université IRD, JEAI WARM, Unité de Modélisation Mathématiques et Informatique des Systèmes Complexes, UMMISCO, 93143 Bondy, France

Abstract. Traffic signal control (TSC) is an important problem that has been interested by many researchers and urban managers. Simulating and optimizing TSC for real-time control system is investigated recently with development by the Internet of things (IoT). The new model integrating Multi-agent system, geographic information system (GIS), and reinforcement learning to optimize TSC is proposed in this paper. The proposed simulation is minimizing total waiting time. Moreover, the simulation is applied into Ba Dinh ward, Hanoi, Vietnam for a case study.

Keywords: Traffic signal control · Simulation
Reinforcement learning · Multi-agent · Optimization

1 Introduction

Normally, traffic signals (in the form of red, yellow and green lights) resolve the conflict of the traffic flows reaching a junction from different directions [1]. Because of the many benefits from effective traffic signal bring to the transportation systems, such as less congestion, lower polluted air environment, reductions in the costs of wasted time, fuel and the stop time of transportations. Thus, Traffic signal control (TSC) problems have attracted many researchers and urban managers [1–7].

In the literature for TSC, there have been some proposed approaches that are separated into two types: microscopic models and macroscopic ones. Macroscopic models are often traditional control paradigms and microscopic models are adaptive control paradigms. Some of macroscopic models are petri-net models [3,4] predictive control model, dynamic programming models (DPM) [2]. Most traffic signal controllers based on macroscopic models are applied on an isolated junction. Moreover, they do not take into account the changing of dynamic road

H. Unger et al. (Eds.): IC2IT 2018, AISC 769, pp. 145–154, 2019.
https://doi.org/10.1007/978-3-319-93692-5_15

network with different behaviors and interactions between vehicles and the different parts of the traffic network. To avoid the disadvantage of macroscopic model, the microscopic model tends to be more adaptive to the current traffic conditions. In the road network dynamics, they can take into account changes of many traffic components such as number of vehicles, number of vehicle types, accidents, rush hours. Some microscopic models are mainly based on artificial intelligence models (AI) such as machine learning (ML) model [6,7], Markov decision process [5], Reinforcement learning-based traffic signal controller [8,9]. The disadvantage of microscopic are cost computation, need big data.

One of macroscopic models that are model predictive control framework using mixed integer programming. This model has been applied for network wide TSC [10]. The computation time required by the mixed integer programming for optimizing all signals reduces. However, the disadvantage of this model is not taking into account the difference of behaviors of vehicles and the interaction of them and traffic signals.

Multi-agent system (MAS) is used to apply to many complex systems, including TSC [9]. The advantage of this model is that it describes naturally the real system with each agent interact the other agents and heterogeneous environment. Concretely, it can take into account different behaviors of each vehicle, light, junction of the transport system and heterogeneous environment even Geographic information system (GIS) environment. The disadvantage of this model is the costly computation time on executing many experiments to find the solution of optimization problem [9].

In the paper [1], authors considered model predictive control (MPC) for Markov decision processes. To optimize a system, they defined value functions based on the sum of rewards given by the performance function at decision step. They considered using experience to incrementally learn value functions over time. To speed up solving problem, reinforcement-learning models are proposed like temporal-difference learning the agent incorporates experience built up through interaction with the system. Combining MPC and Reinforcement-learning models are used in many areas including TSC. The original model, the probabilities is estimated based on the frequentist interpretation of probability the number of vehicles transition states. The required data are limited because using real data, the changing strategy to control, it is difficult to get data for systematic learning. Thus, in this paper MAS, GIS and RL are integrated to simulate many experiments. The data from many different strategies and observation help decision maker can choose the most effective strategy.

This paper addresses the optimization problem of real-time traffic control that includes minimizing trip waiting time. It means maximizing flow rate, satisfying green waves for platoons traveling on main roads. The new hybrid-based model integrating GIS, MAS and RL is investigated to solve this problem for mixed traffic flows (Fig. 1).

The paper is structured into four sections. The introduction and related work of urban traffic signal controllers are presented in Sect. 1. The methodology about the multi-agent system, GIS and Reinforcement learning model and proposed

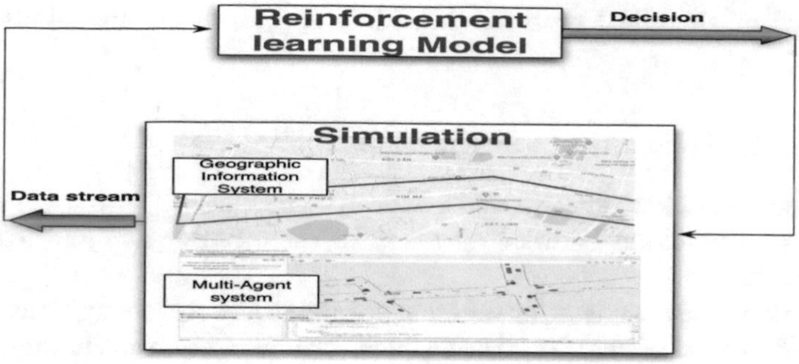

Fig. 1. Hybrid-based model integrating Geographic information system, Multi-agent system and Reinforcement learning.

model is shown in Sect. 2. Section 3 presents the experiments and results are applied our new proposed model. Finally, Sect. 4 concludes the paper, discusses about the advantage/disadvantage of our proposed model and proposes some future works.

2 Methodology

2.1 Geographic Information System for Traffic Signal Control

A geographic information system (GIS) is a computer system for capturing, storing, checking, and displaying data related to positions on Earth's surface. GIS visualize many different kinds of data on one map [11]. GIS is heterogeneous environment that people easily see, analyze, and understand patterns and relationships. Using GIS for the traffic signal control problem, the same map could include sites such that boundary of research area, road system, signal control light system, buildings, alert signs, motors, cars, buses, trucks.

GIS can use any information that includes location. The location can be expressed in many different ways, such as latitude and longitude, address. In traffic signal problem, behaviors of vehicles within space requires some sort of geographical information system (GIS) in order to identify spatial relationships. By linking multi-agent system and GIS we have the ability to model the emergence of phenomena through individual interactions over space and time.

2.2 Multi-agent System

The signal control problem is modelled by multi-agent system that is considered separate intersection, which did not show much of the effect of intersections on an overall network traffic system. In a single intersection traffic signal control model, a signal controller of an intersection is simple. However, the road network, traffic signal controller are complex systems, the interactions of vehicles and signal of different intersection. Concretely, the traffic signal control of the road network is modelled by the ODD protocol [12]. Main species in our model are detailed as follows:

Purposes. The purposes of the model are

- Give some scenarios for GIS and traffic system.
- Observer interaction between agents with the other agents and agents with environment.
- Simulation to create data for different scenarios about waiting time, the number of stop time of vehicles, travel time of vehicles, etc. that finding the optimal scenario.

Entities, State Variable and Scale. In the multi-agent system, all things will be considered as agents such as roads, vehicles, signal control, etc. Each agent has its own behavior and attribution. In addition, each agent interacts with the other agents and the environment.

(a) Geometric agents

Geometric agents are built based on the real data GIS. These agents contain the geometric data such as roads, start waiting, end points, and traffic lanes, and thus have all geometric properties (i.e.: position, length, etc.) and do not possess any behaviors.

(b) Vehicles

Represent the means of traffic systems. Vehicle agents have attributes: running lanes, current velocity. In addition, each vehicle agent has its own type motor, car, bus, etc. that has different size, capacity, the rulers of moving.

The main behaviors of the vehicle include: velocity, direction of movement, identification of control signals, pause, etc.

(c) Lights

The traffic vehicle system will pass through intersection based on the color of the signal lights (if signal light is green, vehicles can go; if the signal light is yellow or red, vehicles must stop).

The number of signal lights depends on intersections, at the intersection of two roads having two-way street, the number of lights is usually 4 or 8, while if the intersection is between a two-way street and a one-way street there are usually only three lights.

The main behavior of the signal lights is to adjust the color according to the current state of the intersection.

(d) Intersection

Each intersection belongs some lights. The road network has a number of intersections in the system, each intersection will control the corresponding signal lights. Therefore, it indirectly controls.

As the main controlling agent, intersections are the most interesting actors in the model. The optimization of the problem is also achieved by the intersections' behavior.

The behaviors of an intersection will change depending on the design scenario, the two most important behaviors is counting the time of the current state and the change of state.

(e) States variables

The two common state variables in this model are the phase indicator of the intersection and the color of the signal. In addition, for some scenarios developed during simulation, different traffic volume (during low, mid, rush traffic hours) is also taken into consideration.

(f) Scales

Spatial scale is defined by GIS data. The spatial scale includes giving intersection. Depending on the scenarios, the time scale may change, the small scale is second and big-time scale of the model and simulation is some hours or a full day.

Process Overview and Scheduling. Vehicles occur at random points on the system, then moves to the specified destination and exits the system. The movement of the vehicle is shown in Fig. 2. Intersection: intersections have fixed locations, represented by five intersections on the traffic system. Depending on the scenario that simulates actual data or performed optimally by the Q-learning algorithm, the process of the intersection will be the process shown in Figs. 3 or 4.

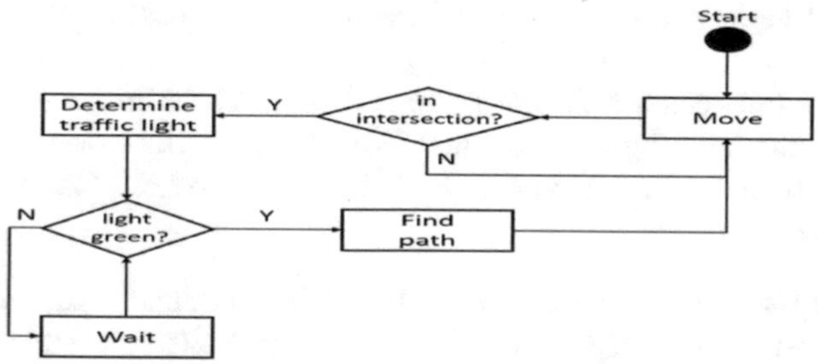

Fig. 2. Process and scheduling of vehicles.

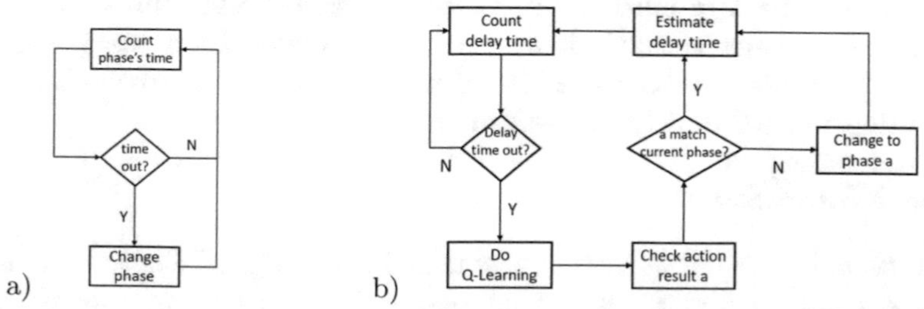

Fig. 3. (a) Process and scheduling of interaction in real-life control. (b) Process and scheduling of interaction which is controlled by Q-learning.

Lights: they will get information about the current state of the intersection corresponding to it to adjust the color accordingly. The process of adjusting the light color is shown in Fig. 4.

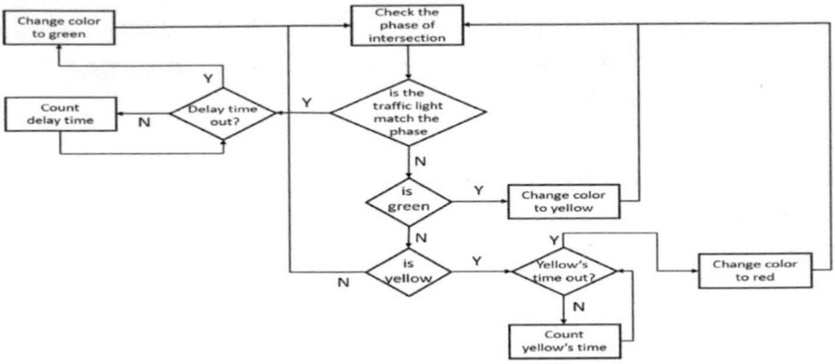

Fig. 4. Process and scheduling of signal control.

Design Concepts. *(a) Basic principles.* Basics principles designed for the model are listed below.

- Movement of vehicles in the system is based on ideal traffic conditions.
- Traffic light signals work exactly following the controlling of the intersection agents.
- Conflictions between traffic flows are not allowed at intersections under the control of the traffic light controllers (for example, the flows of vehicles from Kim Ma to Kim Ma and from Nguyen Chi Thanh to Lieu Giai are not allowed at the same time).

(b) Adaptation. In the scenario applied reinforcement learning algorithms, intersection agents frequently store the number of halting vehicles in the intersection under the form of state vectors at predetermined moments. Hence, we have a wiser selection of actions when a state is homogeneous with states before.

(c) Objective. Goal of a vehicle is moving to the destination with the shortest path. In the scenario applied reinforcement learning algorithms, the purpose of intersection agents is maximizing the reward (reward - a scalar value that calculated from state vectors). This purpose is equivalent to minimizing the total waiting time of vehicles in each intersection.

(d) Interaction

Direct Interaction

- Interaction between intersection agents and traffic lights: The color of a traffic light is changed directly based on phase characteristic of the intersection agent.
- Interaction between traffic lights and vehicles: vehicles are allowed to move through an intersection if and only if the corresponding light turns into green.

- Interaction between environment and vehicles in the scenario that has time domain is a day (or one-day scenario): when environment is changing through time in different milestones such as low traffic, average, and high traffic hours, the number of vehicles on the roads is changing correspondingly.
- Interaction between vehicles and intersection agents in the scenario applied reinforcement learning algorithms: because state vectors are computed based on the number of halting vehicles in an intersection, actions of an intersection agent is induced based on moving vehicles on it.

Indirect Interaction

- Interaction between intersection agents and vehicles: this interaction is performed indirectly by traffic light agents.
- Interaction between environment and intersection agents in one-day scenario: since environment agent affects the vehicle flows capacity, it affects indirectly actions of intersections.

(e) Stochasticity. Vehicle flow in the system at each moment is a Markov random process with a predetermined expected value. Intersection agents choose random actions when there is insufficient information about states in the past.

Initialization. To integrate Multi-Agent system, GIS and reinforcement learning, GAMA platform is used. GAMA platform is the object-oriented paradigm that the notion of class is used to supply a specification for objects, agents in GAMA are specified by their species, which provide them with a set of attributes, actions, behaviors and also specifies properties of their population, for instance its topology or schedule [13]. Further more, agent in GAMA can also inherit its properties from another species, creating a relationship similar to specialization in object-oriented design.

When initializing interconnected intersection system, each intersection has determined traffic lights, and number of traffic lights vary based on the characteristics of their corresponding intersection. When initial time of an intersection is 0 AM, average arrival vehicle flow is distributed based on parameters of low traffic time. Number of loops that correspond to a second is 50, so one hour equals 180000 loops, and one day equals 43200000 loops.

Input

- Geographic data (e.g.: positions, shapes, etc.).
- Average vehicle capacity data in the system at high, average, and low traffic hours.
- Motor/car/bus ratios.
- Traffic light time durations in practice.

Submodels (Q-learning). Q-learning is a reinforcement learning method that studies how to optimize a cumulative long-run reward. Specifically, this method will set a target function to calculate reward based on the state of the dynamic environment depending on the action of the agents. Q-learning will give a action decision for new strategy of the system. Concretely, Q-learning will select an action for that optimizes the reward function [8].

In this paper, the agent in Q-learning is an intersection that take from MAS and GIS (Fig. 1). In addition, the reward function is waiting time of intersection that is total waiting time of vehicles at this intersection that is calculated by MAS and GIS model. The main action is traffic signal of interaction that is the retention time of the current light state and the length of signal traffic. The maximum queue at the moving lanes is a state that decides the action of each intersection to minimize the waiting time of whole system.

3 Experiments and Results

3.1 Input Data

- GIS data of Kim Ma - Nguyen Thai Hoc route, with a total of 5 intersections, where two intersections that intersect between two two-way roads, the other three intersections that intersect between one-way road with two-way road (Fig. 1).
- Average flow capacity per minute and ratios between number of car/bus/motor in 13 points in the system.
- Time durations of traffic lights at intersections in the systems in real-life.

3.2 Implementation Scenarios

Scenario 1: Traffic lights are controlled based on practical database; traffic flows that move into the system follow Poisson process with parameters which are real average number of arriving flows per minute and one-hour duration of total consideration time.

Scenario 2: Traffic lights are controlled by Q-learning algorithm; traffic flows that move into the system follow Poisson process with parameters which are real average number of arriving flows per minute and one-hour duration of total consideration time.

3.3 Results

After running 30 experiment times for each scenario, the experiment data are collected and analyzed. Based on purpose of TSC problem, the waiting time data is especially considered.

At each intersection, the simulation calculates waiting time of a intersection by sum waiting time of all vehicles at this intersection. Then, simulation calculates the total waiting time of all intersection of the traffic system.

Therefore, with 30 experiment times for each scenario, we are able to collect data and then analyze to give our results.

Two data sets for experiment 1 and 2 are analyzed respectively:

Experiment 1's result summarized:

```
  Min. 1st Qu.  Median    Mean 3rd Qu.    Max.
102.4   102.5   103.9   104.1   105.2   106.5
```

Experiment 2's result summarized:

```
 Min. 1st Qu.  Median    Mean 3rd Qu.    Max.
82.95  84.81   85.93   85.86   86.14   90.32
```

Finally, the result comparing average waiting time of two scenarios are described in Fig. 5.

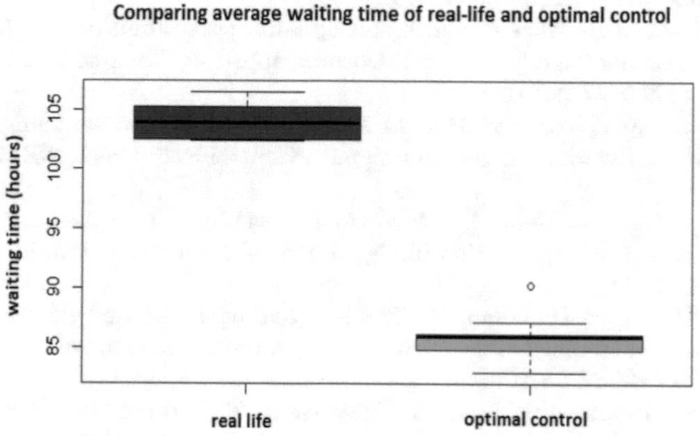

Fig. 5. Box plot for comparing average waiting time of two scenarios.

From the results above, we imply that Q-learning algorithm helps reducing the sum of average waiting time of vehicles in the system from 104.1 h down to 85.9 h (approximately 17.5% in percentage)

4 Conclusion and Discussion

In this paper, a multi-agent system for intersection traffic systems is built and formulated, then integrating GIS, MAS and reinforcement learning is applied to solve the traffic signal optimization problem. The optimal results show effective more than real signal control from 15 to 20%. The results of the paper:

- Obtain real-time traffic mixed vehicle data such as GIS, average vehicle intake, lamp time, etc.
- Successfully build model of traffic system on Kim Ma - Nguyen Thai Hoc street.
- A good combination between the two models is the multi-agent system and the reinforcement learning model.

- Getting initial results when solving a control signal optimization problem by building an integrated model.

Acknowledgment. This research was partially supported by a project of Hanoi University of science and technology, Hanoi, Vietnam and IRD, UPMC Univ Paris 06 UMMISCO.

References

1. Negenborn, R.R., Schutter, B.D., Wiering, M.A., Hellendoorn, H.: Learning-based model predictive control for Markov decision processes. In: Proceedings of the 16th IFAC World Congress, pp. 354–359 (2005)
2. Zhu, F., Ukkusuri, S.V.: Accounting for dynamic speed limit control in a stochastic traffic environment: a reinforcement learning approach. Transp. Res. Part C Emerg. Technol. **41**, 30–47 (2014)
3. dos Santos Soares, M., Vrancken, J.: A modular petrinet to modeling and scenario analysis of a network of road traffic signals. Control Eng. Pract. **20**(11), 1183–1194 (2012)
4. Chen, F., Wang, L., Jiang, B., Wen, C.: A novel hybrid petrinet model for urban intersection and its application in signal control strategy. J. Frankl. Inst. **351**(8), 4357–4380 (2014)
5. Xu, Y., Xi, Y., Li, D., Zhou, Z.: Traffic signal control based on Markov decision process. In: Proceedings of the 14th IFAC Symposium on Control in Transportation Systems, pp. 67–72 (2016)
6. Kamal, M., Imura, J., Ohata, A., Hayakawa, T., Aihara, K.: Control of traffic signals in a model predictive control framework. In: Proceedings of the 13th IFAC Symposium on Control in Transportation Systems, pp. 221–226 (2012)
7. Polson, N.G., Sokolov, V.O.: Deep learning for short-term traffic flow prediction. Transp. Res. Part C Emerg. Technol. **79**, 1–17 (2017)
8. Mannion, P., Duggan, J., Howley, E.: Parallel reinforcement learning for traffic signal control. Procedia Comput. Sci. **52**, 956–961 (2015)
9. Das, S., Bhattacharyya, B.K.: Optimization of municipal solid waste collection and transportation routes. Waste Manag. **43**, 9–18 (2015)
10. Kamal, M., Imura, J., Ohata, A., Hayakawa, T., Aihara, K.: Traffic signal control in an MPC framework using mixed integer programming. In: Proceedings of the 7th IFAC Symposium on Advances in Automotive Control, pp. 645–650 (2013)
11. Kotikov, J.: GIS-modeling of multimodal complex road network and its traffic organization. Transp. Res. Procedia **20**, 340–346 (2017)
12. Grimm, V., Berger, U., DeAngelis, D.L., Polhill, J.G., Railsback, S.F.: The odd protocol: a review and first update. Ecol. Model. **221**(23), 2760–2768 (2010)
13. Grignard, A., Taillandier, P., Gaudou, B., Vo, D.A., Huynh, N.Q., Drogoul, A.: GAMA 1.6: advancing the art of complex agent-based modeling and simulation. In: Boella, G., Elkind, E., Savarimuthu, B.T.R., Dignum, F., Purvis, M.K. (eds.) PRIMA 2013: Principles and Practice of Multi-Agent Systems, pp. 117–131. Springer, Heidelberg (2013)

Image Processing

Digital Image Watermarking Based on Fuzzy Image Filter

K. M. Sadrul Ula(✉), Thitiporn Pramoun, Surapont Toomnark, and Thumrongrat Amornraksa

Multimedia Communication Laboratory, Computer Engineering Department, Faculty of Engineering, King Mongkut's University of Technology Thonburi, Bangkok, Thailand
kmsadrulula@gmail.com, thitiporn1016@gmail.com, surapont@gmail.com, t_amornraksa@cpe.kmutt.ac.th

Abstract. This paper presents an image watermarking method based on fuzzy image filter. In the embedding process, a unique Gaussian noise-like watermark is added into the blue color component from the *RGB* color space of original image. The value of each embedding bit is secured by the use of a secret key-based stream cipher. In the extraction process, a blind detection approach is used so that the original image is not required. The decreasing weight fuzzy filter with moving average value (DWMAV) is considered and applied to the watermarked component to remove the added watermark so that the original blue color component can be estimated. The extracted watermark is finally obtained by subtracting the estimated blue color component from the watermarked one. For performance comparison purpose, the weighted Peak Signal-to-Noise Ratio (w*PSNR*) is used for evaluating the quality of watermarked image, the normal correlation (*NC*) for the accuracy of extracted watermark, and the Stirmark benchmark for the robustness of embedded watermark. Experimental results confirm superiority of the proposed watermarking method compared to the two similar previous methods.

Keywords: Digital image watermarking · Fuzzy logic · Fuzzy image filter Gaussian noise · Decreasing weight fuzzy filter with moving average value

1 Introduction

Digital watermarking is used to protect unauthorized usage and distribution of digital media like images, videos. It is also used to identify the originator and verify authenticity or integrity of data. Everyday a tremendous amount of digital content is adding into cyberspace. Security risk is increasing day by day. Attackers are developing new attacking techniques e.g. self-similarity, line removal attack, etc. To protect attacks and to discourage attackers, watermark is introduced and it ensures to survive intact without affecting the quality of content during formation and transmission through a channel. Hence, digital watermarking has been remaining a very popular topic for researchers over last couple of years, and numerous amount of techniques already developed by them. Image watermarking consists of two processes, watermark

H. Unger et al. (Eds.): IC2IT 2018, AISC 769, pp. 157–166, 2019.
https://doi.org/10.1007/978-3-319-93692-5_16

embedding and retrieval processes. First one involves adding owner information known as watermark into host image. Second one is to retrieve the embedded watermark from the watermarked image. Two image domains are available, i.e. frequency domain [1] and spatial domain [2]. In frequency domain, we embed a watermark into coefficients of transformed image. On the other hand, spatial domain is to embed a watermark into host image directly by modifying image pixels.

Plenty amount of previous work was done on image watermarking. As we focused on fuzzy filtering in this paper, we first considered some existing fuzzy image filters. Ville [3] proposed a new technique for filtering narrow-tailed and medium narrow-tailed noises by a fuzzy filter. Two important features were introduced. First one was a filter which estimated a "fuzzy derivative" in order to be less sensitive to local variations due to image structures such as edges. The second one was the membership functions which were adapted accordingly to the noise level to perform "fuzzy smoothing". Kwan [4] introduced 9 fuzzy filters for noise reduction in noisy images. All these fuzzy filters applied a weighted membership function to an image within a window to determine the center value. The filters were MED (Median filter), MAV (Moving average value filter), GMED (Gaussian fuzzy filter with median), TMED (Symmetrical triangular fuzzy filter with median), ATMED (Asymmetrical triangular fuzzy filter with median), GMAV (Gaussian fuzzy filter with moving average value), TMAV (Symmetric triangular fuzzy filter with moving average value), TMAV (Asymmetric triangular fuzzy filter with moving average value), DWMAV (Decreasing weight fuzzy filter with moving average value). Marudhachalam [5] developed various fuzzy hybrid filtering techniques to remove random noise from medical images, using topological approach. Each of these fuzzy filters, which applied a weighted membership function to an image within the 8-neighbours of a point, was simple and easy to implement. Rahman [6] developed a modified fuzzy filter for Gaussian noise reduction in digital images. It was a modified Gaussian fuzzy filter with max center (GFMC) designed for a statistical noise which usually possessed zero mean Gaussian distribution characteristic.

We now considered some existing image watermarking methods. Sridevi [7] proposed digital image watermarking using genetic algorithm in DWT and SVD transform. In this method, genetic algorithm was used to find the best positions for embedding the watermark to obtain the optimal solution and to achieve robustness and fidelity. Discrete wavelet transform (DWT) was applied to original image because of its spatial multi-resolution nature, and Singular value decomposition (SVD) was also applied on the selected middle band. A strong sensitivity of digital image watermarking scheme for noise disturbance was proposed by Rosiyadi [8]. It was based on the singular value decomposition (SVD) method by the texture masking in accordance with the human visual system (HVS) characteristics. This scheme composed of two main steps, calculating the weighted value of enhanced noise sensitivity by the texture masking model to obtain the strong sensitivity, and modifying the singular value of host image and the singular value of watermark by the SVD operation. Bansal [9] proposed watermarking method where the features of DWT and Discrete Cosine Transformation (DCT) were combined. In embedding process, they used red color component of image to apply the DWT. Later they implemented DCT on the 8×8 block of image. Hurrah [10] proposed a watermarking scheme for color image based on hybrid transform domain. It was developed to resist dual attacks which may be combination of signal processing and geometric attacks. They used a new interblock differencing method in DCT domain to embed a single watermark in all the three color components, i.e. RGB, of the color image. Sudibyo [11] proposed watermarking method using Chinese remainder theorem based on Haar Wavelet Transform (HWT). It can provide security to copyright because of its insertion technique which indirectly encrypts the copyright itself. Thongkor [12] proposed a spatial domain image watermarking method based on regularized filter. In their method, a watermark image was embedded into a host color image directly by modifying the blue color component. The watermark strength was controlled by two factors, i.e. a constant value and the luminance within a local embedding area. The prediction of the original host image was obtained from the watermarked image by using the regularized filter, so that the embedded watermark could be blindly recovered by subtracting the predicted image from the watermarked image. Lastly, Lai [13] proposed an image registration algorithm which works against the local geometric attack. An amplitude modulation watermarking scheme was used to embed and retrieve the watermark. The watermark bits work as implicit anchors to help correct local geometric distortions. They used an automatically multi-scaled neighborhood-matching approach so that it can be used to quickly and efficiently estimate local displacement.

In this paper, we are proposing an image watermarking method based on fuzzy image filter called DWMAV. We consider a decreasing weight fuzzy filter with moving average value within a window as a center value because it can be easily implemented and helps to suppress low, medium and high levels noise [4]. This value is used to find the general window function. A set of experiments is carried out to evaluate performance of our watermarking method. Brief introduction to fuzzy image filter is given in the next section. The proposed image watermarking method implementing the DWMAV and its operation are described in Sect. 3. The experimental results are presented and the conclusion is shown in Sects. 4 and 5, respectively.

2 Background on Fuzzy Image Filter

The basic idea of this filter is to average a pixel using other pixel values from its neighborhood, but at the same time, to preserve the important image structures such as edges. Since images can contain a verity of noises, e.g. uniform noise, impulse (or salt and pepper), Gaussian noise, etc., various linear and nonlinear methods are introduced and applied. However linear filters tend to blur edges and reduce other fine image detail. In this case fuzzy techniques are introduced for detail-preserving smoothing of noisy images. Regarding Gaussian noise reduction, fuzzy filters have superior performance over linear filters because it can manage the vagueness and ambiguity efficiently [14]. In other words, fuzzy filters can preserve edge and image detail properly. That is why they are suitable to remove Gaussian noise.

Noise smoothing and edge enhancement are in intrinsically conflicting processes. Therefore, a single filter may not be suitable for filtering of different parts of an image. Depending on local context, filtering should vary from pixel to pixel. Some portions of an image the local conditions can be assessed only vaguely, that is why it is immensely difficult to set the conditions under which a certain filter should be selected. Hence, the capability of reasoning with vague and uncertain information is a desired requirement, and this suggests the use of fuzzy logic for image filtering. Several fuzzy image filters for noise reduction already have been developed, e.g. the well-known FIRE (Fuzzy Inference Ruled by Else-action) filters, Data dependent fuzzy filters, Hybrid filters, Fuzzy Vector Directional Filters, Fuzzy median filters and Fuzzy stack filters [14]. If we consider *x(i, j)* as the input of a 2-dimensional fuzzy image filter, then output of this filter is defined as [4]:

$$y(i,j) = \frac{\sum_{(r,s)\in A} F[x(i+r,j+s)].\, x(i+r,j+s)}{\sum_{(r,s)\in A} F[x(i+r,j+s)]}, \tag{1}$$

where the general window function is $F[x(i,j)]$ and the area of window is A. The range of r and s for a square window of dimensions $N \times N$, are: $-R \leq r \leq R$ and $-S \leq s \leq S$, where $2R+1 = 2S+1 = N$.

3 Proposed Image Watermarking Method

In this embedding process, first we do permutation on black and white watermark image $I_w(i,j) \in \{0,1\}$ using Gaussian distribution. Thereafter, the result is XORed with the pseudo-random bit stream generated from a secrete key to improve the balance of 1 s and 0 s around a pixel (i,j). Next, the 0 bits are converted into -1 in order that the watermark to be embedded becomes $W(i,j) \in \{-1,1\}$. Later, we adjust the watermark bit's value using scaling factors so that we can control the strength of the watermark for the entire host image. Finally, the watermark embedding is performed by modifying the image pixel in the blue color channel $B(i,j)$, in point-wise modification. The modifications of the image pixel $B'(i,j)$ are either additive or subtractive, depending on $W(i,j)$, and proportional to the modification of luminance of the embedding pixel $L(i,j) = 0.299R(i,j) + 0.587G(i,j) + 0.114B(i,j)$. The watermark embedding process can be given by the following equation.

$$B'(i,j) = B(i,j) + W(i.j)S_t L'(i,j), \tag{2}$$

where S_t is the signal strength used to control the overall quality of watermarked image, $L'(i,j)$ is the modification of luminance obtained from a Gaussian pixel weighted mask [15] which gives more weight at the center pixel in the filtering area in accordance with Gaussian distribution. L' was proved to help increase the watermark extraction performance in [15]. The block diagram of the embedding process is shown below (Fig. 1).

In extraction process, we use the DWMAV to estimate the original blue component. The DWMAV is considered and used because it has the superior performance while comparing the mean square error (MSE) of the original and filter noisy image for various levels of noises in different window sizes. [4] Center value chosen as the decreasing weight fuzzy filter with moving average value within a window is defined as

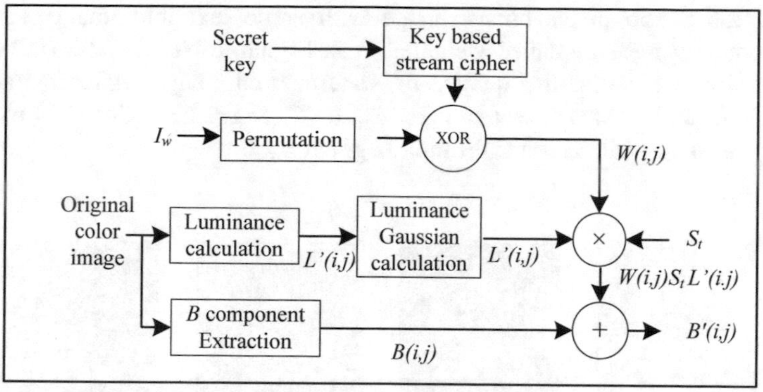

Fig. 1. The embedding process

$$B''(i,j) = \frac{\sum_{(r,s)\in A} F[B'(i+r,j+s)] \cdot B'(i+r,j+s)}{\sum_{(r,s)\in A} F[B'(i+r,j+s)]}, \tag{3}$$

$$F[B'(i+r,j+s)] = 1 - \frac{\max(|r|,|s|)}{\max(|R|,|S|)+t}, \tag{4}$$

where $F[B'(i+r,j+s)]$ is the general window function, square window of dimensions $N \times N$ where N is the width, and the threshold value is t that determines the height of the decreasing triangular shape weighted function at $|r| = R$ and/or $|s| = S$. Normally, $t = 1$, 2 and 3 provides a varying degree of finite impulse response (FIR) filtering performance. The block diagram of the extraction process is shown below (Fig. 2).

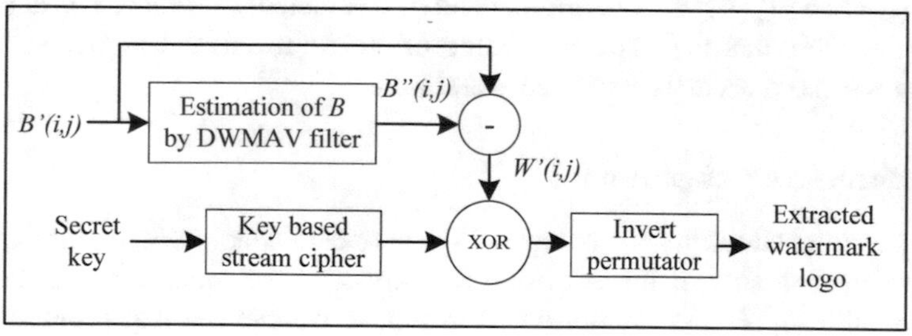

Fig. 2. The extraction process

4 Experimental Results

As seen in Fig. 3, in the experiments, we used 6 standard 256 × 256 pixels color images with different characteristics as host images. The images *'tiffany'* and *'splash'* were used as low level of detail image, *'pepper'* and *'lena'* used as medium level of detail images. For high level, we used *'barbara'* and *'boats'* [16]. As watermark image, we used 256 × 256 pixels binary image with some text into image. A common objective quality measure called weighted Peak Signal-to-Noise Ratio (*wPSNR*) [17] was used for evaluating the quality of watermarked image. After extracting the watermark, its quality was measured in terms of *NC* (Normalized Correlation) and used for comparison as well. The *NC* formula is given by

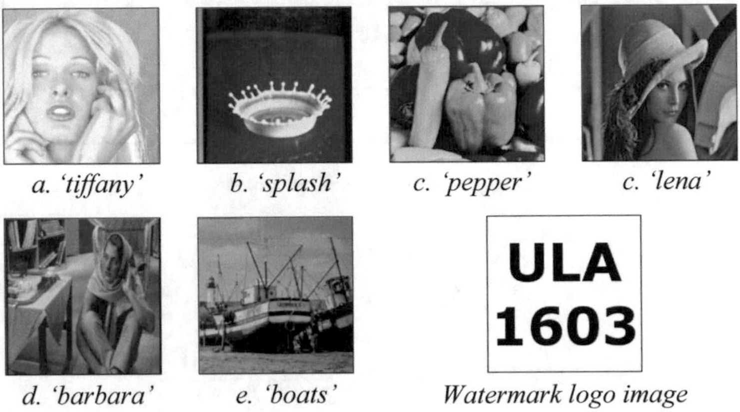

Fig. 3. All test images (original) and watermark image (original)

$$NC = \frac{\sum_{i=1}^{M}\sum_{j=1}^{N} I_w(i,j) I'_w(i,j)}{\sqrt{\sum_{i=1}^{M}\sum_{j=1}^{N} I_w(i,j)^2}\sqrt{\sum_{i=1}^{M}\sum_{j=1}^{N} I'_w(i,j)^2}}. \tag{5}$$

where $I_w(I,J)$ and $I'_w(i,j)$ are the original and the extracted watermark bits at pixel (i,j) respectively. Note that the maximum value of *NC* is 1, and the higher value of *NC* represents the more accurate extracted watermark.

4.1 Performance Comparison

For the experimental results, we compared the extracted watermark in terms of average *NC* for different quality of images using the proposed watermarking method and the previous methods [12, 13]. As illustrated in Fig. 4, we can see that proposed method was more superior to the previous ones. Figure 5 shows results of some watermarked image *'lena'* and its corresponding extracted watermark from these three methods using *wPSNR* of 30 ± 0.01 dB and 40 ± 0.01 dB. Note that the threshold shown in the figures was set by using the maximum *NC* from 1000 different versions of extracted watermarks from every test image. Accordingly, the maximum *NC* of 0.681 was obtained from the image *'splash'*, and thus used as the threshold to validate the extracted watermark.

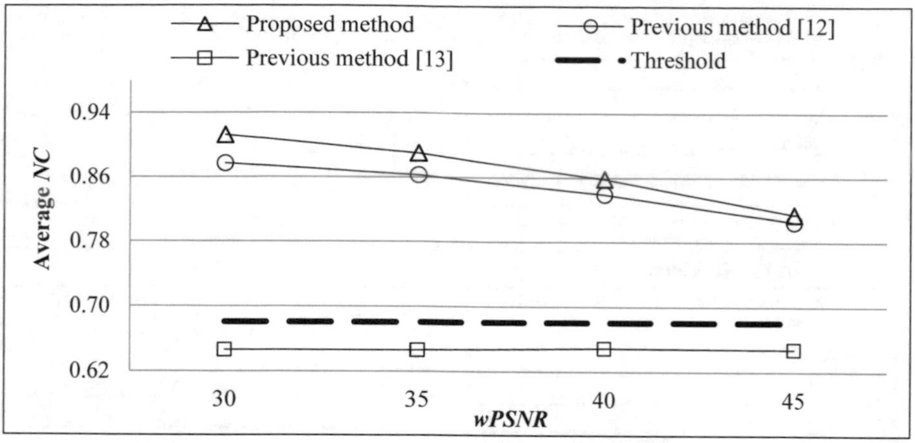

Fig. 4. Performance comparison of the three methods

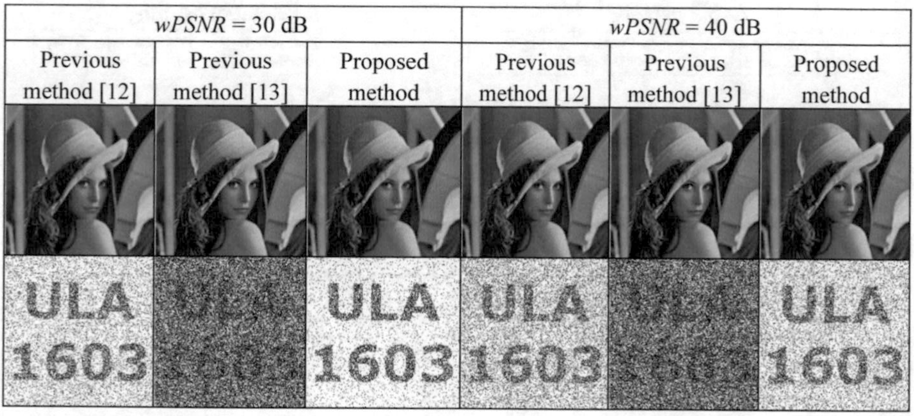

Fig. 5. Examples of the watermarked image *'lena'* and its corresponding extracted watermark

4.2 Robustness Against Attack

Next, we assessed the robustness of the embedded watermark by applying different types of attack at various strengths to watermarked image, obtained from the three methods at *wPSNR* of 30 ± 0.1 dB. Eight types of attacks from Stirmark benchmark [18] were used in the experiments. They were JPEG compression, rescaling, additive noise, rotation/scaling, line removal, self-similarity, rotation/cropping, and rescaling, as indicated in Fig. 6*(a)–(h)*, respectively.

From all the results obtained, it is obvious that our proposed method was more robust against attacks than the previous ones. For JPEG compression test, the proposed method was able to preserve better quality and average *NC*, compared to other methods. Note that the performance from all three methods was equivalent at JPEG quality of 70% and below. That is, average *NC* of all methods was lower than the

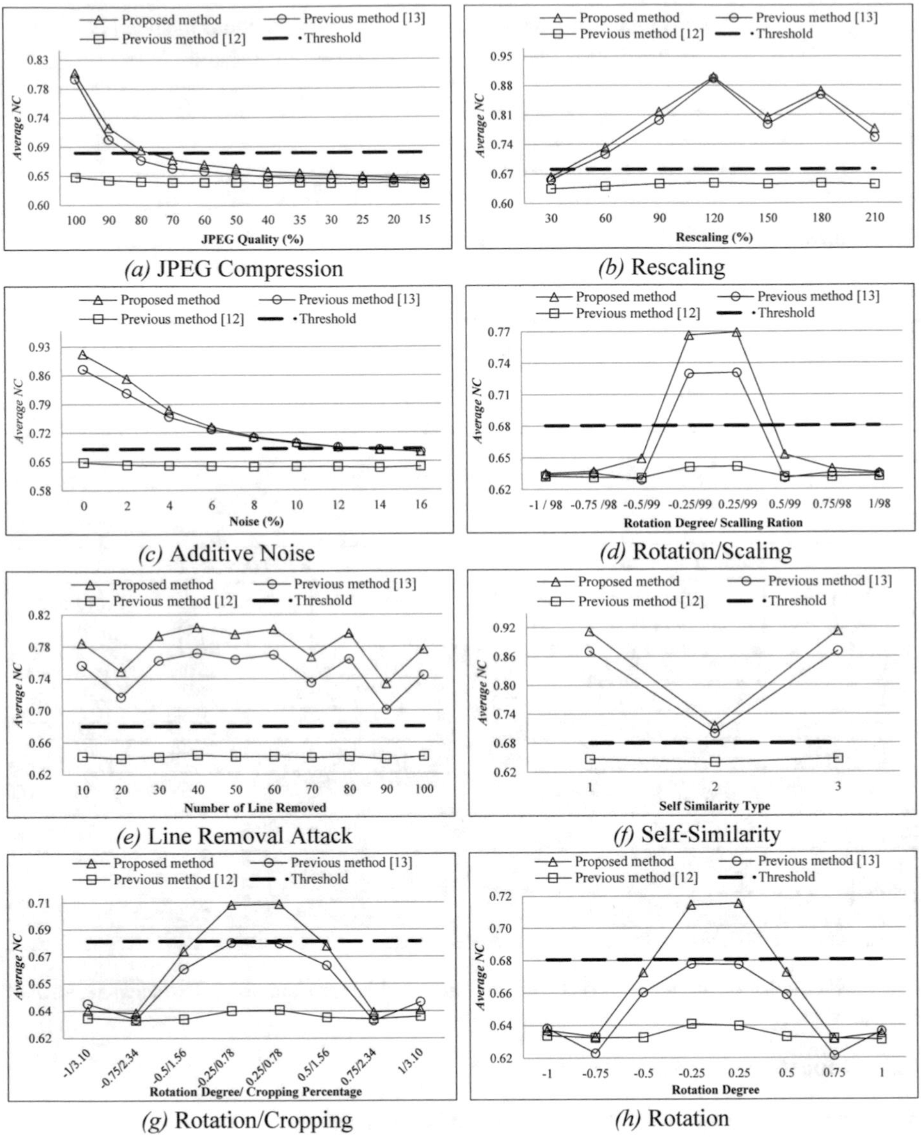

Fig. 6. Robustness comparison against various attacks

threshold. Regarding rescaling attack, the improved performance was dropped but still better than the other two methods. The proposed method was able to achieve superior performance for noise test as well. There was noticeable performance for rotation degree/scaling rotation at $\pm 0.25°$. When we consider number of line removal attack, the proposed method was clearly ahead than the previous methods. In self-similarity type test, average *NC* was reduced at type 2 but still better than others. For rotation degree/cropping percentage and rotation degree test, average *NC* at ±0.25° was still above the threshold, while the previous methods were failed below the threshold.

5 Conclusion

In this paper, an image watermarking method using the DWMAV filters was presented. In our method, the watermark was embedded in the blue component of original image and the DWMAV was used to estimate the original image in the extraction process. It can restrain better the low, medium and high level of Gaussian noise to make our watermarking method efficient. The performance of the proposed method in terms of *NC* at equivalent *wPSNR* was experimentally improved compared to the previous methods [12, 13], and so was the robustness of the embedded watermark.

References

1. Lee, S.J., Tung, S.H.: A survey of watermarking techniques applied to multimedia. In: IEEE International Symposium on Industrial Electronics, Pusan, South Korea, pp. 272–277 (2001)
2. Cox, I.J., Miller, M.L., Bloom, JA.: Digital Watermarking and Steganography. Morgan Kaufmann Publishers, San Franciso (2002)
3. Ville, D.V.D., Nachtegael, M., Weken, D.V.D., Kerre, E.E., Philips, W., Lemahieu, I.: Noise reduction by fuzzy image filtering. IEEE Trans. Fuzzy Syst. **11**(4), 429–436 (2003)
4. Kwan, H.K.: Fuzzy filters for noisy image filtering. In: International Symposium on Circuits and Systems, Bangkok, Thailand (2003)
5. Marudhachalam, R., Ilango, G.: Fuzzy hybrid filtering techniques for removal of random noise from medical images. Int. J. Comput. Appl. **38**(1), 15–18 (2012)
6. Rahman, T., Haque, M.R., Rozario, L.J., Uddin, M.S.: Gaussian noise reduction in digital images using a modified fuzzy filter. In: 17th International Conference on Computer and Information Technology, Dhaka, pp. 217–222 (2014)
7. Sridevi, T., Fatima, S.S.: Digital image watermarking using genetic algorithm in DWT and SVD transform. In: 3rd International Conference on Computational Intelligence and Information Technology, pp. 485–490 (2013)
8. Rosiyadi, D., Lestriandoko, N.H., Fryantoni, D.: A strong sensitivity of digital image watermarking scheme for noise disturbance. In: 2nd International Conference on Technology, Informatics, Management, Engineering & Environment 2014, Bandung, pp. 7–9 (2014)
9. Bansal, D., Mathuria, M.: Color image dual watermarking using DCT and DWT combine approach. In: International Conference on Trends in Electronics and Informatics, Tirunelveli, pp. 630–634 (2017)
10. Hurrah, N.N., Loan, N.A., Parah, S.A., Sheikh, J.A.: A transform domain based robust color image watermarking scheme for single and dual attacks. In: 4th International Conference on Image Information Processing, Shimla, India, pp. 1–5 (2017)
11. Sudibyo, U., Eranisa, F., Rachmawanto, E.H., Setiadi, D.R.I.M., Sari, C.A.: A secure image watermarking using chinese remainder theorem based on haar wavelet transform. In: 4th International Conference on Information Technology, Computer, and Electrical Engineering, Semarang, pp. 208–212 (2017)

12. Thongkor, K., Supasirisun, P., Amornraksa, T.: Digital image watermarking based on regularized filter. In: 14th IAPR International Conference on Machine Vision Applications, Tokyo (2015)
13. Lai, C.H., Wu, J.L.: Robust image watermarking against local geometric attacks using multiscale block matching method. J. Vis. Commun. Image Represent. **20**(6), 377–388 (2009)
14. Russo, F.: Nonlinear fuzzy filters: an overview. In: 8th European Signal Processing Conference, Trieste, Italy, pp. 1–4 (1996)
15. Amornraksa, T., Janthawongwilai, K.: Enhanced images watermarking based on amplitude modulation. Image Vis. Comput. **24**(2), 111–119 (2006)
16. Petitcolas, F.: Watermarking schemes evaluation. IEEE Signal Process. Mag. **17**(5), 58–64 (2000)
17. Abdel-Aziz, B., Chouinard, J.-Y.: On perceptual quality of watermarked images – an experimental approach. In: Kalker, T., Cox, I., Ro, Y.M. (eds.) IWDW 2003. LNCS, vol. 2939, pp. 277–288. Springer, Heidelberg (2004)
18. Yaghmaee, F., Jamzad, M.: Estimating watermarking capacity in gray scale images based on image complexity. EURASIP J. Adv. Signal Process. **2010**(8), 1–9 (2010)

Image Watermarking Against Vintage and Retro Photo Effects

Piyanart Chotikawanid(✉), Thitiporn Pramoun, Pipat Supasirisun, and Thumrongrat Amornraksa

Multimedia Communication Laboratory, Department of Computer Engineering, Faculty of Engineering, King Mongkut's University of Technology Thonburi, Bangkok, Thailand

piyanart.cho@gmail.com, thitiporn1016@gmail.com, {pipat, t_amornraksa}@cpe.kmutt.ac.th

Abstract. Vintage and retro photo effects from various photo and camera software are popular. They are typically obtained from various image processing techniques. However when such effects are applied to watermarked image, they can partially or fully damage watermark information inside the watermarked image. In this paper, we propose an image watermarking method in spatial domain in which the embedded watermark can survive vintage and retro photo effects. Technically, the watermark is embedded by modifying the luminance component in *YCbCr* color space of a host color image. For watermark extraction, a mean filter is utilized to predict original embedding component from the watermarked components, and each watermark bit is blindly recovered by differentiating both components. The watermarked image's quality is evaluated by weighted peak signal-to-noise ratio, while the extracted watermark's accuracy is evaluated by bit correction rate. The experimental results under various percentages of vintage and retro photo effects show that our proposed method was superior to the previous method.

Keywords: Luminance component · Image watermarking · YCbCr color space Vintage and retro photo effects

1 Introduction

Because of limitless communications in digital network, digital media, e.g. image and video, is easy to be shared, copied and edited without permission from the real owner. Images are popularly redistributed in social networks, e.g. Instragram, Facebook. While the image is redistributed in such networks, several image processing techniques from many applications, e.g. Camera360, Line camera, can be applied to them. The most popular effects frequently applied to digital color images are vintage and retro photo effects [1]. These effects can introduce distortions to the images leading to some difficulties for verifying the real owner of the images. To verify their ownership, image watermarking methods are utilized. Image watermarking is performed by embedding owner information into an image before releasing it to the public. The embedded information can later be utilized to verify the real image owner. The image to be

H. Unger et al. (Eds.): IC2IT 2018, AISC 769, pp. 167–176, 2019.
https://doi.org/10.1007/978-3-319-93692-5_17

embedded and the owner information are usually known as "host image" and "watermark", respectively.

Image watermarking method may be categorized into two types based on its working domain: spatial and frequency domain. Image watermarking in frequency domain can conceptually survive most compression techniques, e.g. JPEG compression, while image watermarking in spatial domain can conceptually survive most geometrical attack, e.g. rotation, scaling, and translation. Image watermarking methods in spatial domain were for instances those proposed in [2–4]. In [2], the blue color component from the *RGB* color space was proposed to embed watermark. Positions for embedding watermark were estimated by using standard deviation and average pixel values around the embedding pixel. The same algorithm was utilized to determine positions for extracting watermark. In [3], the blue color component was also utilized for embedding watermark. In embedding process, the blue color component was first divided into four parts, where each part was utilized to embed the same watermark. For watermark extraction, the original watermark was required to compare with watermarked image in order to extract the watermark. Image watermarking method in frequency domain were for instances those proposed in [5–7]. In [7], the luminance component (*Y*) of the *YCbCr* color space was proposed to embed watermark. The *Y* component was first divided into blocks of 8×8 pixels followed by the Discrete Cosine Transform (DCT). The DC coefficient was selected to embed watermark. In extraction process, the same coefficient was selected to extract the watermark. However, most of this method can embed a smaller size of watermark, compared to the size of host image. Observably, the image watermarking methods proposed in the past have no consideration for vintage and retro photo effects.

In this paper, an image watermarking method designed to be robust against most vintage and retro photo effects is proposed. In the proposed embedding method, the watermark is embedded into the *Y* color component in the *YCbCr* color space in order to avoid distortions introduced by most vintage and retro photo effects. Since the watermark is embedded into the *Y* component the same way as a random noise, mean filter is utilized to remove such noise in order to predict original *Y* component. Finally, the watermark is obtained from a difference between the embedded component and the predicted original *Y* component. Our proposed method experimentally shows better performance in terms of Bit Correction Rate (*BCR*) under both attack and non-attack situations.

2 Theories

2.1 Vintage and Retro Photo Effects

Vintage and retro photo effects are performed by adding or creating antique charm to any modern images. The effects are so popular especially in many applications for Android and iOS, e.g. Instagram, Camera360, Line camera and so on [1]. Most of these effects were developed by applying three types of color effects: Vignette, VignetteEffect and SepiaTone. The Vignette is to reduce the brightness of an image at the periphery, the VignetteEffect is to modify the brightness of an image around the periphery of a specified region, and the SepiaTone is to map the colors of an image to various shades of brown [8]. As a result, color of the input images is modified and adjusted in such a way that the output image appears in a retro tone [9]. Obviously, color information within the input image is more distorted from vintage and retro photo effects than brightness or luminance information, and thus the watermark should then be focusing on the luminance component of the image.

2.2 *YCbCr* Color Space

YCbCr is a common color space utilized in digital color image and video processing. The representation of this color space is suitable for getting rid of some redundant color information in many image and video compression standards e.g. JPEG and MPEG4. The color space is represented by three components i.e. Luminance (*Y*), and two chrominance components (*Cr* and *Cb*). The transformation equation of the *RGB* color space to *YCbCr* one is given below.

$$\begin{bmatrix} Y \\ Cb \\ Cr \end{bmatrix} = \begin{bmatrix} 0.257 & 0.504 & 0.098 \\ -0.148 & -0.291 & 0.439 \\ 0.439 & -0.368 & -0.071 \end{bmatrix} \begin{bmatrix} R \\ G \\ B \end{bmatrix} + \begin{bmatrix} 16 \\ 128 \\ 128 \end{bmatrix} \quad (1)$$

3 Proposed Method

Our proposed watermarking method consists of 2 processes: watermark embedding and watermark extraction. In the embedding process, the luminance (*Y*) component obtained from the *YCbCr* color space is utilized to carry a watermark because as previously mentioned vintage and retro photo effects less distorts the *Y* component value than the chrominance ones, and the embedded watermark will be more robust against such effects.

For preparing watermark, a binary watermark image, I_w, is permuted to change positions of bit '0' and '1' to obtain permuted watermark, I_{w_p}. The I_{w_p} is then XORed with pseudo random bit generation for balancing number of bits '0' and '1'. Bit 0s of the XORing result are changed into −1, so that the watermark to be embedded become as $w \in \{-1, 1\}$. The watermark embedding is achieved by modifying the *Y* component in accordance with *w* and watermark strength, *s*. The watermarked component, I'_y, is determined by

$$I'_y = I_y + s.w \quad (2)$$

where *s* or scaling parameter, is utilized to adjust the strength of *w*. The result is lastly combined with the original color components, I_{Cr} and I_{Cb}, and retransformed to the *RGB* color space to obtain the watermarked image, I'_{im}.

In the extraction process, the watermarked image is first transformed to *YCbCr* color space and the watermark component, I'_y, is utilized to extract watermark. The embedded watermark is assumed to be noise-like signal added to original image, mean filter, $M(x,y)$ is thus utilized to remove this signal from the watermarked image to predict the original Y component, I''_y. The filtering step is given below.

$$I''_y(x,y) = \sum_{m=-1}^{1} \sum_{n=-1}^{1} M(m,n) * I'_y(x+m, y+n), \tag{3}$$

where m and n are coordinates of the filter, and $*$ is convolution operation. The I''_y is then subtracted from I'_y to acquire extracted watermark, w'.

$$w' = I'_y - I''_y \tag{4}$$

Figure 1 shows block diagram of our proposed watermarking method.

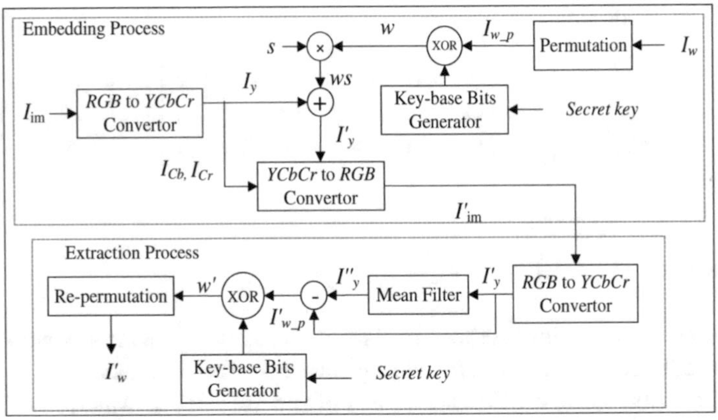

Fig. 1. Block diagram of the proposed watermarking method

4 Experimental Settings and Results

The performance of the proposed method was tested on six standard 256 × 256 pixels color images containing different characteristics i.e. 'lena', 'barbara' and 'peppers', 'watch', 'boats', and 'house' The test images 'babara' and 'boats' are representatives of high level of image complexity, 'watch' and 'lena' are representatives of medium level of image complexity and 'house' and 'peppers' are representatives low level of image complexity [10]. All the test images are demonstrated in Fig. 2. The watermark is a binary image with the same size as the host images (see Fig. 4d).

To measure the quality of watermarked images, a modified version of Peak Signal-to-Noise Ratio (*PSNR*) called weighted Peak Signal-to-Noise Ratio (*wPSNR*) [11] was utilized. The *wPSNR* of a watermarked image is calculated by

$$wPSNR = 10\,log_{10}\left(\frac{255}{RMSE \times NVF}\right), \tag{5}$$

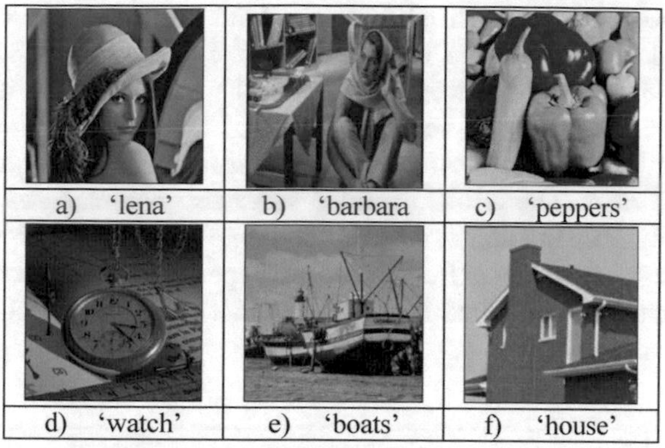

Fig. 2. Test images

where *NVF* is the noise visibility function and *RMSE* is root mean square error as can be obtained by

$$RMSE = \sqrt{\frac{\sum_{z=1}^{K}\sum_{x=1}^{N}\sum_{y=1}^{M}\left(I_{im}(z,x,y) - I'_{im}(z,x,y)\right)^2}{K \times M \times N}} \tag{6}$$

where *K*, *M* and *N* represent the image plane, the numbers of row and column in the images, respectively. The *NVF* value represents weighted based on human perception such that highly textured areas result 0 and flat areas result 1.

The quality of extracted watermark was evaluated by a measurement called Bit Correction Rate (*BCR*) [12]. *BCR* is performed by measuring the percentage of similarity between the original watermark and extracted watermark [12]. The *BCR* of an extracted watermark is calculated by

$$BCR = \frac{1}{M_w \times N_w} \times \sum_{i=1}^{M_w}\sum_{j=1}^{N_w} \overline{I_w(i,j) \oplus I'_w(i,j)} \times 100\% \tag{7}$$

where M_w and N_w are the number of row and column of watermark image, $\oplus$ and – are Exclusive OR and inverse operations, respectively. The perfect quality of extracted watermark in terms of *BCR* is 100%.

For performance comparison, the previous method [2] was selected and compared because it could embed a watermark with the same size as host image, and also achieved a blind detection property. In other words, this particular method was chosen because it can embed watermark bits into every image pixel and its extraction process needs no original image. First of all, in the experiments, we confirmed the reliability of the extracted watermark from both methods by defining a threshold. The threshold was utilized to identify the existence of valid watermark. The threshold was obtained by performing two processes: to determine *BCR* of extracted watermark with 1000 incorrect secrete keys, and *BCR* from non-watermarked images. The highest *BCR* value from the two processes was utilized as a threshold. The highest *BCR* values of the proposed method and the previous method [2] were 50.21% and 49.42%, respectively. Hence, *BCR* of 51% and 50% were utilized as the thresholds for the proposed and previous watermarking methods, respectively. We then evaluated the performance of both methods at various qualities of watermarked image, i.e. *wPSNR* 25–45 dB, in terms of average *BCR*. The results obtained were also compared as demonstrated in Fig. 3. Apparently from the figure, the proposed outperformed the previous one for the entire range. The *BCRs* obtained from the two comparing methods at *wPSNR* of 35 dB for each test image are presented in Table 1, while some graphical results from the experiments are shown in Fig. 4.

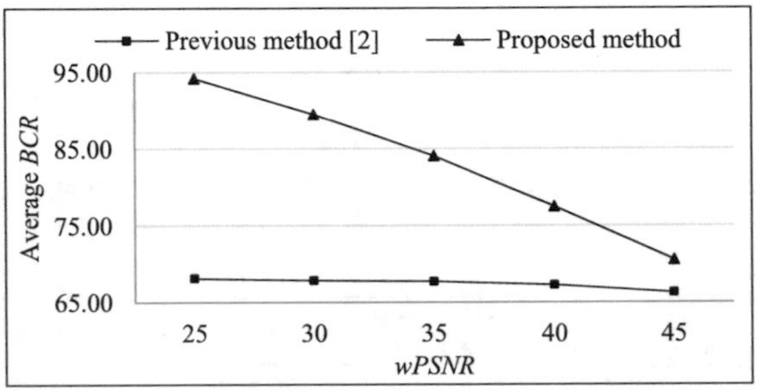

Fig. 3. Performance comparison at various *wPSNR* values

Next, we evaluated robustness of the embedded watermark against vintage and retro photo effects. To apply such effects to the watermarked images, we utilized two popular software: first a commercial software available on mobile applications with iOS 11.2 called "Line camera", and second a web-based photo editor called "LunaPic" [13]. In the experiments, Line camera was utilized to provide 0–100 percentage of vintage and retro photo effect to I'. Note that the result from Line camera was in compressed and rescaled format. This was similar to applying three attacks to the watermarked image: vintage and retro photo effect, JPEG compression, and scaling 250%.

To observe the sole effect of vintage and retro photo effects, we created such effects using LunaPic with 0–100 percentages. Note that this website introduced vintage and retro photo effects without applying compression and scaling. Figure 5 shows graph of the average *BCR* of extracted watermark at *wPSNR* of 35 dB between the proposed method and the previous method [2] under various percentages of vintage and retro photo effect from LunaPic. Some examples of the watermarked image with vintage and retro photo effects from LunaPic are shown in Fig. 6.

Figure 7 shows graphical result of the average *BCR* of extracted watermark at *wPSNR* of 35 dB between the two comparing methods under Line camera at various vintage and retro photo effect percentages, while some results from Line camera and its extracted watermark are shown in Fig. 8.

Accordingly, from the resultant average *BCR* in Figs. 5 and 7, apart from obtaining more robustness, judged from the higher value of *BCR*, the proposed method was less

Table 1. Comparison of *BCR* between the two watermarking methods at *wPSNR* of 35 dB.

Method	Test image					
	'lena'	'barbara'	'peppers'	'watch'	'boats'	'house'
Previous method [2]	69.17	60.91	68.00	73.14	69.13	65.91
Proposed method	**86.84**	**81.14**	**87.85**	**83.36**	**80.72**	**84.52**

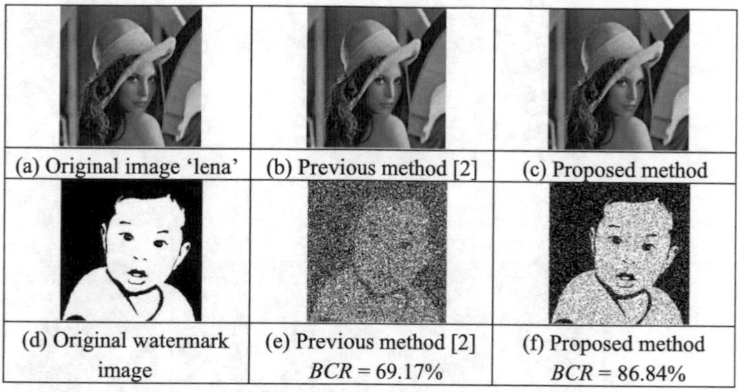

Fig. 4. Examples of original image 'lena', original watermark image, watermarked images and extracted watermark from the two watermarking methods

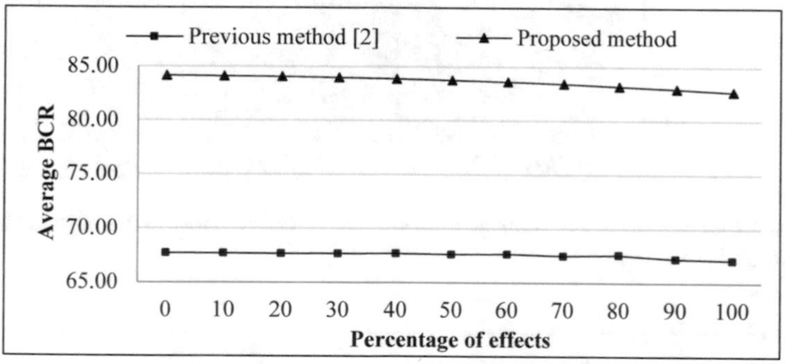

Fig. 5. Robustness of embedded watermark against *LunaPic*

affected from vintage and retro photo effects than the previous method [2]. Note that the *BCR* values of extracted watermark from both methods were higher than the thresholds. It is worth noting that embedding a watermark in the luminance component will more noticeably affect the human visual system (HVS) than that in the color components because changes in the luminance component are more sensitive to human eye. Thus, the strength of watermark must be carefully adjusted to avoid any visual quality reduction of watermarked image.

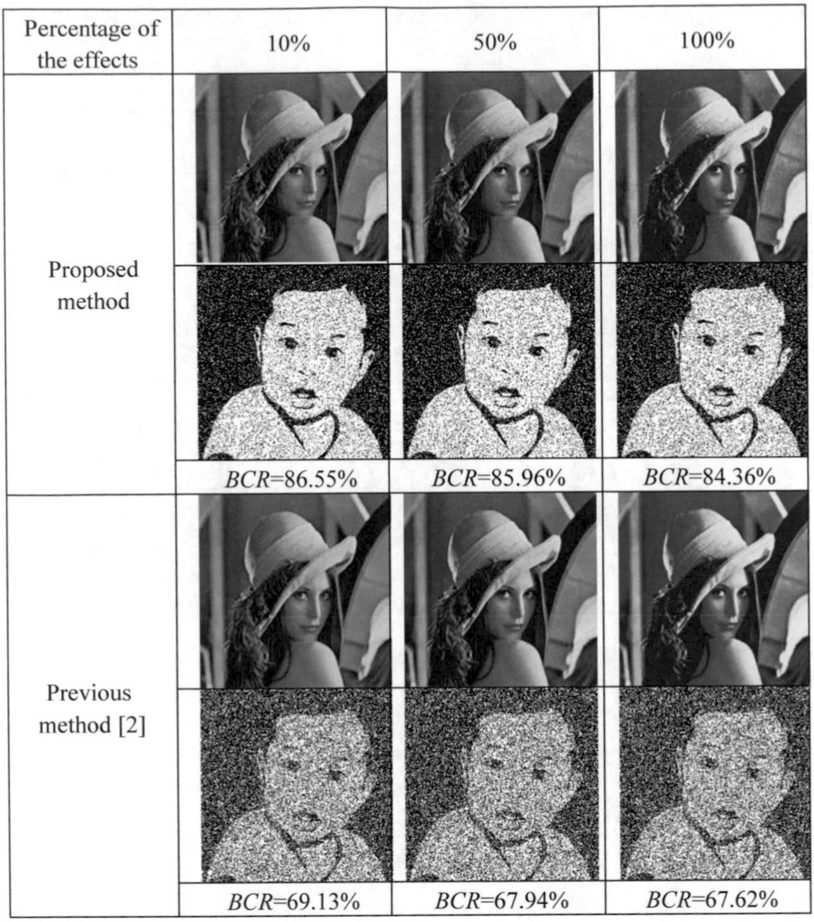
Percentage of the effects
10%
50%
100%
Proposed method
BCR=86.55%
BCR=85.96%
BCR=84.36%
Previous method [2]
BCR=69.13%
BCR=67.94%
BCR=67.62%

Fig. 6. Examples of watermarked image with vintage and retro photo effects created by *LunaPic* and its extracted watermark

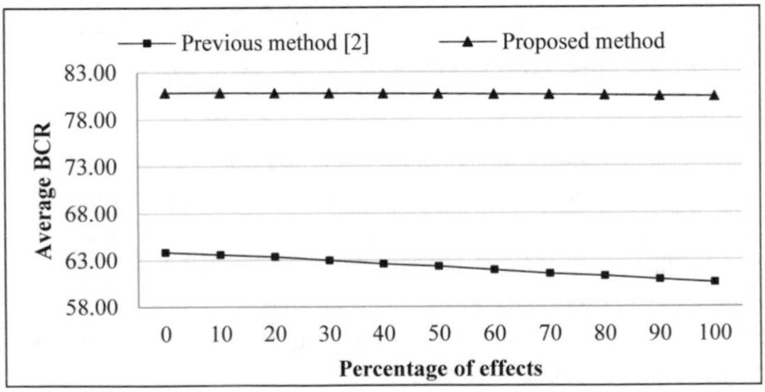

Fig. 7. Robustness of embedded watermark against *Line camera*

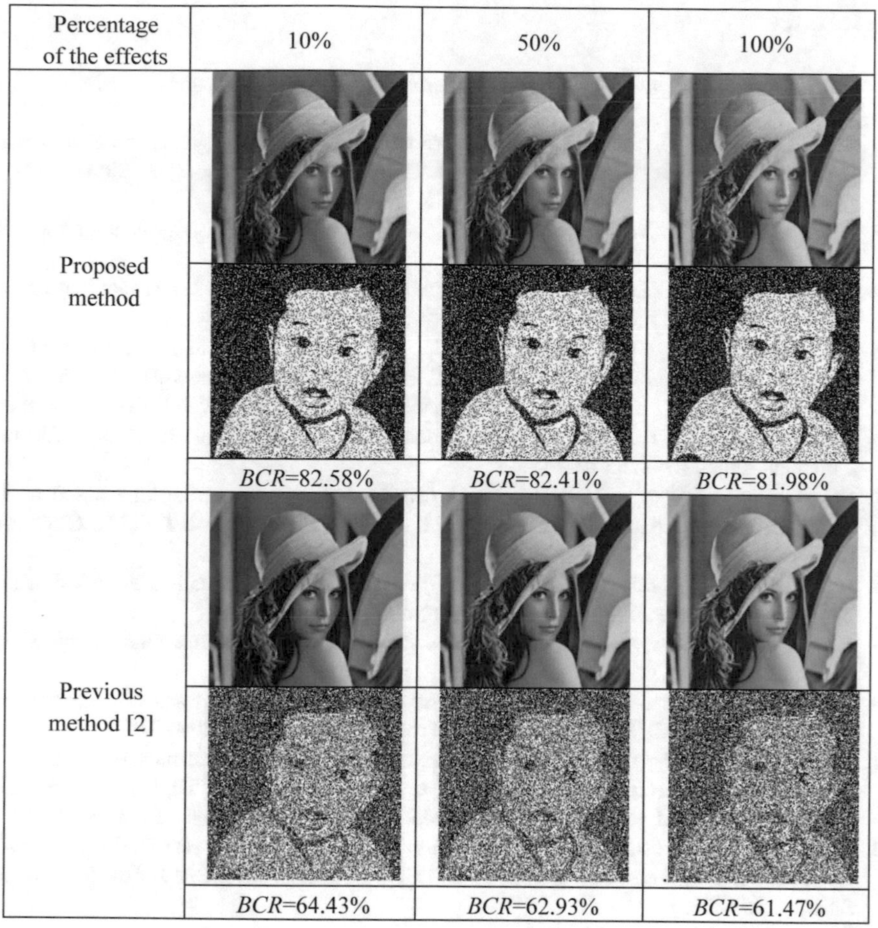

Fig. 8. Examples of watermarked image with vintage and retro photo effects created by *Line camera* and its extracted watermark

5 Conclusion

We have presented an image watermarking method with the consideration of vintage and retro photo effects. In the proposed method, the *Y* component in *YCbCr* color space was utilized to embed the watermark, while a mean filter was utilized in the watermark extraction to achieve a blind detection property. We have shown experimentally that, at equivalent quality of the watermarked image, the embedded watermark from our proposed method was more robust against vintage and retro photo effects than the previous method [2].

References

1. Sasso, F.: Super easy vintage look to photos in pixelmator. http://abduzeedo.com/node/72210
2. Lai, C.-H., Wu, J.-L.: Robust image watermarking against local geometric attacks using multiscale block matching method. J. Vis. Commun. Image Represent. **20**(6), 377–388 (2009)
3. Nasir, I., Weng, Y., Jiang, J., Ipson, S.: Multiple spatial watermarking technique in color images. SIViP **4**(2), 145–154 (2010)
4. Li, L.-D., Guo, B.-L.: Localized image watermarking in spatial domain resistant to geometric attacks. AEU Int. J. Electron. Commun. **63**(2), 123–131 (2009)
5. Hernandez-Guzman, V., Cruz-Ramos, C., Nakano-Miyatake, M., Perez-Meana, H.: Watermarking algorithm based on the DWT. IEEE Latin Am. Trans. **4**(4), 257–267 (2006)
6. Das, C., Panigrahi, S., Sharma, V.K., Mahapatra, K.K.: A novel blind robust image watermarking in DCT domain using inter-block coefficient correlation. AEU Int. J. Electron. Commun. **68**(3), 244–253 (2014)
7. Su, Q., Niu, Y., Wang, Q., Sheng, G.: A blind color image watermarking based on DC component in the spatial domain. Optik Int. J. Light Electron Opt. **124**(23), 6255–6260 (2013)
8. Apple.inc. https://developer.apple.com/library/content/documentation/GraphicsImaging/Reference/CoreImageFilterReference/index.html
9. Pixelmator. http://www.pixelmator.com/mac/tutorials/adjusting-colors/vintage-photo-look-in-pixelmator/
10. Yaghmaee, F., Jamzad, M.: Estimating watermarking capacity in gray scale images based on image complexity. EURASIP J. Adv. Sig. Process. **2010**(8), 851920 (2010)
11. Abdel-Aziz, B., Chouinard, J.-Y.: On perceptual quality of watermarked images – an experimental approach. In: Kalker, T., Cox, I., Ro, Y.M. (eds.) Digital Watermarking: Second International Workshop, IWDW 2003, pp. 277–288. Springer, Heidelberg (2004)
12. Benoraira, A., Benmahammed, K., Boucenna, N.: Blind image watermarking technique based on differential embedding in DWT and DCT domains. EURASIP J. Adv. Sig. Process. **2015**(1), 55 (2015)
13. LunaPic. https://www.lunapic.com/

Performance Analysis of Lossless Compression Algorithms on Medical Images

I Made Alan Priyatna, Media A. Ayu(✉), and Teddy Mantoro

Faculty of Engineering and Technology,
Sampoerna University, Jakarta, Indonesia
alanpriyatna@gmail.com,
{media.ayu, teddy.mantoro}@sampoernuniversity.ac.id

Abstract. Medical images play importance role in medical science. Through medical images, doctor can do more accurate diagnoses and treatment for patients. However, medical images consume large data size; therefore, data compression is necessary to be applied in medical images. This study presents a comparison of lossless compression techniques: LOCO-I, LZW, and Lossless JPEG algorithms which tested on 5 modalities of medical images. All the algorithms are theoretically and practically a lossless image compression which has MSE equal to 0. Moreover, LZW offers higher compression ratio and faster decompression process but slower compression process than the other two algorithms. Lossless JPEG has the lowest compression ratio and requires more time both to compress and decompress the images. Therefore, in general, in this study, LZW is the best algorithm to implement in compressing medical images.

Keywords: Medical imaging · Lossless image compression · LOCO-I
LZW · Lossless JPEG

1 Introduction

Nowadays with the advances of technologies, a giant volume of data produced, transferred and stored every day. Consequently, data compression is needed to reduce the actual size of data such as text, image, audio, video or any type of digital data to achieve small memory usage and allow faster data transmission. Data compression is defined as a method which transforms or changes from one representation of data to another new compressed representation by reducing the number of bits where the delivered information stays the same. Therefore, by compressing the data, it may reduce the use of storage and faster data transmission [1, 2].

Lossless and lossy compression are two major family techniques of data compression. Lossless compression is a technique where there is no loss of information occurs during compressing the data. This technique is used when the original data is very important, that losing any details may give different information. On the other hand, Lossy compression is a technique by removing nonessential information. Decompressed file from this technique is not identical with the original data [3]. Those two major techniques have each advantages and disadvantages when the technique implemented during the compression. In lossless compression, there will be no loss of

H. Unger et al. (Eds.): IC2IT 2018, AISC 769, pp. 177–186, 2019.
https://doi.org/10.1007/978-3-319-93692-5_18

quality; the compressed data will be as good as the original data. However, the size of the compressed data is also only slightly decreased. The most advantage of using lossy compression is the significant reduction of the size of the original data, which also means it will give lower quality compare to the original data [4].

Data compression currently is widely used in many applications including in medical imaging technology. Medical imaging is defined as a technique to make images of part of human body to do more accurate diagnoses and appropriate treatment decisions [5, 7]. Medical image can be send over the internet to provide patient faster diagnoses and appropriate treatment. One of the challenges in medical imaging is the size of the medical image which is usually very large. In 2011, U.S. hospital produces at least 27,000 terabytes a year for medical images. That number increase 20% from 2005 due to the improvement in medical image technology such as increasing pixel resolution and faster processing [6]. For this reason, it is important to implement data compression on medical images to decrease the memory usage which indirectly decrease the cost and provide faster transmission rate.

Since there is no standard of data compression on medical images, it becomes interesting topic in research to determine the most fit compression algorithms on medical images. Therefore, this study focuses on analyzing the performance of lossless compression algorithms on medical images. Remaining part of this paper is organized as Sect. 2 discussing the related works, Sect. 3 the methodology used in this study. In the Sect. 4 we will show the result of the study. Lastly, conclude the study in Sect. 5.

2 Literature Review

Medical imaging is a technique to produce or process an image of human body which use to do more accurate diagnoses and appropriate treatment decision [5, 7]. To produce a good medical image with high throughput acquisition and high resolution of image is required large size of memory usually in term of megabytes or even gigabytes [8, 9]. For this reason, image compression is undoubtedly needed to compress the medical image into smaller size, indirectly give faster transmission rate.

Both lossless and lossy image compression techniques can apply on compression of medical image. However, lossy compression cannot reconstruct compressed image to be the same as the original image, which means some information of medical image will loss and can lead to incorrect diagnoses [10]. Thus, lossless compression is commonly used in medical image compression to warrant no loss of information happening during the process of compression [11]. Huffman Coding, Run Length Encoding (RLE), Arithmetic encoding, Lossless JPEG, LZ family algorithm are well-known lossless image compression for medical image compression [12].

In past decades, several researches have been done for purposing, analyzing and comparing one algorithm with other algorithm to find most fit algorithm for medical image compression. Some researchers proposed compression techniques for medical image compression based on the region of interest (ROI) of the medical image. The work by [13] divided the image into two parts: ROI and non-ROI parts by setting the ROI value by them self. The non-ROI parts are then encoded using their proposed encoding mechanism and then combine with ROI parts. Then next step in their technique is compressed the image with LZW lossless compression. As the result, authors show the LZW lossless compression has better performance than other LZ family such as LZ77 and LZ78 compression techniques. On the other case, [11] proposed a mixed compression both lossless and lossy compression on the medical image. The ROI part compressed losslessly and RONI part compressed using lossy compression algorithm.

Other work by [14] compared TMW, CALIC and LOCO-I techniques. Those algorithms tested on 5 medical image modalities with total 10 images and the result shows that LOCO-I has the best performance among other. [15] have done comparing between various standard image compression such as Lossless JPEG, JPEG – LS, JPEG 2000, PNG and CALIC. They found that JPEG-LS based LOCO – I has the best performance. One more research done in [16] comparing LOCO-I, JPEG2000, PNG and CALIC lossless compression techniques. From 380 thermal images tested for each technique, LOCO – I has the best result.

The previous works show that they focused on testing the algorithms on several modalities only and concerned only on the compression performance, compression ratio and compression speed are the indicators of deciding which algorithm has better performance among others. However, the decompression performance also an important part of the whole performance of compression algorithm. Thus, this study discusses and compare the quality of lossless image compression for medical images, the compression, decompression and the effect of the compression on transmission rate. Lossless JPEG, LZW and LOCO – I lossless compression techniques are discussed and compared. The reason those techniques are chosen because Lossless JPEG is the simple, efficient and become the first standard techniques from JPEG compression's family for medical image compression [17]. Moreover, based on the literature, LZW and LOCO – I techniques show better performance among other techniques.

3 Research Methodology

Figure 1 shows the schematic flow of this research. The literature review done to gain more understanding about the topic, review what other researchers have done, and the position of this research. The lossless image compression techniques, study materials, data collection and the analysis will be discussed further in this section.

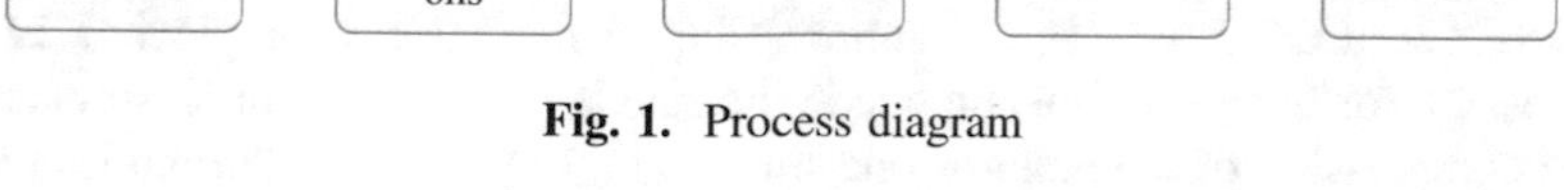

Fig. 1. Process diagram

3.1 Lossless Image Compression

Lossless JPEG. The Lossless JPEG is based on Differential Pulse Code Modulation (DPCM). This compression technique encodes the image based on the principle of predictive coding. JPEG standard has defined eight Prediction Selection Value (PSV) as shown in Fig. 2. PSV 0 is for differential coding, PSV 1, 2, 3 are one dimensional predictors and PSV 4,5,6,7 are two dimensional predictors. Each pixel (for example: X in Fig. 2) value is predicted by the neighborhood pixel's (A, B, C) value using the prediction model.

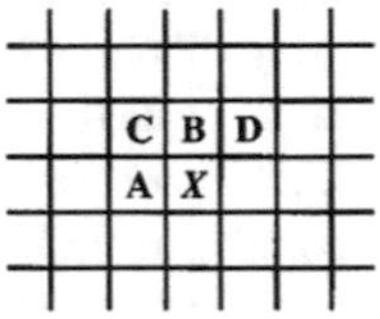

Selection	Predictor
0	None
1	A
2	B
3	C
4	A+B-C
5	A+(B-C)/2
6	B+(A-C)/2
7	(A+B)/2

Fig. 2. Lossless JPEG predictor configurations

For the predictor value of most upper left corner pixel of the image is calculated by 2P-Pt-1 where P is the pixel precision bits and by default, Pt the point transform parameter can be equal as 0. First row of image except the first pixel, the predictor value of each pixel is calculated by using selection 1 and selection 2 for the first column. For all other pixels, one of the eight selection options need to choose to calculate the predict value. Then the difference between the original pixel and the calculated predict value is called as prediction error and this value will further used in encoding process. While calculating the prediction error, module 216 is used, therefore, the prediction error value range of −32767 to 32768 for 16-bit input.

The next step is known as source modeling or encoding. The prediction error then encoded using either Huffman or arithmetic coding. Because of the simplicity of Huffman coding, this technique is widely used. Instead of encoding every single prediction error values using Huffman coding, modified Huffman code (HMC) approach where the prediction error values are first represented as a pair of symbol (CATEGORY of prediction error, MAGNITUDE or the value of prediction error). Assign the encoded CATEGORY symbol and appended it with the additional CATEGORY bit number which represent the MAGNITUDE to get the codedword.

To decode the encoded bits stream, simply just extract the Huffman code (the CATEGORY symbol) first than followed by the prediction error (the MAGNITUDE symbol). Add the prediction error with the predicted value to get the original pixel back.

LOCO-I. The LOCO – I is an abbreviation from LOw COmplexity, LOssless COmpression for Image. This technique is the standard technique for lossless and near lossless compression of continuous-tone image [1]. LOCO – I algorithm known as a simple and efficient algorithm since uses adaptive predictor, context modeling and simple encoding process: Rice-Golomb coding or Run Length Coding [17]. Figure 3 shows the encoding process of LOCO – I compression technique. Whereas the process divided into two parts: context modeling and encoding the image.

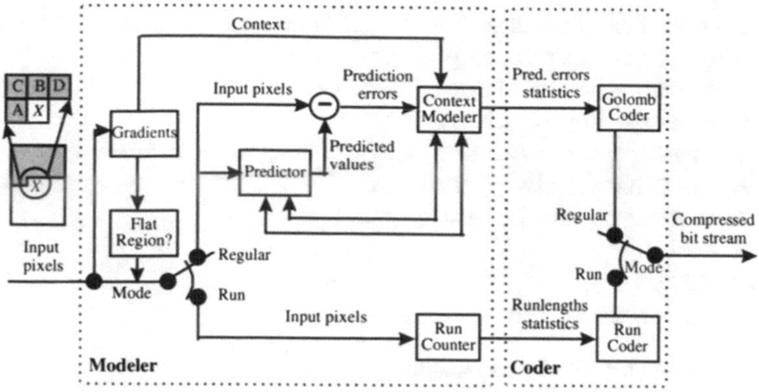

Fig. 3. Block diagram of LOCO-I encoder

In LOCO – I, context is used to predict pixel x and determine what mode use to encode the pixel x: regular mode or run mode. Run mode is chosen if the context suggests that pixel x, y, z… are likely identical. On the other hand, regular mode is selected to encode the current pixel.

In regular mode, the variation of Median Adaptive Prediction is used to predict the value of pixel x by the mean of the b, c and a pixels [2]. Then the predicted value x is subtracted by the original value pixel x to obtain the prediction error. The prediction error then quantized and mapped to certain range and ready for next step: encoding. The mapped prediction error is encoding by using Rice – Golomb Code which is ideal for the integers n which are distributed geometrically [1] The code consists of two parts, those are unary code for the part of n/m and the binary representation of n/m.

In Run mode, the encoder starts at the current pixel x and find the longest run of pixels which is identical to the context pixel a. In this mode, only the length of the run needs to be encode, the pixels in the runs not encoded since those pixels are identical to context pixel a. This works since context pixel a is already known by the encoder. LOCO – I technique a symmetric compression method. Therefore, decoder need to reverse the process of encoding process to decode the encoded bit stream.

LZW. Not like other two algorithms, LZW is a dictionary-based compression technique whereas save the group of symbols from input string and convert it into codes. This is compression since the codes is requires less space than the strings they replace.

LZW starts with initialize a standard character set of 256 characters (8 bits for 1 character) for its dictionary. Then reads the next characters in input string. If the new substring not in dictionary, add it to dictionary and encodes it depending on its indexing in the dictionary. If the new substring already in dictionary, LZW then reads next character and concatenate with the previous substring.

```
table initialization with single character strings
P = get the first input character
WHILE not end of input stream
  C = next input character
  IF P+C is in the string table THEN P = P+C
  ELSE Output the code for P
  Add P+C to the string table
  P=C
END WHILE
output code for P
```

[Pseudocode for LZW compression]

LZW creates identical dictionary during decompression with dictionary on compression process with condition that both compression and decompression process start with the same initial dictionary (e.g. all 256 ASCII characters) [18]. The dictionary then updates for each character in the encoded string stream. To decode, LZW read the code in the encoded string stream and translating back to the original string based on the dictionary.

```
table initialization with single character strings
prev = first input code;
output translation of prev
WHILE not end of input stream
  now = next input code
  IF now is not in the string table THEN
    S = translation of prev
    S = S + C
  ELSE    S = translation of now
  OUTPUT S
  C = first character of S
  insert prev + C to string table
  prev = now
END WHILE
```

[Pseudocode for LZW decompression]

3.2 Materials and Tools Used in the Study

The sample images taken from http://www.cancerimageingarchive.net/ website with format ".dcm" format (DICOM format). The compression algorithms are tested on 5 modalities of the medical images such as computed radiography (CR), computed tomography (CT), magnetic resonance (MR), nuclear medicine (NM), and "other" modality (OT), where each modality has special characteristics of image. Five medical images of each modalities are used in the testing and analyzing of the performance of

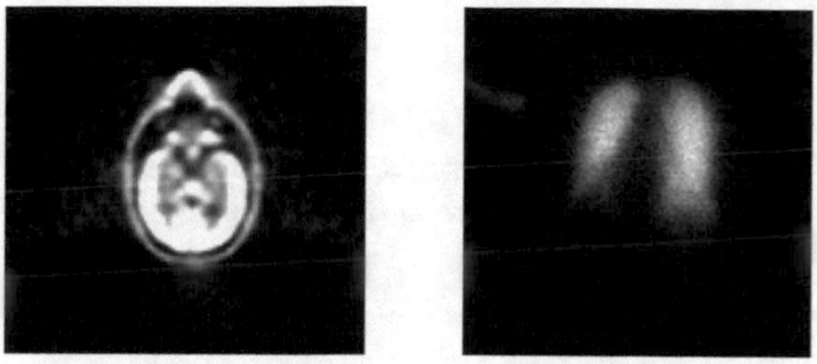

a. PT1 b. NM1

Fig. 4. Images example and its name

compression algorithms for each modality, which means the algorithms are going to be tested on 25 medical images. The images named as abbreviation of the modality and followed by the number 1 to 5. Figure 4 shows the example of images and its name.

3.3 Data Collection and Analysis

Since this study used lossless compression technique, the quality performance of the technique can be evaluated using only compression ratio [19, 20]. Here, compression ratio is a parameter in data compression to know the ratio between compressed and original image size. Compression Ratio (CR) can be calculated using the following formula 1:

$$CR = \frac{Original\ Image\ Size}{Compressed\ Image\ Size} \tag{1}$$

The formula shows that higher compression ratio indicates better compression efficiency.

To evaluate the quality of the decompressed image, Mean Square Error (MSE) is calculated. The goal of MSE is to measure or provide quantitative scores of two signals/image which describe how similar the signal to each other, the degree of error or distortion among the 2 images (original image and decompressed image). MSE can be obtained by calculating the average of squared intensity of two or more (inputs) images as shown in MSE's formula 2.

$$MSE = \frac{1}{NM}\sum\nolimits_{m=0}^{M-1}\sum\nolimits_{n=0}^{N-1}\left(x_{j,k} - x'_{j,k}\right)^2 \tag{2}$$

Where NM is the image size, $x_{j,k}$ indicates the jk-th pixel value of compressed image and $x'_{j,k}$ indicates the jk-th pixel value of original image. The smaller the MSE value the more identical the decompressed image to the original image.

This study also uses compression and decompression time to determine which lossless image compression technique has better performance among others.

4 Result and Discussion

The algorithms are implement in MATLAB® code and tested on 25 medical images with 5 modalities and in the same environment of the system. The compression ratio, compression and decompression time are show in Figs. 5, 6 and 7.

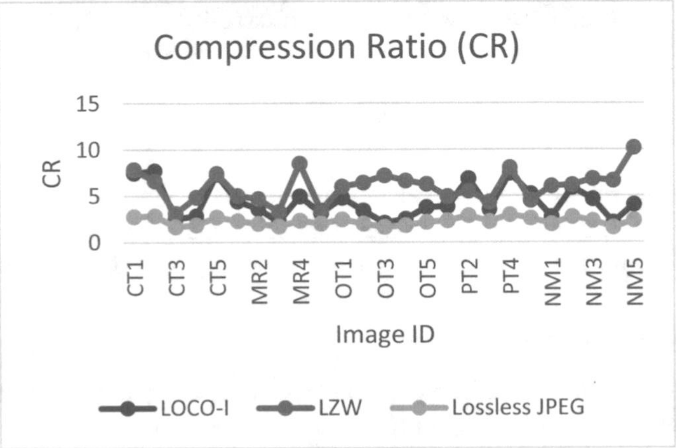

Fig. 5. Compression time

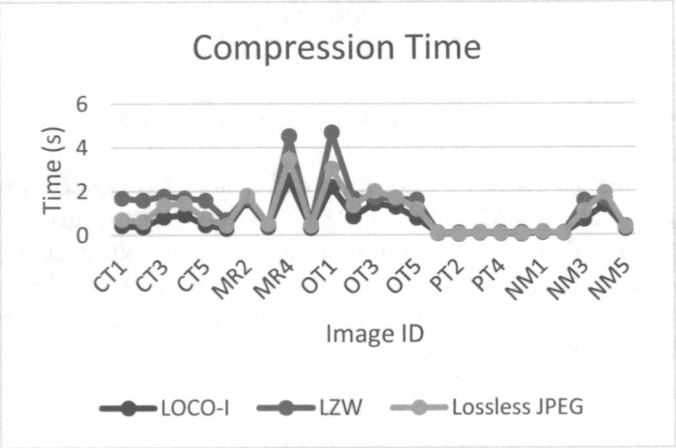

Fig. 6. Compression time

Figure 5 shows that LZW lossless compression mostly has higher compression ratio than the other two lossless compression: LOCO-I and Lossless JPEG. In average, LZW has compression ratio around 6.00, LOCO-I has 4.35 and the smallest compression ratio, Lossless JPEG with 2.23. Thus, LZW 38% higher than LOCO-I and

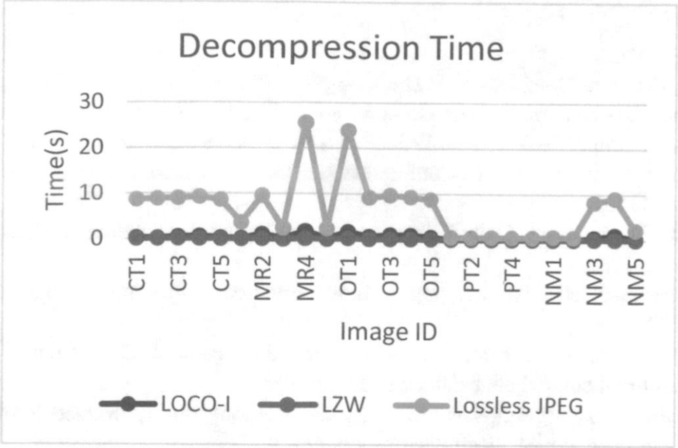

Fig. 7. Decompression time

169% higher than Lossless JPEG. The higher compression ratio means the smaller time require to transfer the medical images for diagnose or any purposes.

Figures 6 and 7 shows the compression and decompression time for all algorithms. LZW which has the highest compression ratio in general, required longer time to compress the images than the other two lossless compression. In contrast to compression time, LZW performs a faster decompression time than LOCO-I and Lossless JPEG. LOCO-I faster in compression process but not in decompression process compare with LZW. Lossless JPEG require more time both to compress and decompress the image.

The MSE has calculated for all lossless compression algorithm. The value of MSE for each image dataset is 0. That value means the decompressed image is identical to the original image. No data loss during the compression process. Therefore, the algorithms are theoretically and practically a lossless image compression and appropriate to use in medical images.

5 Conclusion

This study compared and analyzed the performance of three lossless image compression on medical images: LOCO-I, LZW, and Lossless JPEG. The algorithms tested on 5 modalities of medical images with total image 25 images. The algorithms are lossless image compression theoretically and practically also lossless by showing the value of MSE is equal to 0. As a result, LZW offer higher compression ratio and decompression time than LOCO-I and Lossless JPEG. However, it requires more time to compress the images. Lossless JPEG has lower performance in any aspect, lower compression ratio, slowest algorithm in both compression and decompression process. Since the purpose of compression is to get smallest possible data size or largest possible compression ratio, therefore, LZW offers better performance than LOCO-I and Lossless JPEG on medical images.

References

1. Salomon, D.: Data Compression - The Complete Reference. Springer, London (2007)
2. Sayood, K.: Introduction to Data Compression, 4th edn. Elsevier Ltd., Waltham (2012)
3. Yong, K.K., Chua, M.W., Ho, W.K.: CUDA lossless data compression algorithms: a comparative study. In: IEEE Conference on Open Systems (ICOS), pp. 7–12. IEEE, Langkawi (2016)
4. Pu, I.M.: Fundamental Data Compression. Butterworth-Heinemann, Elsevier, London (2006)
5. Innovatemedtec: Medical Imaging. Innovatemedtec. http://innovatemedtec.com/digital-health/medical-imaging
6. Panner, M.: What's next for the health-care data center? Data Cent. J. http://www.datacenterjournal.com/whats-healthcare-data-center
7. FDA, Medical Imaging. https://www.fda.gov/RadiationEmittingProducts/RadiationEmittingProducsandProcedures/MedicalImaging/default.html
8. Banalagay, R., Covington, K.J., Wilkes, D., Landman, B.A.: Resource estimation in high performance medical image computing. Neuroinformatics **12**(4), 563–573 (2014)
9. Lancaster, J.L., Hasegawa, B.: Fundamental Mathematics. CRC Press, Boca Raton (2016)
10. Annadurai, S., Geetha, P.: Efficient secured lossless compression of medical images - using modified runlength coding for character representation. In: IEEE Indicon 2005 Conference, pp. 14–18. IEEE, Chennai (2005)
11. Reddy, B.V., Reddy, P., Kumar, P.S., Reddy, A.S.: Lossless compression of medical images for better diagnosis. In: 2016 IEEE 6th International Conference on Advanced Computing, pp. 404–408 (2016)
12. Tajallipour, R., Wahid, K.: Efficient implementation of adaptive LZW algorithm for medical image compression. In: 14th International Conference on Computer and Information Technology (ICCIT 2011), Dhaka, Bangladesh (2011)
13. Singh, S., Pandey, P.: Enhanced LZW technique for medical image compression. In: 2016 International Conference on Computing for Sustainable Global Development (INDIACom), pp. 1080–1084 (2016)
14. Krivijarvi, J., Ojala, T., Kaukoranta, T., Kuba, A., Nyul, L., Nevaleinen, O.: A comparison of lossless compression methods for medical images. Comput. Med. Imaging Graph. **4**(22), 323–339 (1998)
15. Ukrit, F.M., Umamageswai, A.: A survey on lossless compression for medical images. Int. J. Comput. Appl. **31**(8), 47–50 (2011)
16. Schaefer, G., Starosolski, R., Zhu, S.Y.: An evaluation of lossless compression algorithms for medical infrared images. In: Proceedings of the 2005 IEEE Engineering in Medicine and Biology 27th Annual Conference, pp. 1673–1676. IEEE, Shanghai (2005)
17. Kim, Y., Horii, S.: Handbook of Medical Imaging. SPIE Press, Washington (2000)
18. LZW Data Compression. https://www2.cs.duke.edu/csed/curious/compression/lzw.html
19. Mrak, M., Grgic, S., Grgic, M.: Picture quality measures in image compression systems. In: The IEEE Region 8 EUROCON 2003. Computer as a Tool, pp. 233–237. IEEE, Ljubljana (2003)
20. Eskicioglu, A., Fisher, P.: Image quality measures and their performance. IEEE Trans. Commun. **43**(12), 2959–2965 (1995)

Brain Wave Pattern Recognition of Two-Task Imagination by Using Single-Electrode EEG

Sararat Wannajam[1] and Wachirawut Thamviset[2](✉)

[1] Department of Information Technology, Faculty of Science, Khon Kaen University, Khon Kaen, Thailand
sararat.w@kkumail.com

[2] Department of Computer Science, Faculty of Science, Khon Kaen University, Khon Kaen, Thailand
twachi@kku.ac.th

Abstract. This research aims to develop a method using an Electroencephalography (EEG) based brain-computer interface technology for enabling people to answer a true/false question by imagining to an answer in their mind; especially, for people with disabilities who cannot communicate with normal ways. Since, a single-electrode EEG is accurately method to recognize the internal state of mind. The acquisition of EEG signals on frontal electrode, which is pasted on the forehead between the left eye and hair (Fp1) from eight subjects for classification of mental task into two classes. The mental tasks were induced by showing images with symbol or text to allow the users reply a true/false question by imagination to an answer. The event related potential (ERPs) features were determined from the processed EEG signals for two classes of mental task. The acquired brain wave data were processed by the OpenViBE platform. Then, the classification was performed using LibSVM and Multiple-Layers Perceptron (MLP), while the parameters were differently validated with 10-Fold Cross Validation methods, and were compared for selection the highest accuracy of classification. The results show that the proposed method achieves 91% on maximum in accuracy.

Keywords: Pattern recognition · Single-electrode EEG · Multilayer perceptron Support vector machines

1 Introduction

Social interaction activities are necessary for all human beings. In general, a true/false question is an efficiently way to communicate that it is very easy for normal people to tell the answer questionnaire such as speech, body language or writing on paper. However, it is difficult to be possible for people with disabilities. In the past several year, there is considerable research on the use of the electroencephalogram (EEG) signals and extract useful information out of them [1]. The most commonly used technology is called BCI. Brain computer interface (BCI) is systems build a communication bridge between human brain and the external world eliminating the need for

H. Unger et al. (Eds.): IC2IT 2018, AISC 769, pp. 187–196, 2019.
https://doi.org/10.1007/978-3-319-93692-5_19

typical information delivery methods. They manage the sending of messages from human brains and decoding their silent thoughts [2]. For example, the invention of the device for paralysis patients to transfer commands to the mechanical arm or wheel chair that can help them able to take care of themselves, come back to a normal life and makes the quality of life better. In additional, BCI technology had been used in other application such as video game control, mouse movement control [3, 4], etc. Furthermore, there are many research works on EEG; for example, Tan [5] used a cheap EEG device to record the brain wave in order to classify mental states of different learners by predicting the mental states of children and adults. This indicated the benefits of using the time structure of EEG signal to help identify the mental states. Meanwhile, Zhang [6] used the BCI technology to recognize the brain wave patterns during imagining Chinese vocabulary by using the Support Vector Machine, which as a result contributed to the average accuracy of classification. According to such problems, there has been an attempt to improve the brain wave recognition and classification methods to be more accurate. After the previously mentioned research works, it can be seen that the development of BCI in other aspects is also of interest.

In this research, conducted to record the EEG signals, which come from the brain activity while doing some workload, called the Event-Related Potential (ERPs) which of brain electrical waves will be found in cognitive processes, such as reflection to a specific image or character [7]. The hypothesis of the research is the brain workload during imagining can be classified at least into two classes and can be used to response the answer of true/false question. For constructing a classification model, brain signals had been recorded from volunteers by using single electrode EEG device, called NeuroSky Mindset. During recording process, each volunteer was given a set of incidents, which involve the true/false symbols that the recorded signals can be formed to two classes: (0) signals of imagine to "False" symbol and (1) signals of imagine to "True". Then, the signals were processed to pre-processing since the EEG data contained interfering signals, so was used pre-processed to distinguish the features representing the mental workload from the signals. After the noise reduction from the signal, essential features from the brain signals were extracted. For feature extraction from EEG signals use technique like Discrete Wavelet Transform (DWT), and signal average by performed in OpenViBE software. After that, in classification method the support vector machines (SVM) technique and artificial neural network with multilayer perception (MLP) are used to train the classification models with the extracted features because classifiers are mostly used in BCI design [8] and the value of accuracy is quite high. Both constructed SVM and MLP models were evaluated and compared for selection the highest accuracy of classification by performed in WEKA software. The framework of answer classification process is shown in Fig. 1. This research work is divided into sections as follows: Sect. 2 discusses the related theories, Sect. 3 explains the research methodology, Sect. 4 reveals the findings of this research, Sect. 5 summarizes and discusses the findings as well as suggestions, and Sect. 6 contains references in this research work.

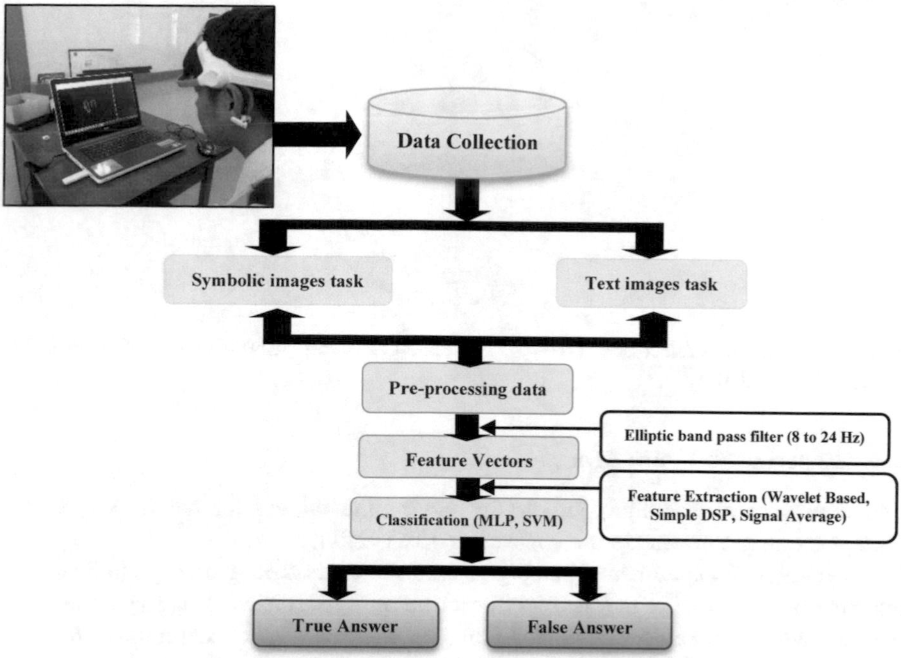

Fig. 1. Framework of answer classification process

2 Data Collection

The Data Collection starts with collecting the EEG signals from eight subject were students. Which there an average age of 19.6 years. The subjects were divided into 6 male and 2 female with normal body and agree to the experiment.

3 Methodology

3.1 Signal Acquisition

In this research, a set of the single electrode device called NeuroSky Mindset because this device is easy to use and cheap. A 11channel used for recording the EEG signal and later these signals are collected from OpenViBE software via Bluetooth. An electroencephalogram (EEG) was recorded from reference position horizontally called Fp1 at located between the left eye and hair while another electrode is placed on the ears. The electrodes were placed according to the International Electrode (10–20) Placement System [11, 12] as shown in Fig. 2. The output data of device returns the raw EEG data sampled at 512 samples per second.

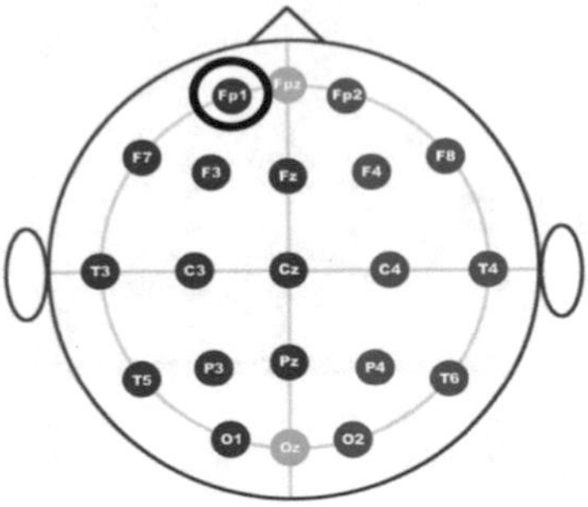

Fig. 2. International Electrode (10–20) Placement System (Source: Multi-Channel EEG (BCI) Devices, 2015).

3.2 Experimental Paradigm

The experimental paradigm judges the environmental conditional in sequence to prompt the user to generate the controlling EEG activity.

Two different experimental paradigms are followed. One is using symbolic cue and the other is text cue. The task is to perform true or in true answer imagery according to the cue. Subjects are evidently guide to imagine the visualization sensation of true/false answer. The order of cues is random. The experiment consists of several runs with 20 trials each after each. Each trial lasts 16 s. The whole experiment took approximately 4 min 49 s for each subject.

The methods of visual cue as follows, symbols and texts are used to display the imagination tasks to be performed. The steps of the trials is illustrated shown in Figs. 3 and 4. After trial begin, the subjects will have to wait for the target images for 5 s, at there was 1 beep, an answer are displayed on the right screen for make specific conditional. Such as, the true symbolic images are displayed, so will have to imagine images as seen etc. Then, the beginning of the trial are fixation cross (+) were displayed in the middle of the screen in order to stimulate. The subjects are required to stare their gaze and attention at the center of fixation cross. The cue images of the imagination task appears 3 s after the cross is made visible. When, the cue images disappears, the subject instructed to perform relax. Blinking and swallowing are permitted only during.

3.3 Pre-processing

Since the data obtained from EEG signals contained interfering signals, because of blinks, eye movements or head movements. The frequency filter was used to distinguish the features representing the mental workload from the signals. The temporal filtering is a fundamental IIR filter which is one of the significant innovations of the discrete time processing system. The IIR filter consists of the equation of output signals which are present input signals, past input signals and past output signals which can be written in difference equation [11] as shown in Eq. 1.

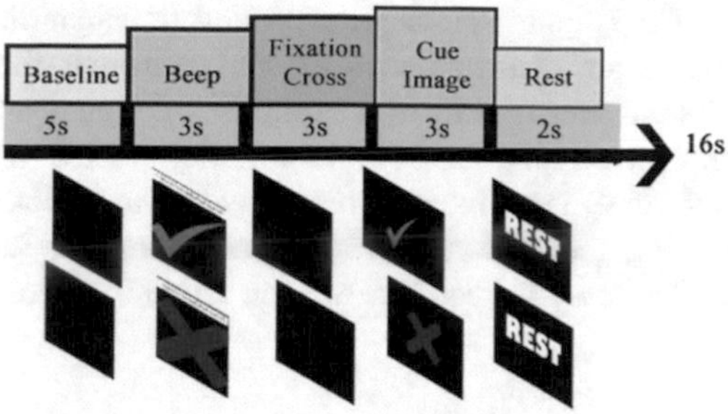

Fig. 3. Symbolic images cue training

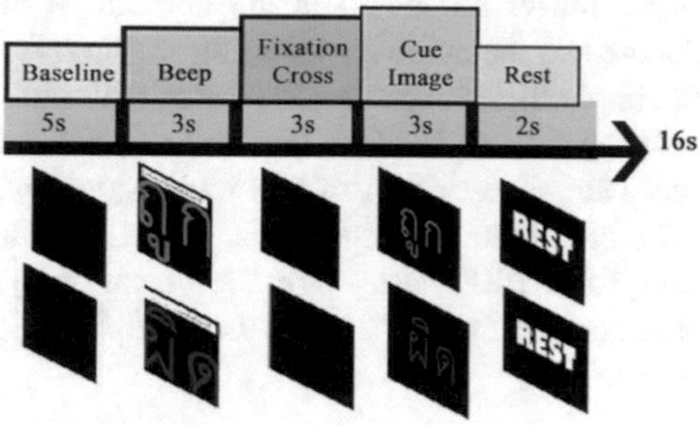

Fig. 4. Text images cue training

$$y(n) - \sum_{k=1}^{N} a_k y(n-k) = \sum_{k=0}^{M} b_k x(n-k) \tag{1}$$

where x (n) is the sequence of the input signal and the Y (n) is the sequence of signals get stable output, N M is the number of samples of output and input values. This can be used to calculate the output signal to each other. Thus, the frequency filter adjusted the signal to be in the range of 8–24 Hz, because the frequency of notice was within the range of Alpha (α) and Beta (β) band [12, 13]. Furthermore, The Butterworth filter is a type of IIR filter designed to have a frequency response as flat as possible in the passband [14, 15]. The signals were reduced with sampling rate from 512 Hz to 320 Hz which adjustment reduced the size of sampling which made it easy for storing and processing to further extract the features of data [16].

3.4 Feature Extraction

The signals are processed to extract distinct features. For this purpose, the ERPs (event related potential) attributes were extracted and multidimensional feature extraction is carried out. i.e. Features are extracted from both time domain for tasks classification using OpenViBE software. The techniques used are Discrete Wavelet Transform, Simple DSP and Signal average. Wavelet, which is called a mathematical microscope for analyzing signals, has the ability to analyze signal which is localized in time domain or frequency domain. EEG signal's nonstationary and transient characteristics make it difficult to extract the exact characteristics of EEG through the ordinary spectrum analysis methods. The wavelet transform decomposes a signal into a set of functions obtained by shifting and dilating one single function called mother wavelet [17]. Daubechies Wavelet 4 (db4) is selected as the wavelet base. Db4 is applied as it was appropriate for analyzing the discontinuous wavelets in practice and it resulted in proper time accuracy due to its D4 coefficients correspond to central beta band [18] as shown in Eq. 2.

$$\mathrm{DWT}(m,n)=\frac{1}{\sqrt{a_0^m}}\sum\nolimits_k f(k)\Psi\left[\frac{n-kb_0a_0^m}{a_0^m}\right] \tag{2}$$

where m, n, k represent integer number n is the amount of data m is the number indicating the scale change k is the number indicating shifting. Additionally, extraction of feature also use the simple digital signals processing (Simple DSP) used to apply a mathematical formulae to each sample of an incoming signal and output. In here, the resulting signal use the natural logarithm of the input signal plus one [19], and the mean obtained from the calculation of input data, so will get a total of data dim is 100×4 feature vectors for a ERPs tasks. For the visual analysis a Power Spectral Density Estimate plot is acquire from 4 features Low Alpha, High Alpha, Low Beta, and High Beta shown in Fig. 5.

3.5 Analysis and Classification

After feature extraction will get feature matrix is normalized in order to have feature in a same range. Machine learning algorithms are essential tools for classification SVM and MLP classifiers are observed.

Support Vector Machines Technique (SVM). Support Vector Machines was developed with the aim for pattern recognition. The principle of Support Vector Machines is to find the coefficient of equation in order to classify input data entering the system so that the system can learn and choose the optimal separating hyper plane [20] by using Kernel function to determine the features and variables which are used to determine the multi-dimensional plane called feature. For feature selection, the set of features used to describe in another case (e.g. expectation line) are called vectors. Therefore, the purpose of Support Vector Machines to benefit multi-dimensional plane which classifies the groups of vectors, one of which is the group of variables on one side of the plan and the other of which is on another plane. The vectors on the multi-dimensional plane are called support vectors as shown in Fig. 6. Moreover, the data classification with Support Vector Machine often relies on linear SVM because this type of classification results in successful outcome applicable to a number of BCI issues. The linear SVM is more complicated in terms of kernel consisting of data mapping in other areas by increasing dimensions of data. The kernel function often used in BCI research is Gaussian method or Radial Basis Function (RBF). The kernels can be calculated as shown in Eqs. 3 and 4, respectively [8].

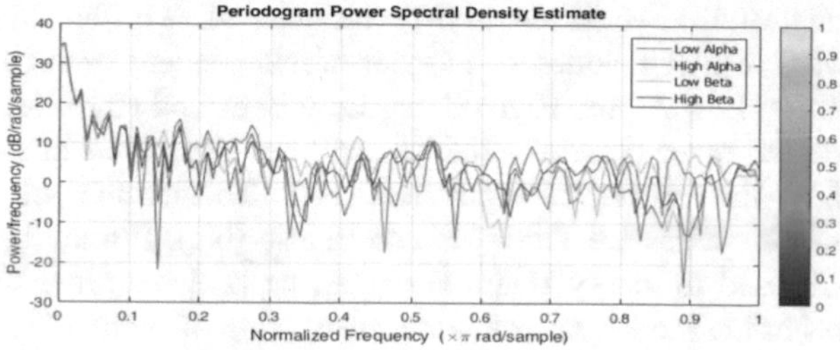

Fig. 5. Power Spectral Density Estimate of features

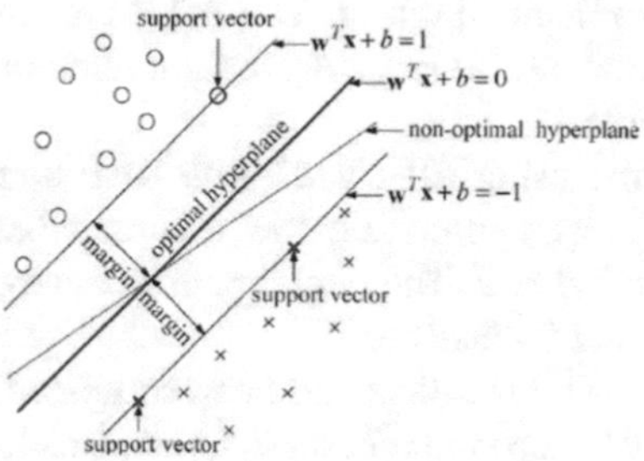

Fig. 6. Optimal Separating Hyper plane (Source: Ajay Unagar, 2017).

$$\mathrm{K}(x, x_i) = (x^T x_i + 1)^P \tag{3}$$

$$\mathrm{K}(x, x_i) = \exp(-y\|x - x_i\|^2) \tag{4}$$

Multi-Layer Perceptron Technique (MLP). The function of artificial neural network is to imitate the function of human brain through the learning mechanism by utilizing past examples during the neural network exercise with backpropagation. It consists of input data meaning the data to be processed, weight meaning the specific value for each input datum in order to differentiate the input data, summation function being the total of input data and weight in each layer used to summarize the relationship among all input data. Transformation function is the calculation of function of artificial neural network which shows the result of function in this element called "Activation Function". The transformation function can be divided into 3 types: linear, threshold and sigmoid. Input data are actual results presented to the multilayer perceptron users [21].

4 Result and Discussion

The classification of two task in EEG-data during symbolic images and text images using machine learning algorithms has been proven here. The classification was performed in Weka. Training and testing was to performance the accuracy of classification on all subjects by making different combinations of eight attributes extracted. Firstly, the classifier using Multi-Layer Perceptron Technique (MLP) is defined by the structure of a neural network, which includes the number of independent features as input nodes. There are 4 features: Low Alpha, High Alpha, Low Beta, and High Beta. Additional, it is also alongside the other control variables are Start time, End time, Event Id, and Event duration, so the number of input node is 8. The number of node output as number of answer 2 classes are True (Yes) and False (No). While, the number of the hidden layer will used to theory Baum-Haussle rule [22] for order to determine the number of nodes in the hidden layer so, the most value of hidden layer have number about 80 node. Moreover, the structure of a neural network can determining the learning rate and momentum value. Therefore, in here will done change the learning rate from 0.1–0.5, momentum from 0.5–0.9 and the number of hidden layer until the accuracy value have performance highest. The MLP classifier using training time as 5,000 epochs and the model performance evaluation method by k value of k-fold was set as 10 as shown in Table 1.

Secondary, the classifier using LIBSVM[1] with RBF kernel it will modify gamma value to determine the most appropriate the accuracy best of accuracy observed, reaching 91% as shown in Table 2. The model performance evaluation method 10-fold Cross-validation same as MLP classifier.

Finally, the results of both classifiers had been compared. As a result, the accuracy analysis found that the SVM technique best performance and the recognition rate is as high as 91% and the highest accuracy MLP technique is 80% respectively. Meanwhile, the comparing result of tasks recognition rate is as high performance with symbolic images task as both models work well. Therefore, show that the symbolic images task appropriate with SVM technique most, so for compare the performance for the classification of two tasks during symbolic images task and text images task for two classes (True/False) in EEG-data using machine learning algorithm the results obtained are shown in Fig. 7.

5 Conclusion and Future Work

This article presented the classification of workload by two patterns of images which were symbolic images and texts by extracting the features of ERPs from EEG signals. According to the table showing the accuracy value and the comparison of accuracy between two workloads and two classification techniques, the workload of two - class images with Support Vector Machines resulted in the highest accuracy being 91%. Our detection propose that it is possible to translate the true or false answer to the given problem with diagnose the ERPs response at the time of stimulus presentation. Therefore in the future, in order to reach more accuracy, the scope of the device should be expanded or more various tools should be used to increase the number of electrodes to detect signals. Moreover, we can test the same model control some application for instruction interactive between student and teacher.

---

[1] https://weka.wikispaces.com/LibSVM.

Table 1. The recognition accuracy in % of the MLP classifier

ERPs tasks	Multilayer Perceptron (MLP)					
	Hidden layer	Leaning rate/Momentum				
		0.1/0.5	0.2/0.6	0.3/0.7	0.4/0.8	0.5/0.9
Symbolic images task	20	74%	**80%**	68%	66%	56%
	40	72%	77%	74%	65%	58%
	60	76%	74%	70%	63%	56%
	80	69%	72%	72%	61%	53%
Text images task	20	78%	76%	74%	63%	56%
	40	75%	77%	73%	58%	55%
	60	75%	76%	78%	61%	57%
	80	74%	**79%**	74%	59%	51%

Table 2. The recognition accuracy in % of the SVM classifier.

ERPs tasks	Support Vector Machines (SVM)		
	Gamma (γ)	10-fold Cross-validation test accuracy (%)	Mean absolute error
Symbolic images task	0.1	**91%**	0.09
	0.3	78%	0.22
	0.5	76%	0.24
Text images task	0.1	**89%**	0.06
	0.3	77%	0.23
	0.5	72%	0.28

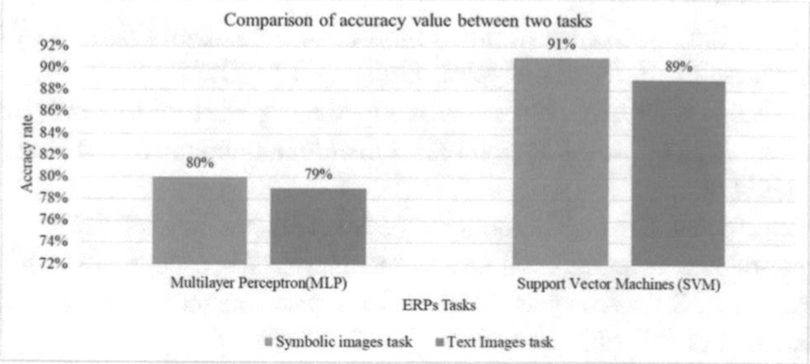

Fig. 7. Comparison of EEG signals between Symbolic images task and Text images task of values obtained from the accuracy values of extracted features of the testing data from two classes of both algorithm.

References

1. Siuly, Y.L., Wen, P.: Classification of EEG signals using sampling techniques and least square support vector machines. In: Wen, P., Li, Y., Polkowski, L., Yao, Y., Tsumoto, S., Wang, G. (eds.) Rough Sets and Knowledge Technology, vol. 5589, pp. 375–382. Springer, Heidelberg (2009)
2. Abdulkader, S.N., Atia, A., Mostafa, M.-S.M.: Brain computer interfacing: applications and challenges. Egypt. Inform. J. **16**(2), 213–230 (2015)
3. Tanajak, J.: BCI Technology to follow, COMPASS MAGAZINE. http://www.compasscm.com/view/133
4. Liu, N.-H., Chiang, C.-Y., Chu, H.-C.: Recognizing the degree of human attention using EEG signals from mobile sensors. Sensors **13**(8), 10273–10286 (2013)
5. Tan, B.H.: Using a low-cost EEG sensor to detect mental states. School of Computer Science, Computer Science Department, Carnegie Mellon University, Pittsburgh, PA (2012)
6. Zhang, X., Li, Y., Peng, X.: Brain wave recognition of word imagination based on support vector machines. TELKOMNIKA Telecommun. Comput. Electron. Control **14**(3A), 277–281 (2016)
7. Siripornpanich, V.: Evaluation of attention using electroencephalography and application in children with attention deficit hyperactivity disorder. Res. Cent. Neurosci. Inst. Mol. Biosci. Mahidol Univ. **20**(1), 4–12 (2013)
8. Lotte, F., Congedo, M., Lécuyer, A., Lamarche, F., Arnaldi, B.: A review of classification algorithms for EEG-based brain–computer interfaces. J. Neural Eng. **4**, 1–24 (2007)
9. 10–20 system (EEG) - an overview|Science Direct Topics. http://www.sciencedirect.com/topics/biochemistry-genetics-and-molecular-biology/10-20-system-eeg
10. Phrasin, A.: Analysis and classification of brainwave signals from P300 speller paradigm. King Mongkut's Institute of Technology North Bangkok (2006)
11. Benjangkaprasert, C.: Variable step-size adaptive algorithms for IIR notch filter and ITS applications. King Mongkut's Institute of Technology Ladkrabang (2006)
12. Ergenoglu, T., Demiralp, T., Bayraktaroglu, Z., Ergen, M., Beydagi, H., Uresin, Y.: Alpha rhythm of the EEG modulates visual detection performance in humans. Cogn. Brain. Res. **20**(3), 376–383 (2004)
13. De Blasio, F.M., Barry, R.J.: Prestimulus alpha and beta determinants of ERP responses in the Go/NoGo task. Int. J. Psychophysiol. **89**(1), 9–17 (2013)
14. Butterworth filter – Wikipedia. https://en.wikipedia.org/wiki/Butterworth_filter
15. Documentation: Home. http://openvibe.inria.fr//documentation/2.0.0/Doc_BoxAlgorithm_TemporalFilter.html
16. Daltrozzo, J., Conway, C.M.: Neurocognitive mechanisms of statistical-sequential learning: what do event-related potentials tell us? Front. Hum. Neurosci. **8**, 437 (2014)
17. Sivakami, A., Devi, S.S.: Analysis of EEG for motor imagery based classification of hand activities. Int. J. Biomed. Eng. Sci. **2**(3), 11–22 (2015)
18. Ngaopitakkul, A.: Wavelet transform and neural networks for fault-classification in transmission lines. Ladkrabang Eng. J. **20**(1), 49–54 (2003)
19. Documentation: Home. http://openvibe.inria.fr//documentation/2.0.0/Doc_BoxAlgorithm_SimpleDSP.html
20. Donporntan, K.S.N.: Adaboost SVM-based technique for image classification. In: The National Conference on Computer Information Technologies, Bangkok, pp. 16–20 (2010)
21. Ammaruekarat, P.: A comparative efficiency of feature selection and neural network classification. In: 5th National Conference on Computing and Information Technology (NCCIT 2009), Bangkok (2009)
22. Baum, E.B., Haussler, D.: What size net gives valid generalization? Neural Comput. **1**(1), 151–160 (1989)

Isarn Dharma Handwritten Character Recognition Using Neural Network and Support Vector Machine

Saowaluk Thaiklang and Pusadee Seresangtakul(✉)

Natural Language and Speech Processing Laboratory,
Department of Computer Science, Khon Kaen University, Khon Kaen, Thailand
saowaluk.th@kkumail.com, pusadee@kku.ac.th

Abstract. In the last decade, handwritten character recognition has become one of the most attractive and challenging research areas in field of image processing and pattern recognition. In this paper, we proposed handwritten character recognition for the Isarn Dharma character. We collected the character images by scanning ancient palm leaf manuscripts. The feature extraction techniques including zoning, projection histogram, and histogram of oriented gradient (HOG) were used to extract the feature vectors. ANN and SVM were used as classifiers in character recognition, and five-fold validation was used to evaluate the recognition results. The experiment result demonstrated that SVM classifier outperformed the other methods in all feature extractions. The recognition accuracy rate through the application of HOG was outstanding, and proved slightly better than HOG applied with zoning. This study further expresses that the gradient feature like HOG significantly outperformed the statistical features, such as zoning and projection histogram.

Keywords: Handwritten character recognition · Feature extraction
Neural network · Support vector machine · Isarn Dharma character

1 Introduction

Handwritten character recognition has become one of the most attractive and challenging research areas in the field of image processing and pattern recognition in the last decade. Recent advancements in technology and heightened expectations in human-computer interaction requiring automatic and semi-automatic system in various applications [1] have led to the development of the handwritten character recognition system, which is an automatic conversion of characters from images into texts. The system is generally classified into either online character recognition or offline character recognition, depending on the type of input data [2, 3]. Online character recognition involves the identification of characters while they are being written, and are captured by special forms of hardware like smart pens or tablets, which are sensitive devices capable of measuring both pressure and velocity. On the other hand, offline character recognition analyzes characters written on many kinds of documents, including contemporary and historical manuscripts derived from several sources, such as scanners and digital cameras, and are then converted into a computer readable format [3].

H. Unger et al. (Eds.): IC2IT 2018, AISC 769, pp. 197–205, 2019.
https://doi.org/10.1007/978-3-319-93692-5_20

For a variety of applications based upon current demands, such as document reading, mail sorting, and postal address recognition, offline handwritten recognition continues to grow with the continuous advancement of new technologies that strive to improve the accuracy and performance of their recognition.

In general, handwritten character recognition is divided into three main processes, including pre-processing, feature extraction, and classification (recognition). Firstly, pre-processing comprises several image processing operations, such as binarization, noise elimination, and normalization, which aim to prepare and produce a character image appropriate for feature extraction [4]. Secondly, feature extraction extracts the representative information from the character images into a set of features called the feature vector that can characterize each class. The widely used feature extraction techniques are template matching, projection histogram, zoning, gradient features, etc. [5]. Lastly, classification is the main decision process in character recognition which uses feature vectors as inputs to classify distinguishing class. A number of classifiers was employed; including Hidden Markov Model (HMM), Support Vector Machines (SVM), Artificial Neural Networks (ANN), k-Nearest Neighbors (k-NN), and others [6].

Research in handwritten character recognition research has been carried out in various kinds of handwritten character scripts in English, Devnagari, Thai, Bangla, Latin, and with basic numerals [7–10]. However, research in the Isarn Dharma character has been minimal. This character format is considered an ancient language, which was used to record the history, tradition, culture, ritual and pharmacopoeias inscribed on palm leaf. In order to preserve the ancient language and maintain access to the valuable information present in these character scripts, it is crucial to develop an Isarn Dharma character recognition system [11, 12].

This study proposed the handwritten character recognition for Isarn Dharma scripts written on palm leaf documents. These documents are often degraded, owing to long duration of deterioration and poor maintenance, which can also decrease the efficiency of recognition. The pre-processing process functions to make the handwritten character image appropriate for feature extraction and recognition. In our proposed, we compared two classification approaches, namely ANN and SVM, and evaluated the performances of several feature extraction techniques, including zoning, projection histogram, histogram of oriented gradient (HOG), as well as various combinations of each.

The paper is organized as follows; we present related literature in Sect. 2. Our proposed methodology in outlined in Sect. 3. The experiment is demonstrated and discussed in Sect. 4. Lastly, our conclusions with our inclination towards future works are given in Sect. 5.

2 Related Literature

Research in handwritten character recognition over the past few years has generated a wide variety of techniques, or evolving existing ones, capable of providing higher recognition accuracy.

Surinta et al. [10] proposed the use of local gradient feature descriptors, namely the Scale Invariant Feature Transform Descriptor (SIFTD) and the Histogram of Oriented Gradients (HOG), for handwritten character recognition. The kNN and SVM algorithms were employed as classifiers. Three different language scripts, namely Thai, Bangla, and Latin, consisting of both handwritten characters and digits, were used within the dataset. Upon comparison, the SVM significantly outperformed the kNN classifier. Moreover, the SVM within the SIFTD method obtained very high recognition accuracies.

Boukharouba and Bennia [13] proposed a handwritten recognition system for Persian numerals, in which a chain code histogram combined with transition information in the vertical and horizontal directions of each character image were utilized as a feature set to extract the features of the numeral images. This classification system was based on SVM. The experiment results showed that although the dimensionality of the proposed feature set space was smaller than those presented in previous works, the performance of system was better.

Tanvir et al. [14] evaluated three approaches including SVM, ANN, and kNN for optical character recognition (OCR) from handwritten manuscript. These approaches were compared in the proposed system and their performances were evaluated against two feature extraction techniques, namely Local Binary Pattern (LBP) and Zoning. The evaluation revealed that the LBP feature extracting technique with the SVM classifier provided the most optimum results.

Kesiman et al. [15] investigated and evaluated the performance of the combination of feature extraction methods for character recognition of Balinese script written on palm leaf manuscripts. There were ten feature extraction methods following projection histogram, celled projection, distance profile, crossing, zoning, moments, some directional gradient based features, Kirsch Directional Edges, Neighborhood Pixels Weights (NPW), and the combination of NPW features applied on Kirsch Directional Edges images, with HOG features and zoning. Two classifiers including SVM, and kNN, were employed in experiments. They also compared the results with a convolutional neural network based system, in which feature extraction was not applied. The study determined that recognition rates can be significantly increased by applying NPW features on four-directional Kirsch edge images. Furthermore, the use of NPW on Kirsch features in combination with HOG features and zoning can increase the recognition rates superior to those using a convolutional neural network.

Research in Isarn Dhamma handwritten character recognition was presented by Wijitjareon and Seresangtakul [16], and then by Khakham et al. [17]. Wijitjareon and Seresangtakul presented an Isarn Dhamma handwritten recognition using a back-propagation neural network. Zone base, projection histogram, as well as a combination of zone base and projection histogram feature sets were applied to extract the Isarn Dharma character features. Evaluations of the recognition rates involved a dataset of character based upon 900 images from 36 classes. The experimental results indicated that the combination of both features (zone base and projection histogram) performed with higher efficiency compared to the individual features. Khakham et al. then developed the application for Isan Dhamma handwritten character recognition utilizing Functional Trees as a classifier. Both global feature and grid feature extraction methods were presented. Based on their experiments, the average accuracy rate of the system, which evaluated with 300 handwritten character images from five classes, was considered satisfactory.

3 Proposed Methodology

In this section we present the proposed Isarn Dharma handwritten character recognition which is divided into three main processes including pre-processing process, feature extraction process, and recognition process as shown in Fig. 1.

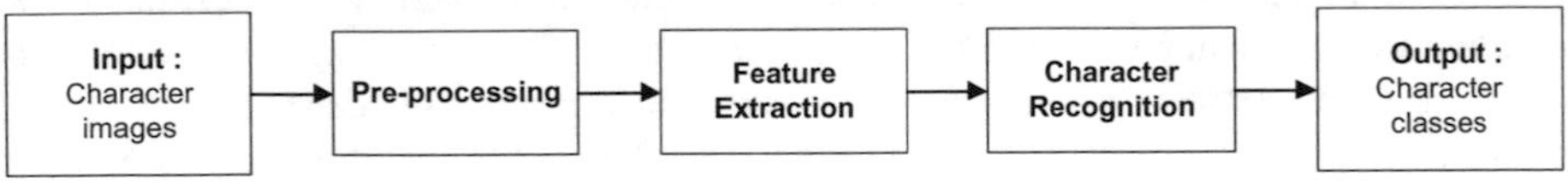

Fig. 1. Isarn Dharma handwritten character recognition

3.1 Pre-processing Process

Pre-processing, one of the most important aspects in character recognition, is responsible to improve the quality of the character images, transform them into binary images, the filtration of impurities, and normalization; as illustrated in Fig. 2. We collected the images by scanning ancient palm leaf manuscripts, and then segmented each character into pieces of the character image, Fig. 2(a). Then, binarization, which can be described as the process converting the RGB image to both a grayscale image and binary image occurs; Fig. 2(b) and (c), respectively. As noise can occur after binarization, usually salt-and-pepper noise, a median filter, appropriate to reduce salt-and-pepper noise, is applied; Fig. 2(d). In the next step, each character image is cropped and rescaled into a format of 64 $\times$ 64 pixels, Fig. 2(e). Lastly, a morphological operation is utilized with the binary image to retrieve its skeleton, Fig. 2(f).

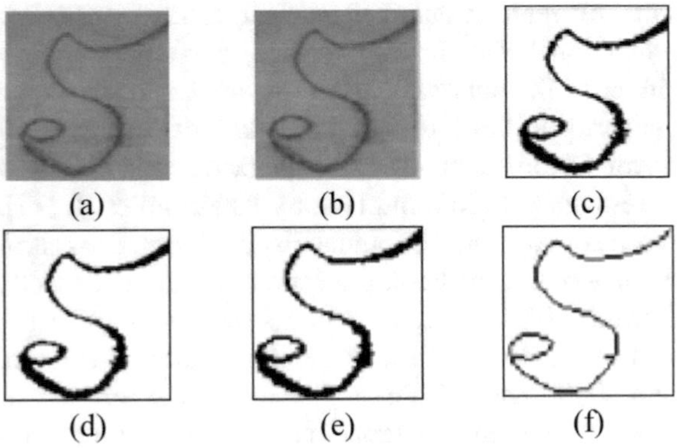

Fig. 2. Example of pre-processing process: (a) original image, (b) grayscale image, (c) binary image, (d) noise removal image, (e) rescaled image, (f) skeleton image

3.2 Feature Extraction Process

Feature extraction is an important process, as an effective method can significantly increase accuracy rate and reduce the misclassification. This process is used to extract information from the binary image and into a set of features, called feature vector. In this process, three feature extraction techniques including zoning, projection histogram, and histogram of oriented gradient (HOG) are applied to extract crucial features necessary for recognition, as described below.

Zoning. Zoning is the method that computes the density of a zone. The character image is divided into small zones, which can be square, horizontal, vertical, diagonal (left and right) [15]. In this study, we utilized the square zone method, which divided a binary image into 64 zones, each of which was 8 × 8 pixels in size. The 64 attributes from 64 zones are represented in Fig. 3.

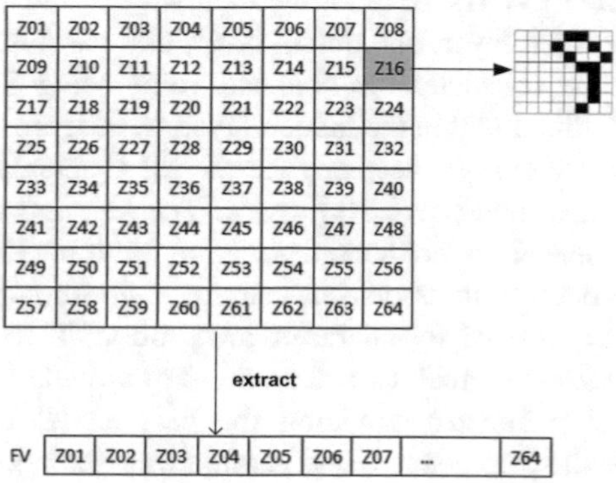

Fig. 3. Division indicating the densities of the zoning feature.

Projection Histogram. The projection histogram of a binary character may result in a horizontal zone or vertical zone, in that the direction of the projection histogram may be vertical or horizontal. Each projection histogram is divided into 32 zones, encompassing a horizontal area of 2 × 64 pixels, and a vertical area of 64 × 2. In total, there are 64 attributes from 32 horizontal projection and 32 vertical projection zones.

Histogram of Oriented Gradient (HOG). The HOG descriptor is generally defined as the distribution of the local intensity gradient of an image, which is computed by applying its derivatives in both horizontal and vertical directions from small connected regions, called cells [10, 15]. In each cell, the histogram of the directed gradient is generated by determining the gradient direction of each pixel into certain range of orientation bin β. In our experiments, we computed the HOG feature from a binary image with a cell size of eight pixels, combined with nine different orientations to produce a 1764 attributes within the HOG feature vectors (Fig. 4).

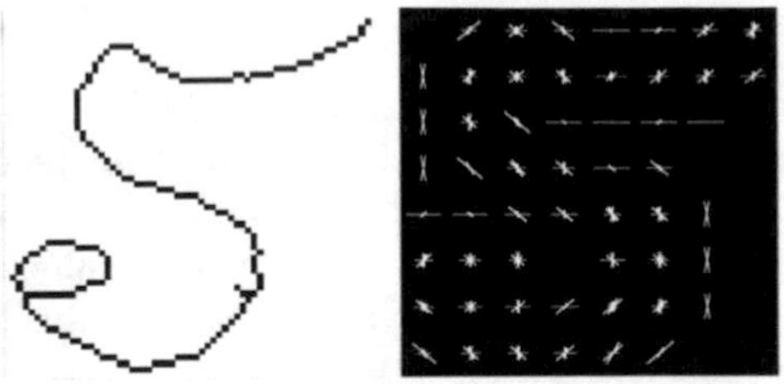

Fig. 4. HOG features for binary image

3.3 Recognition Process

In this process, the feature vectors of characters were used as input data to recognize the character classes with two classifiers namely, Artificial Neural Network (ANN) and Support Vector Machine (SVM). ANN, a well-known classification in pattern recognition, consists of one input layer, one output layer, and various hidden layers. In this study, each hidden node calculates the obtained value using Sigmoid function, and number of nodes is modified within the hidden layer, in order to determine the number of nodes that will provide the best accuracy for the MLP neural network; which, after adjusting the hidden layer node, was 100 nodes. The learning algorithm used in the network was back-propagation gradient descent algorithm. The SVM, which is a supervised learning models with associated learning algorithms analyze data used for classification, were also applied for character recognition. Throughout our study, we adjusted various parameters (kernel, C, and gamma) to provide the best accuracy. Post adjustment, the kernel parameter providing the best model was the Radial Basic Function (RBF), while the parameters C and gamma were varied, depending on dataset.

4 Experimental Results

Our experiment consisted of input data of the Isarn Dharma handwritten alphabetic character images scanned from palm leafs. There were 36 alphabets; each containing 25 alphabet images, written by several writers, as demonstrated in Fig. 5. A total of 900 character images of different size and brightness were used for classification. After classification, the original input images underwent the preprocessing process, each of which produced a binary image, with a size of 64×64 pixels.

The binary images were then subjected to feature extraction through several methods, including zone base, projection histogram, and HOG. They, individually and in various combinations, formed the resulting feature dimensions. Our two classification methods like ANN and SVM were then applied to train the recognition model. For recognition evaluation, k-fold validation, in which k was set as five, was utilizes to split the dataset into sets of training data and testing data. Table 1 shows the comparison of recognition accuracies of different feature extraction techniques with two classification methods, ANN and SVM.

The accuracy rates reveal that the SVM classification method achieved greater accuracy than ANN. In comparison with works in the past [16] our Isarn Dharma

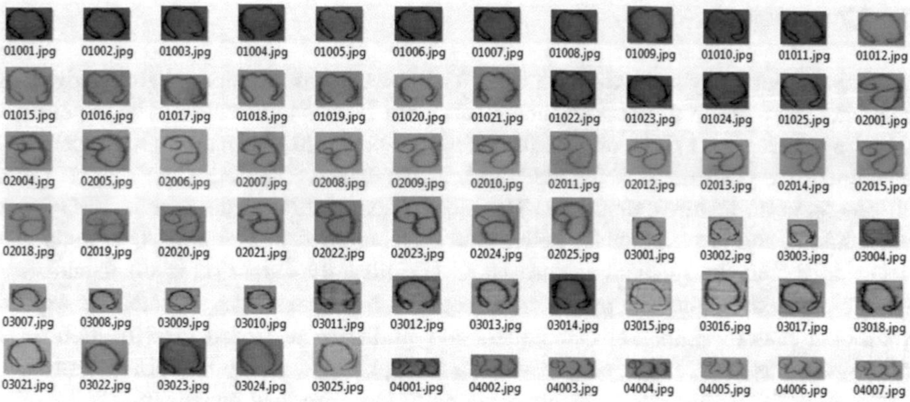

Fig. 5. A selection of the Isarn Dharma handwritten character images.

Table 1. Comparison of recognition accuracies of different feature extraction techniques using ANN and SVM.

No.	Features	Feature dimensions	Recognition accuracy (%)	
			ANN	SVM
1	Zone	64	91.11	95.22
2	Horizontal+Vertical	64	82.44	88.22
3	Zone+Horizontal	96	90.44	94.78
4	Zone+Vertical	96	89.56	94.22
5	Zone+Horizontal+Vertical	128	90.44	94.56
6	HOG	1764	93.44	96.56
7	HOG+Zone	1828	92.22	95.33
8	HOG+Horizontal+Vertical	1828	86.44	88.67
9	HOG+Zone+Horizontal	1860	91.56	94.67
10	HOG+Zone+Vertical	1860	90.74	94.22
11	HOG+Zone+Horizontal +Vertical	1892	91.56	94.67

handwritten character recognition system, which incorporates feature extraction zoning, horizontal histogram projection, vertical histogram projection, and various combinations of each; offered significantly greater performance. Moreover, HOG feature extraction provides the greatest accuracy rate. The additional HOG feature also increased performance, yielding even higher accuracy rates.

5 Conclusions

We present our study and experiments on feature extraction techniques for handwritten character recognition of palm leaf inscribed Isarn Dharma characters. The experiment relied on 13 feature extraction sets for training model and evaluation. Our experiment contained two classifiers, ANN and SVM, in which the SVM classifier outperformed all others, in all feature extractions. The recognition accuracy rate applying HOG was outstanding, and proved slightly better than HOG applied with zoning. Our study also determined that the gradient feature HOG significantly outperforms other statistical features, such as zoning and projection histogram. In future works, we intend to explore additional feature extraction techniques, and utilize more varied classification algorithms, such as kNN, DNN, etc., to achieve the optimal accuracy rate of recognition of recognition for Isarn Dharma handwritten palm leaf inscribed characters.

Acknowledgements. We wish to gratefully acknowledge the support received through a scholarship from the Ministry of Science and Technology, Thailand.

References

1. Pradeep, J., Srinivasan, E., Himavathi, S.: Diagonal based feature extraction for hand-written character recognition system using neural network. In: 2011 3rd International Conference on Electronics Computer Technology, pp. 364–368. IEEE, Kanyakumari, India (2011)
2. Pal, U., Jayadevan, R., Sharma, N.: Handwriting recognition in indian regional scripts: a survey of offline techniques. ACM Trans. Asian Lang. Inf. Process. TALIP **11**(1), 1–1:35 (2012). ACM
3. Arif, M., Hassan, H., Nasien, D., Haron, H.: A review on feature extraction and feature selection for handwritten character recognition. Int. J. Adv. Comput. Sci. Appl. 6 (2015)
4. Ketcham, M., Yimyam, W., Chumuang, N.: Segmentation of overlapping Isan Dhamma character on palm leaf manuscript's with neural network. In: 2016 Recent Advances in Information and Communication Technology, pp. 55–65. Springer, Cham (2016)
5. Kumar, G., Bhatia, P.K.: A detailed review of feature extraction in image processing systems. In: 2014 Fourth International Conference on Advanced Computing Communication Technologies, pp. 5–12. IEEE, Rohtak, India (2014)
6. Parvez, M.T., Mahmoud, S.A.: Offline Arabic handwritten text recognition: a survey. ACM Comput. Surv. **45**, 23:1–23:35 (2013)
7. Ouchtati, S., Mohamed, R., Bedda, M.: A set of features extraction methods for the recognition of the isolated handwritten digits. Int. J. Comput. Commun. Eng. **3**, 349–355 (2014)
8. Choudhary, A., Rishi, R., Ahlawat, S.: Off-line handwritten character recognition using features extracted from binarization technique. AASRI Procedia. **4**, 306–312 (2013)
9. Arora, S., Bhattacharjee, D., Nasipuri, M., Malik, L., Kundu, M., Basu, D.K.: Performance Comparison of SVM and ANN for Handwritten Devnagari Character Recognition (2010). ArXiv10065902 Cs
10. Surinta, O., Karaaba, M.F., Schomaker, L.R.B., Wiering, M.A.: Recognition of hand-written characters using local gradient feature descriptors. Eng. Appl. Artif. Intell. **45**, 405–414 (2015)

11. Lakkhawannakun, P., Seresangtakul, P.: Isarn dharma alphabets to Thai language translation by ATNs. In: Jeong H., Obaidat, S.M., Yen, N., Park, J. (eds.) Advances in Computer Science and its Applications. Lecture Notes in Electrical Engineering, vol. 2779, pp. 775–782. Springer, Heidelberg (2014)
12. Phaiboon, N., Seresangtakul, P.: Isarn Dharma Alphabets lexicon for natural language processing. In: 2017 9th International Conference on Knowledge and Smart Technology (KST), pp. 211–215. IEEE, Chonburi, Thailand (2017)
13. Boukharouba, A., Bennia, A.: Novel feature extraction technique for the recognition of handwritten digits. Appl. Comput. Inform. **13**, 19–26 (2017)
14. Tanvir, S.H., Khan, T.A., Yamin, A.B.: Evaluation of optical character recognition algorithms and feature extraction techniques. In: 2016 Sixth International Conference on Innovative Computing Technology (INTECH), pp. 326–331. IEEE, Dublin, Ireland (2016)
15. Kesiman, M.W.A., Prum, S., Burie, J.C., Ogier, J.M.: Study on feature extraction methods for character recognition of Balinese script on palm leaf manuscript images. In: 2016 23rd International Conference on Pattern Recognition (ICPR), pp. 4017–4022. IEEE, Cancun, Mexico (2016)
16. Wijitjareon, A., Seresangtakul, P.: A hybrid features extraction for Isarn Dharma handwritten character recognition using neural network. In: 2012 4th International Conference on Knowledge and Smart Technology (KST), pp. 45–52. IEEE, Chonburi, Thailand (2012). (in Thai)
17. Khakham, P., Chumuang, N., Ketcham, M.: Isan Dhamma handwritten characters recognition system by using functional trees classifier. In: 2015 11th International Conference on Signal-Image Technology Internet-Based Systems (SITIS), pp. 606–612. IEEE, Bangkok, Thailand (2015)

Network and Security

Subset Sum-Based Verifiable Secret Sharing Scheme for Secure Multiparty Computation

Romulo L. Olalia Jr.(✉), Ariel M. Sison, and Ruji P. Medina

Technological Institute of the Philippines, Quezon City, Philippines
romulo.olalia@sancarloscollege.edu.ph,
ariel.sison@eac.edu.ph, ruji.medina@tip.edu.ph

Abstract. Despite the information theoretic security of Shamir Secret Sharing Scheme and the ideality of Verifiable Secret Sharing Scheme in ensuring the honesty of a dealer of the shared secret and the shared secret itself, the detection and removal of an adversary posing as shareholder is still an open problem due to the fact that most of the studies are computationally and communicationally complex. This paper proposes a verifiable secret sharing scheme using a simple subset sum theory in monitoring and removing compromised shareholder in a secure multiparty computation. An analysis shows that the scheme cost minimal computational complexity of O(n) on the worst-case scenario and a variable-length communication cost depending on the length of the subset and the value of *n*.

Keywords: Secret sharing scheme · Subset sum
Secure multiparty computation · Shamir scheme · Imposter · Shareholder
Dealer · Network security

1 Introduction

A network system is secure only if it preserves the confidentiality of the information and proper authentication have been carried out [1]. A private information should remain private and should not be lurking around the network [2]. A part of a private information should not also be broadcast because knowledge of much little information may hint a potential attacker. Basing on the intrusion triangle (opportunity, motive, and means) [3], an opportunity should be at least prevented.

The question of whom to trust is essential especially on the topic of network security and secret sharing [4]. A host, which holds a secret that is connected to a network, can either be trusted or not. Sometimes, a host that was initially trusted might be compromised over a period of time. Adi Shamir introduced a concept on how a certain secret (S) can be securely divided into several pieces and distribute it in such a way that with a given number *(k)* of shares, secret *(S)* can be reconstructed but a combination of $k - 1$ shares will reveal nothing about the secret *(S)*. Although information is theoretically secure, it lies on the issue that the dealer is honest and the distribution happens on a private connection. Verifiable Secret Sharing Scheme (VSSS) [5] addressed this issue wherein it allows dealer *D* to share secrets among n number of parties and each party can confirm that the dealer is honest and that parties are not

H. Unger et al. (Eds.): IC2IT 2018, AISC 769, pp. 209–219, 2019.
https://doi.org/10.1007/978-3-319-93692-5_21

cheating on the process [6]. Despite this, VSSS lack on the discovery and removal of adversary posing as a shareholder in a multiparty computation.

This study modifies the existing and proven verifiable secret sharing scheme while protecting each party from adversaries in a distributed environment using subset sum theory. Specifically, the scheme or protocol will enable each shareholder to constantly monitor if a shareholder is still the original shareholder or just an imposter at the least possible time without disturbing the master server or dealer of the secret *S*. The scheme will also allow the removal of the compromised shareholder and renewal of a new shared secrets, commitments and subsets among trusted shareholder.

2 Related Literature

2.1 Secret Sharing Scheme and Its Problem

Network and information security have a long history of challenges and researches. In the area of secret sharing, Adi Shamir has introduced the concept of threshold (k, n) cryptography where a certain secret (S) can be securely divided into several pieces and distribute it in such a way that with a given number (k) of shares, secret (S) can be reconstructed but combination of $k - 1$ shares will reveal nothing about the secret (S). He called it (k, n) threshold scheme or Secret Sharing Scheme (SSS).

Although his concept is information theoretically-secure, it is vulnerable to cheaters and attackers from inside and out [7]. It also dwells on the assumption that the dealer is honest and the distribution happens in a private connection. However, that is not always the case. If the dealer is dishonest, the secret is compromised and if the participants in the network cheated, the secret cannot be recovered. Answering the problem, Verifiable Secret Sharing Scheme has been introduced [8].

2.2 Verifiable Secret Sharing Scheme

Verifiable secret sharing scheme (VSSS) is based on the work of Shamir only that any party or shareholder holding a shared secret can confirm that the dealer is honest and that shareholders are not cheating on the process. However, if a shared secret has been discovered and in the course of time the culprit had acquired enough number of shares (t), the secret may be disclosed. This type of attack can be offset by refreshing of the shared secret.

To secure the shared secret on a distributed network, constant modification of the shared secret may be done such that Secret S stays the same [9]. This means that the original shared secret will be refreshed at an interval time. However, refreshing the shared secret does not detect and remove an impersonator computer because it will just receive the secret in every refresh. Some researches use part of the shared secret to be exchanged to other shareholders [10, 11], however, sharing a part of the shared secret is like giving an opportunity to the attacker to make an attack and does not detect imposters [12]. Algorithms and theories involved in different researches were also computationally heavy. They use a combination of multiplication and addition of polynomial to secure the shared secret and its communication [13].

2.3 Subset Sum Theory

There is a simple way to achieve both security and low computational and communication cost while taking advantage of the information-theoretic security of Shamir's secret sharing scheme and the verification capability of VSSS and avoiding an imposter to penetrate the network. This is in the use of the theory of subset sub, a semantically secure cryptosystem [14]. The subset sum theory suggests that given a positive integer, an unknown combination from the list should sum up to a given integer *S*. This concept will not be used entirely on this study to make it even more computationally low yet secure. Simply, generate a set of integers which summed up to a given integer *b*. The same additive homomorphic property has been used in addressing privacy protection for collection and merging of data to test its equality [15] ensuring data security in a public key cryptosystem [16], logic optimization, an application on LTE networks [17, 18] and mobile applications and graphics processing units [19].

Although some researches were conducted in solving the problem, it is not enough to address the concern. Therefore, it is a proposition of this research that with the integration of concept from subset sum theory in the information-theoretic secure SSSS and the verification capability of VSSS, an imposter in the network can be detected and eliminated; other conventional attacks can be deterred such as man-in-the-middle and lattice-based attack; collusion between shareholder can be discourage; and the communication and computational cost will be reduced thus having a more secured multi-party and distributed network.

3 Design Architecture

3.1 Concept of the Scheme

The conceptual framework of the study was depicted in Fig. 1. Initially, the (trusted) dealer/master server will generate shared secret *(x, y)* from secret S based on Shamir's secret sharing scheme or the *(k, n)* threshold scheme.

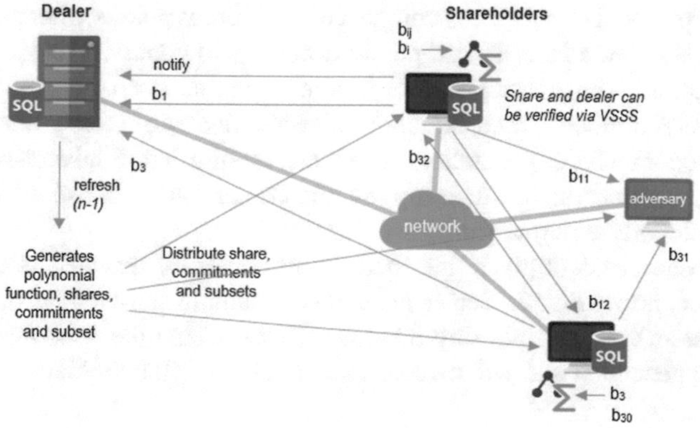

Fig. 1. Concept of the study

The *(k, n)* threshold scheme is based on polynomial interpolation given *k* points in a 2-dimensional plane *(x, y)*. The n is the number shared secrets to be generated and k is the threshold or the number of shared secrets that are needed to reconstruct the secret. The dealer will then generate n number random integer number b_i which is the subset of and sum up to b (1):

$$b = \sum_{i=1}^{n} b_i \tag{1}$$

The generated b_i, and other private and public values will be distributed to the shareholders. Confirmation of the trustworthiness of the dealer and the truthfulness of shared secret will be done through verifiable secret sharing scheme.

Upon receiving the four variables (multivariate), the shared secret (x_i, y_i) will be stored and all shareholders will generate n number of subset b_{ij} whose sum is the shared secret b_i (2).

$$b_i = \sum_{j=1}^{n} b_{ij} \quad \ni b_{ij} \neq b_i \tag{2}$$

Subset $b_{ij} \mid b_{ij} \neq b_{i1}$ will be distributed to all other shareholders leaving behind b_{i0}. The received $n - 1$ number of *bij* will be summed with b_{i1} thus having a new b_i. This will loop until a shareholder misses a b_{ij} from a shareholder. When that happens, a shareholder will notify the dealer. The dealer will then ask all shareholders to provide their respective b_i. Regardless of the round when all b_i are computed, it will be always equal to the original generated number b (1) by the dealer.

If a dealer did not receive b_i from the imposter, the sum will be wrong then the server will know that particular shareholder is an imposter. Thus, re-computes new shared secret using different polynomial removing the imposter $(n - 1)$ from the circle and redistributes the shared secret again including the commitment and subset to the shareholders.

3.2 Overview of the Protocol

The proposed protocol or scheme is composed of different phases wherein entities such as dealer, shareholder, adversary and public domain will generate, receives, shares and holds information. The existing VSSS includes phases such generation of shares, generation of commitments, reconstruction of secret, and verification of share while the proposed protocol includes generation of subsets, monitoring of adversary, notification of the dealer, verification of adversary by the dealer and removal of the adversary within the multiparty environment.

The proposed protocol given that there is no adversary detected in the multiparty computation is shown Fig. 2. Dealer generates a random polynomial of degree $t - 1$ with the secret as the constant using Shamir scheme. Using the generated polynomial, shares will be generated and will be sent to n number of shareholders. Aside from the shares, the Dealer will generate prime p and q such that $q \mid p - 1$ and generator $g \in Z_p^*$. Values of p, q and g are all public. Shareholders may verify the correctness of their respective share and the honesty of the Dealer using verification scheme such as of Feldman scheme. The secret can be re-constructed using Lagrange Interpolation.

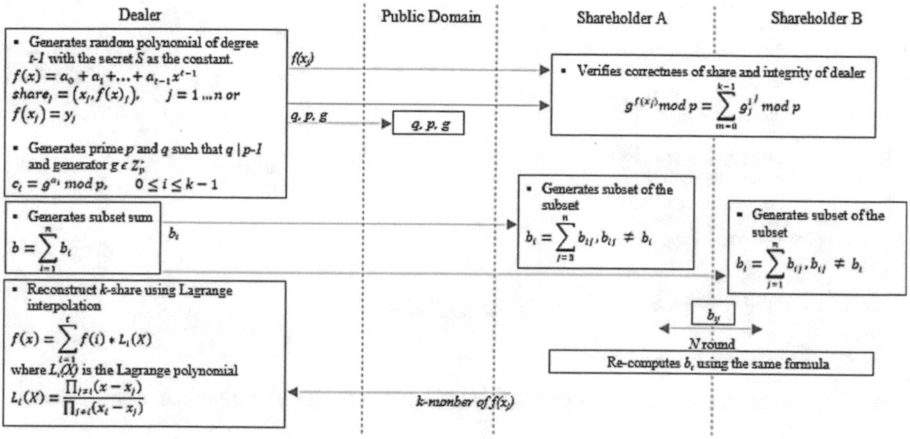

Fig. 2. Subset sum-based VSSS protocol given no adversary in the multipart computation

N-number of subset b_i of a sum b will also be generated by the Dealer and will be distributed to all shareholders. Upon receipt, the shareholder will generate its own subset b_{ij} whose sum is equal to b_i. These subsets will be sent by the shareholder to other shareholders except for b_{i0} which will be retained with it and that any b_{ij} is not equal to b_i.

On the other hand, the proposed protocol given that an adversary poses a shareholder among others is showcased in Fig. 3. Same with the first protocol, Dealer generates shares based on Shamir's scheme, commitments to the polynomial and subsets. These values will be distributed to shareholders and the public domain respectively. An honest shareholder can verify the integrity of the Dealer while the adversary would not because verification may or may not be done and it is assumed that the adversary does not know the algorithm used by other honest shareholders. Dealer can still reconstruct the secret given that the remaining honest shareholders are $\geq$k and $\leq$n.

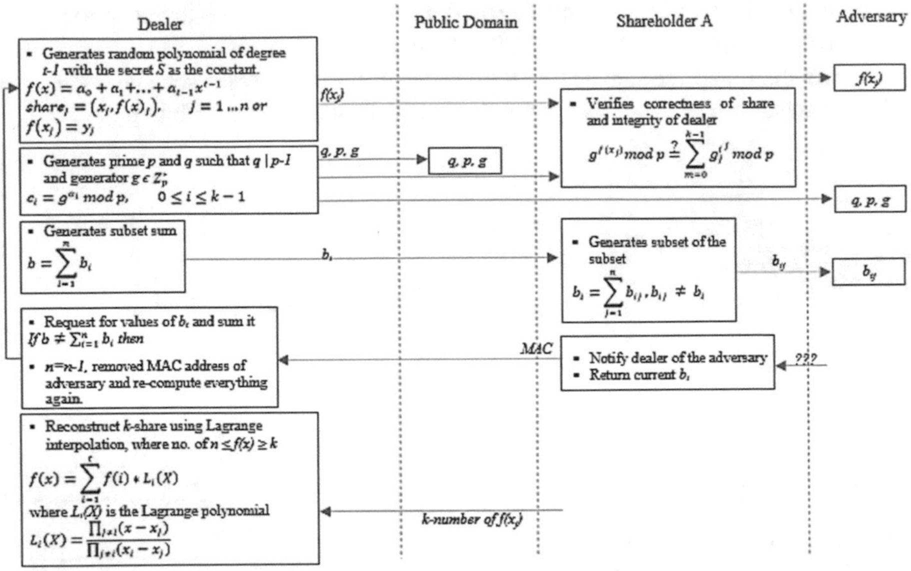

Fig. 3. Subset sum-based VSSS protocol with an adversary in the multiparty computation

4 Evaluation of the Modified Scheme

Computational complexity can be analyzed using an asymptotic notation, particularly Big O notation which uses the worst case scenario. Big O notation provides theoretical estimation of resources needed to solve a particular problem using an algorithm with respect to time and space. This paper analyzes only the contribution of the VSSS.

The protocol includes two basic functions: generation of initial subset numbers and summing it up; and generation of subset numbers based on a given sum. These functions happen on the Dealer and shareholders respectively.

In the generation of subset numbers and summing it up, the following pseudocode was performed:

```
for (i=1; i<=n; i++) {
  bi = 256-bitRandomNo
  b = b + bi
}
```

Analyzing the code, the following notation can be derived (3):

$$f(n) = (c1 + c2) \times n \quad (3)$$

or in Big O notation, it is simply (4):

$$f(n) = O(n) \tag{4}$$

The second function comprises of generating a group of subsets b_{ij} whose sum is equal to b_i. The following code depicts the function:

```
a = bi
for (i=1; i<=n; i++) {
  bij = 256-bitRandomNo(1,a)
  tmp = tmp + bij
  a = a - tmp
  if (i==n)
  bij = bij + a
}
```

The corresponding notation is (5):

$$\begin{gathered} f(n) = c1 + n \times c2 + c3 + c4 + c5 + c6 \\ f(n) = O(n) \end{gathered} \tag{5}$$

Based on the computational complexity analysis, the proposed scheme only need O (n) complexity, at the worst case scenario, when performing the needed functions which is very simple compared to other protocol. This means that when implemented, it will not cost many resources, therefore, making it an efficient addendum to an existing protocol.

On the communication complexity, Dealer needs to send subsets bi to n number of shareholders during the initial stage of the protocol. Moreover, when the shareholder receives their respective bi, they need also to send subsets bij to n − 1 number of shareholders. Finally, when an adversary broke out, shareholders will send back current values of n − 1 number of bi to the Dealer.

When these values were sent over the communication line (256-bit values), the following formula (Fig. 4) can be used to compute for resources *r* that is needed:

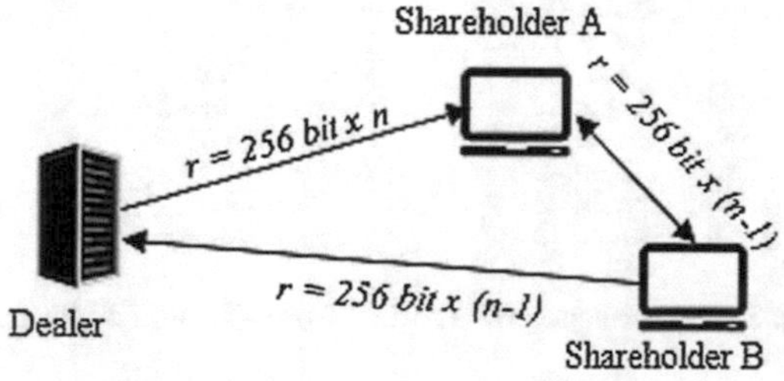

Fig. 4. Communication cost of the scheme

Given, that the number of shareholder n is = 3 and the size of the subsets are 256 bits when the Dealer sends subsets b_i to all shareholders, the communication cost is (6):

$$\mathrm{r} = 256\,\text{bits} \times 3 = 768\,\text{bits} \tag{6}$$

When the shareholders received their respective shares, a new set of subset will be generated and will be distributed to other shareholder except for the one that generated that share. Thus (7):

$$\mathrm{r} = 256\,\text{bits} \times (3 - 1) = 512\,\text{bits/round} \tag{7}$$

Similarly, when an adversary was detected, shares (b_i) will be requested by the Dealer and will travel through the communication line. Thus (8):

$$\mathrm{r} = 256\,\text{bits} \times (3 - 1) = 512\,\text{bits/round} \tag{8}$$

To further analyze the proposed scheme, the generation of the subsets are simulated. Using a computer with a 64-bit AMD A8 Quadcore 2.0 Ghz processor, 6 GB of RAM, AMD Radeon R5 Graphics, and with 64-bit Windows 10, a Javascript program was created. Since Javascript cannot handle large number very well, a Javascript library was used called BigInteger.js. This library is an arbitrary-length library for JS which allows arithmetic operations bypassing the memory and time limitations. Some functions were called from this library. Values in these simulations are in 256-bit range.

To get the actual time spent for each function, the program was executed using Google Chrome 64.0.3282.167 (Official Build) 64-bit Developer Edition utilizing its DevTools. Performance monitor (Fig. 5) was used to record the performance of the Javascript program while running. Functions and its respective time spent in the simulations are shown in Table 1.

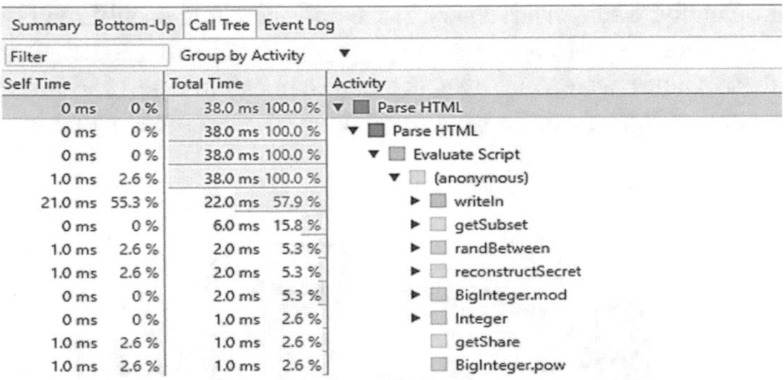
Summary Bottom-Up Call Tree Event Log

Filter Group by Activity

Self Time		Total Time		Activity
0 ms	0 %	38.0 ms	100.0 %	Parse HTML
0 ms	0 %	38.0 ms	100.0 %	Parse HTML
0 ms	0 %	38.0 ms	100.0 %	Evaluate Script
1.0 ms	2.6 %	38.0 ms	100.0 %	(anonymous)
21.0 ms	55.3 %	22.0 ms	57.9 %	writeln
0 ms	0 %	6.0 ms	15.8 %	getSubset
1.0 ms	2.6 %	2.0 ms	5.3 %	randBetween
1.0 ms	2.6 %	2.0 ms	5.3 %	reconstructSecret
0 ms	0 %	2.0 ms	5.3 %	BigInteger.mod
0 ms	0 %	1.0 ms	2.6 %	Integer
1.0 ms	2.6 %	1.0 ms	2.6 %	getShare
1.0 ms	2.6 %	1.0 ms	2.6 %	BigInteger.pow

Fig. 5. Performance monitoring from Chrome's DevTools

Table 1. Time spend of each function used in the protocol

Function	Time spent (in milliseconds)
randBetween()	2.0 ms
BigInteger.add()	1.0 ms
subtractSmall()	1.0 ms
BigInteger.subtract()	2.0 ms
parseValue()	1.0 ms
getSubset()	6.0 ms

Following the pseudocode above, k is set to 3 wherein the program will generate 3 random 256-bit numbers and total them together. The first subset b_i will be taken and from it generate 3 random numbers b_{ij} which sum is equal to the subset.

Based on the performance monitoring of the simulation in Fig. 5 that shows the timeline and the corresponding elapse time per function that was recorded during the execution of the code, the generation of the initial subset values by the dealer took 4.0 milliseconds (ms) wherein it includes randomization function - 2.0 ms, parsing of values from integer to string - 1.0 ms and adding of large numbers - 1.0 ms).

With the generation of the second group of the subset by the shareholder, it only took 10 ms wherein it includes subtracting small numbers - 1.0 ms, subtracting big numbers - 2.0 ms, parsing of values from integer to string - 1.0 ms and finally the predefined function that creates subsets b_{ij} - 6.0 ms. With an overall time of just 14 ms, it can be inferred that the scheme only requires very low computational resources thus can be easily integrated even when using an arbitrary large number.

5 Conclusion and Recommendation

Comparing the proposed subset sum based verifiable secret sharing scheme with the conventional VSSS, the proposed scheme provides a way to monitor and remove an adversary posing as a shareholder that is contrary to the conventional VSSS. This can be done by monitoring a subset value from a given sum and monitoring this value that will be passed by the other shareholders.

Based on the computational complexity analysis, the proposed scheme only need O (n) complexity, at the worst case scenario, when performing the needed functions which is very simple compared to other protocol. This means that when implemented to existing schemes, it will not cost many resource, the additional computational requirement is minimal.

Moreover, the communication complexity assessment of the proposed scheme shows that it will only need a variable-length bandwidth depending on the required length (in number of bits) of the system to be applied to multiply by the n number of shareholders. Furthermore, the simulation of the generation of the subset of the two required major functions (generation of the subset by the dealer and generation of the subset by the shareholder) totals to only 14 ms. This means that with the very low computational resources required of the proposed scheme, it can be easily integrated to the desired system even using a very arbitrary large number.

Thus, it is recommended that future researchers should look at the security analysis of the proposed scheme. It is also encouraged that comparison between other researches on securing multiparty computation be compared with the proposed protocol in terms of performance, memory usage, and bandwidth.

Acknowledgment. The researchers would like to acknowledge the financial support provided by the Commission on Higher Education K to 12 Transition Program PMU, Quezon City, Philippines and San Carlos College, San Carlos City, Pangasinan.

References

1. Thomas, G., Sibi, L., Maneesh, P.: A novel mathematical model for group communication with trusted key generation and distribution using shamir's secret key and USB security. In: International Conference Communication and Signal Processing, pp. 435–438 (2015)
2. Alwen, J., Hirt, M., Maurer, U., Patra, A., Raykov, P.: Anonymous authentication with shared secrets. In: Aranha, D., Menezes, A. (eds.) Progress in Cryptology - LATINCRYPT 2014. LNCS, vol 8895, pp. 219–236. Springer, Cham (2015)
3. Van Ruitenbeek, E., Keefe, K., Sanders, W.H., Muehrcke, C.: Characterizing the behavior of cyber adversaries: the means, motive, and opportunity of cyberattacks. In: Networks, pp. 17–18 (2010)
4. Tavakolifard, M., Almeroth, K.C.: Trust 2.0: Who to believe in the flood of online data? In: Proceedings of International Conference Computer Network Communication, pp. 544–550 (2012)
5. Feldman, P.: A practical scheme for non-interactive verifiable secret sharing. In: 28th Annual Symposium on Foundations of Computer Science (sfcs 1987), pp. 427–437 (1987)
6. Backes, M., Kate, A., Patra, A.: Computational verifiable secret sharing revisited. In: Lecture Notes in Computer Science (Including Subseries Lecture Notes in Artificial Intelligence and Lecture Notes in Bioinformatics). LNCS, vol. 7073, pp. 590–609 (2011)
7. Harn, L., Lin, C., Li, Y.: Fair secret reconstruction in (t, n) secret sharing. J. Inf. Secur. Appl. **23**, 1–7 (2015)
8. Chor, B., Goldwasser, S., Micali, S., Awerbuch, B.: Verifiable secret sharing and achieving simultaneity in the presence of faults. In: 26th Annual Symposium Foundations of Computer Science, pp. 21–23. IEEE, Portland (1985)
9. Nimmy, K.: Novel multi-server authentication protocol using secret sharing. In: International Conference on Data Mining and Advance Computing, pp. 1–6. IEEE, Ernakulam (2016)
10. Shukla, S., Sadashivappa, G.: Secure multi-party computation protocol using asymmetric encryption. In: 10th Asia Conference Information Security, pp. 780–785. IEEE, New Delhi (2014)
11. Wei, L., Ai, J., Liu, S.: A tightly secure multi-party-signature protocol in the plain model. In: 8th International Conference Biomedicine, Engineering and Informatics, pp. 672–677. IEEE, Shenyang (2016)
12. Pakniat, N., Noroozi, M., Eslami, Z.: Reducing multi-secret sharing problem to sharing a single secret based on cellular automata. IACR Cryptology ePrint Archive 642 (2017)
13. Aljumah, F., Soeanu, A., Liu, W.M., Debbabi, M.: Protocols for secure multi-party private function evaluation. In: 2015 1st International Conference Anti-Cybercrime, pp. 1–6. IEEE, Riyadh (2015)
14. Movatadid, L.: CSC 373 - algorithm design, analysis, and complexity - subset sum. In: Summer 2016, pp. 2–5 (2005)

15. Ren, S.Q., Meng, T.H., Yibin, N., Aung, K.M.M.: Privacy-preserved multi-party data merging with secure equality evaluation. In: Proceedings of International Conference Cloud Computing Research and Innovation, ICCCRI 2016, pp. 34–41 (2016)
16. Marwan, M., Kartit, A., Ouahmane, H.: Applying Secure multi-party computation to improve collaboration in healthcare cloud. In: International Conference System Collab, pp. 1–6. IEEE, Casablanca (2016)
17. Chen, K.: Efficient sum-to-one subsets algorithm for logic optimization. In: 29th ACM/IEEE Design Automation Conference, pp. 443–448. IEEE, CA (1992)
18. Zhou, D., Song, W., Ju, P.: Subset-sum based relay selection for multipath TCP in cooperative LTE networks. In: Globecom 2013 - Wireless Network Symposium Subset-Sum, pp. 4735–4740. IEEE, GA (2013)
19. Ristovski, Z., Mishkovski, I., Gramatikov, S., Filiposka, S.: Parallel implementation of the modified subset sum problem in CUDA. In: 22nd Telecommunication Forum Telfor, pp. 923–926. IEEE, Belgrade (2014)

The Semantics Loss Tracker of Firewall Rules

Suchart Khummanee(✉)

Faculty of Informatics, Mahasarakham University, Mahasarakham 44000, Thailand
suchart.k@msu.ac.th
https://www.it.msu.ac.th

Abstract. Frequently, firewall rules are overlapped and duplicated. The problems are usually resolved by merging rules. However, sometimes merged rules lead to the semantics loss. This paper proposed the tracker system for analyzing and alerting the semantics loss of firewall rules while they are being merged, namely SELTracker. SELTracker data structure is built from the Path Selection Tree (PST). PST does only keep all anomaly rules but also maintain normal rules. While firewall rules are being merged, SELTracker analyzes merging rules against PST. Based on the testing results, the proposed system has the ability to effectively detect the semantics loss. Moreover, SELTracker can also detect all other anomalies.

Keywords: Firewall rule · Anomaly · Semantics loss
Path selection tree

1 Introduction

As the modern network connectivity, the security is very important. The firewall is a first security system which is considered to restrict unauthorized user from resources over networks. Firewalls are forced by rules. A rule consists of two parts, the first part is a predicate and another is an action. The predicate refers to conditions of verification. The conditions are usually five fields i.e., the source and destination ip address (0 to $2^{32}-1$ for IPv4 and 0 to $2^{128}-1$ for IPv6), source and destination port (0 to $2^{16}-1$ for WWW = 80), and the protocol (0 to $2^{8}-1$ for TCP = 17 and UDP = 6). The action is either 0 (deny) or 1 (accept) only. Any packet passing through the firewall is verified against the predicate of the first rule. If the result from verifying is true, the packet will be followed by the action of the first rule. On the other hand, if the result is false, the packet will be verified against the next rule in a sequence until no more rules. The number of firewall rules depends on the complication of policies of each organization. Firewall rules are frequently conflict. Al Shaer et al. [1] described the conflict as five cases and called anomalies i.e., the shadowing, correlation, generalization, redundancy and irrelevant. Firewall rule anomalies reduce the firewall performance. The solution which is efficient to resolve the problem is reducing the number of firewall rules

H. Unger et al. (Eds.): IC2IT 2018, AISC 769, pp. 220–231, 2019.
https://doi.org/10.1007/978-3-319-93692-5_22

by merging rules that are overlap and duplicate. However, some merged rules will lost its meaning from original objectives. For example, company A defined firewall rules as follows.

Rule 1 (Boss of company A): boss can access to the internet.
Rule 2 (Employees of company A): employees can access to the internet.

Both rules are overlapping. Conceptually, the administrator should merge two rules (Rule 1 and Rule 2) to be one rule (Rule 1 and Rule 2 are merged to Rule 1) as follows.

Rule 1 (Boss and Employees of company A): all can access to the internet.

After that, the company A launched a new policy which does not allow all employees to surf the internet. Thus, the administrator needs to remove the Rule 1 from the firewall which allows both Boss and Employees to access the internet. As a result, Boss cannot access the internet. In fact, the new policy requires to restrict only employees. This is called ***the semantics loss or lost meaning*** of firewall rules. In addition, there is another way to block employees which most administrators like to do, that is to add a new rule which is not only allowed employees on the top of firewall rules like:

Rule 1 (Block only employees): employees cannot access the internet.
Rule 2 (Allow both Boss and Employees): all can access to the internet.

However, this approach will increase more conflict into the rules. But most admins choose to do it without ignoring the conflicts. As a result, the firewall rules are bulky, incompatible and ineffective. Rule optimization is necessary before deploying them, but such technique will also lead to the semantics loss of rules. This paper aims to build the system which can detect and alert the semantics loss while rules are merging. The rest of the paper: firstly, reviewing related work in Sect. 2, then explaining the concept of the semantics loss and design in Sect. 3, the development of the data structure and evaluation in Sect. 4, and finally in Sect. 5 for the conclusions.

2 Related Work

2.1 Firewall Rule Anomalies

Indeed, the anomaly and conflict problem of the firewall rule are long and ongoing research that have taken place over the past decade and will continue. The famous person who defined the anomaly and conflict is Al-Shaer [1]. He defined five kinds of anomalies, which are shadowing, correlation, generalization, redundancy and irrelevance as follows.

Given sip = source ip address, dip = destination ip address, sp = source port, dp = destination port, and pro is the protocol.

The shadowing is any rules which are obscured by rules in the preceding order. As a result, the rules below are not processing as shown in Fig. 1(a). Let Rx (Rule x) and Ry (Rule y) be any rules in the firewall by x $<$ y, then Ry is shadowed by Rx if all numbers in Ry are in Rx, and there are the different action. The rules based on shadowing anomaly cannot be merged. To resolve shadowing, Al-Shaer recommended swapping between Rx and Ry or removing Ry instead.

The correlation is any rules which are related or overlapped, and there are the different decision. Therefore, some members of the rules below are processed and some are not processed as shown in Fig. 1(b). Rx is correlated with Ry if some of members in Ry are members of Rx and some of Rx are in Ry, and there are the difference action. On the other words, Rx intersects with Ry by different actions. Correlated rules cannot reduce the number of rules because it effects on security and performance. Moreover, it can not switch between Rx and Ry as well.

The generalization is any rules which all members are a subset of other rules, and they have the different decision. Thus, some of the rules below will be not processed as illustrated in Fig. 1(c). Ry is partially covered by Rx if all members in Rx are members on Ry, and they have the different operation. The generalization rules cannot be combined, as with any of the other anomalies mentioned above. Swapping Rx and Ry leads to the shadowing anomaly suddenly.

The redundancy is any rules which are covered by the rules above and, they have the same action as Fig. 1(d). The redundancy can be merged into compact rules. Several researchers proposed the algorithms to resolve this problem as follows. Liu et al. [2] proposed a firewall compressor which compressed firewall rule sets about 52.3% by using dynamic programing against one-dimensional and two-dimensional firewalls. This is consistent with Hu et al. [3], they contributed a solution to resolve overlapping rules using a method called Firewall Anomaly Management Environment (FAME). They claimed that FAME could be higher efficiency and effectiveness. Rule-based merging methods were introduced in several ways. The service-grouping [4] is the best approach in order to reduce the number of rules. They compared merging performance of service-grouping against the traditional and FIREMAN algorithm [12]. They claimed that the total processing time was lower than both algorithms. The heuristic solution proposed (HAF) by Yoon et al. [5] is the effective way to resolve the size of rule sets. The results show that HAF reduces the maximum firewall rule sets about 2 to 5 times by comparing against TTA and FTA. Graph theory is another popular method. Katic et al. [6] proposed a efficiency approach (FIFO) to merge similar rules. They compared the performance with Firewall Policy Advisor [7], they concluded that FIFP was more convenient. The last approach to resolve anomaly and to reduce the size of rule sets is SDD [8]. The SDD applied the concept of the single domain decision to build tree for detecting anomalies and correcting duplicated rules. SDD could eliminate the root of anomalies, but it can also cause the memory problem when rules are large.

The last anomaly problem is irrelevance. It is any rules that are not members of any rules at all (Disjoint). On the other words, any packets do not match any rules in firewall as shown in Fig. 1(e).

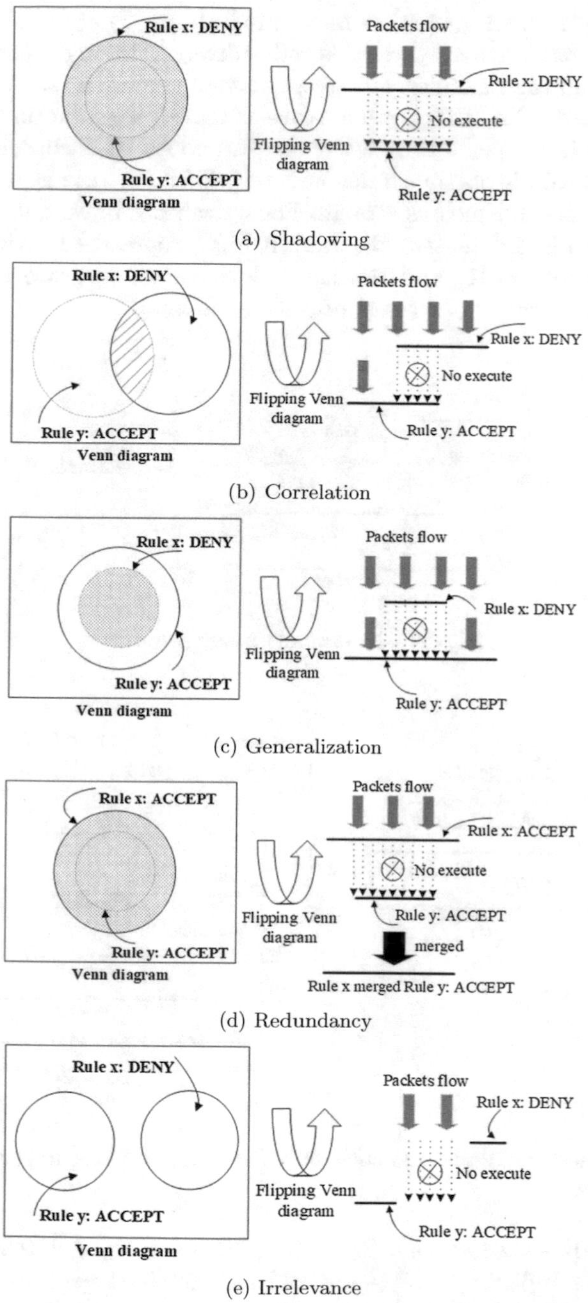

(a) Shadowing

(b) Correlation

(c) Generalization

(d) Redundancy

(e) Irrelevance

Fig. 1. Firewall rule anomalies

2.2 The Semantics Loss of Firewall Rules

Due to the five firewall anomalies mentioned above. There is only one kind of anomalies that can be merged, that is redundancy. Although merging rules will make the firewall rules smaller. However, none of the researchers considered the impact of merged rules. This section explains more the semantics loss of firewall rules and how it happens. Supposing the boundary of each rule is a straight line. The scope of all possible rules is 1 to 100. Each rule is represented in a different linear fashion such as straight lines, dash dot lines, dotted lines and so on as shown in Fig. 2. The straight lines in Fig. 2 represent R_1 (Rule number 1) and R_5. Dot lines are R_2 and R_3. Dash dotted lines indicate R_4, R_6 and R_7 respectively. Each rule has a range of rules as follows.

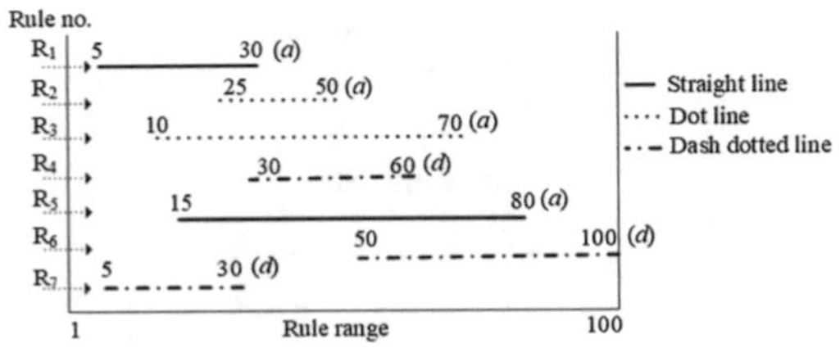

Fig. 2. The semantics loss problem

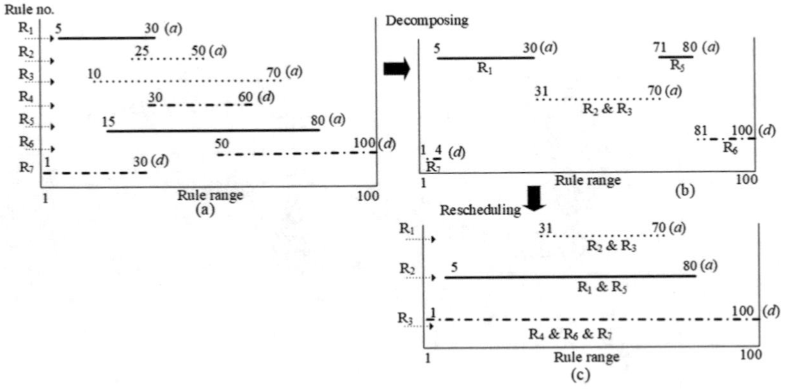

Fig. 3. Reducing duplicated rules by using Firewall Compressor algorithm

$R_1 \in \{5, 30\} \rightarrow a\ (accept)$, $R_2 \in \{25, 50\} \rightarrow a$, $R_3 \in \{10, 70\} \rightarrow a$, $R_4 \in \{30, 60\} \rightarrow d\ (deny)$, $R_5 \in \{15, 80\} \rightarrow a$, $R_6 \in \{50, 100\} \rightarrow d$, $R_{71} \in \{1, 30\} \rightarrow a$

To add a general firewall rule, the latest added rule is always on the top of the firewall rules. For example, R_7 is the first rule of firewall rules, and R_1 is the latest rule. From examples of firewall rules in Fig. 2, when compiling the rules with the Firewall Compressor algorithm [2,9], the results are shown in Fig. 3.

The Firewall Compressor has two major processes, there are decomposing and rescheduling. The decomposing part prunes out rules which are obstructed by rules above as shown in Fig. 3(b). In addition, the rescheduling part proceeds to stretch and adjust the order of rules, as shown in Fig. 3(c). Notice that such Firewall Compressor ignores the order of rule generation and the relationship between rules. This is the reason for arising a new anomaly called the semantics loss in firewall rules. Let's look at Fig. 3(b) again, some members of R_2 to R_7 disappear (Caused by pruning). This means that some meanings have been lost from pruning as well. Practically, for example, R_3 in Fig. 3(a) was created for managers to access the database system ($R3 \in \{10, 70\}$). R_2 was added for staffs to access the storage too ($R2 \in \{25, 50\}$). Once both rules are merged, the new rule (Fig. 3(c)) will be in the range between 31 and 70 (R_1 in Fig. 3(c)). As a result, some members of R_2 and R_3 are lost. In the later period, the company announces a new policy to deny accessing the database for all staffs. Therefore, the administrator is necessary to remove the rule that applies to both managers and staffs (R_1 in Fig. 3(c)). Consequently, the managers cannot access the database anymore. As well as adjusting the schedule, rules are reordered without ignoring the history of the rules, which results in the same lost meaning. For example, R_4 from Fig. 3(a) is swapped and merged together with R_6 and R_7 to be R_3 in Fig. 3(c). Thus, R_4 in Fig. 3(a) will be not activated because it is permanently obscured by the rules above (R_1 and R_2 in Fig. 3(c)).

3 SELTracker Design

3.1 The Contributions

This paper has proposed the algorithm and data structure for detecting and alerting the semantics loss of firewall rules, which is caused by merging rules of the redundancy anomaly, namely SELTracker. Besides, SELTracker can also inspect other anomalies such as shadowing, correlation and generalization.

3.2 Tracking Design

The process of tracking semantics loss as shown in Fig. 4, starting from building a tree structure used to detect rule anomalies, called Path Selection Tree (PST). PST is applied from [10] by enhancing the tracking capability. When new rules are added to the firewall, they will be checked against PST. If new rules are not anomaly, they are added to PST suddenly, and built to the rule list for the firewall engine to continue processing. On the other hand, an alert is sent to the administrator indicating that an error has occurred. For shadowing, correlation and generalization anomaly, the system only warns in the normal alert. But if it is a redundancy anomaly, the system alerts as the semantics loss to an admin. The administrator will make decisions based on his discretion such as deleting rules, switching rules or breaking rules. These actions are beyond the scope of this paper.

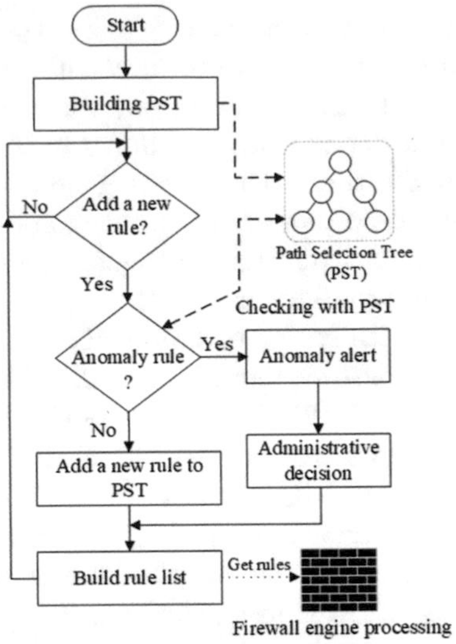

Fig. 4. The tracking process to detect and alert the semantics loss

3.3 PST Design and Development

PST is used as the blueprint for detecting the semantics loss. Therefore, it is necessary to describe the detailed design and operation structure, as detailed below. To make the explanation easy to understand, this paper has set out simple firewall rules to describe the creation and operation in Table 1. Consider the following Table 1, PST will be constructed as follows.

Table 1. Sample firewall rules

Rule no.	sip (source ip address)	dip (dest ip address)	dp (dest port)	pro (protocol)	action
1	5–30	*	80	tcp	*a*
2	10–20	*	80	tcp	*a*
3	1–50	*	70–80	tcp	*d*
4	40–70	20–80	*	tcp	*a*

Note: $sip, dip(*) \in \{1, 100\}$, $dp(*) \in \{1, 100\}$, pro = tcp (port number 17)

The first node built on PST is the root as shown in Fig. 5(a). Next, building the first node in the first level (*sip* level) on the PST (Fig. 5(b)) which maintains a range of source ip addresses from 5 to 30, called *sip* node. Significantly, the *sip* range must be linked to rule no. 1 ({1:{5, 30}}) which is used for tracking anomalies. In the next step, the nodes are hierarchically built from *dip* (Fig. 5(c)), *dp* (Fig. 5(d)), *pro* (Fig. 5(e)) and *action* respectively. Each node is always recorded a data range and a rule number. When the rule no. 1 is built successfully, it is illustrated in Fig. 5(e). The second rule is added on PST in the same way as rule no. 1. According to Fig. 6(a), the *sip*, *dip*, *dp*, *pro* and an *action* of rule no. 2 are a subset of rule no. 1, and they are the same decision. So, the rule no. 2 is built on the same branch tree like the rule no. 1, and every node is necessary to record the scope range in conjunction with the rule no. 1 as shown in Fig. 6(a).

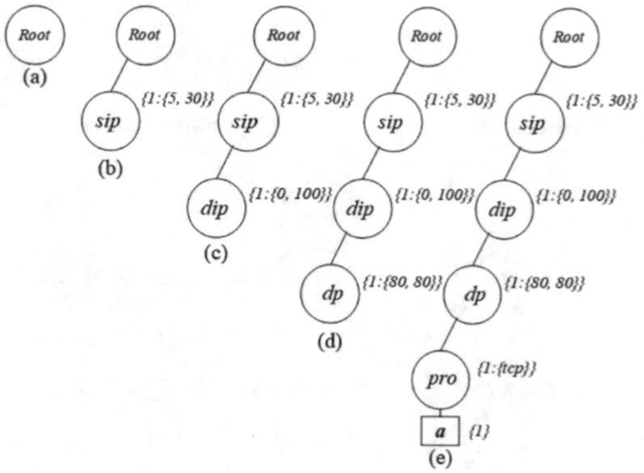

Fig. 5. Creating all nodes on PST of Rule no. 1, and information of each node represents by {rule number: {range of data}}

Inserting the 3rd rule, *sip* of rule no. 3 is the superset of rule no. 1 and rule no. 2. So, the rule no. 3 is divided into two parts: some members are in the first branch in the PST together with the rule no.1 and 2, and some will be built in a new branch as shown Fig. 6(b). Some of the destination ports (*dp*) of the rule no. 3 in the first branch intersect with the rule no. 1 and 2 ({80, 80} - {70, 80} = {70, 79}), Therefore, the algorithm is necessary to create a new branch to maintain other ports that are not the member of the first branch, and to make the decision to be deny (*d*). The final rule (rule no. 4) is built as 3 branches. Each generated branch must always be checked against the rules that were created before. If it is the subnet with the previous rules, the algorithm must remove their members immediately. For the rest of the members, the algorithm builds a new branch on the tree to maintain them as shown in Fig. 6(c).

3.4 Detecting and Alerting for the Semantics Lost

Detecting the semantics lost over PST can be achieved by tracking the merging rules in the tree structure. Table 1 and Fig. 6(c), for example, rule no. 1 and 2 can be merged into the single rule. However, the algorithm will alert to the administrator that the meaning of both rules may be the semantics lost. The decision

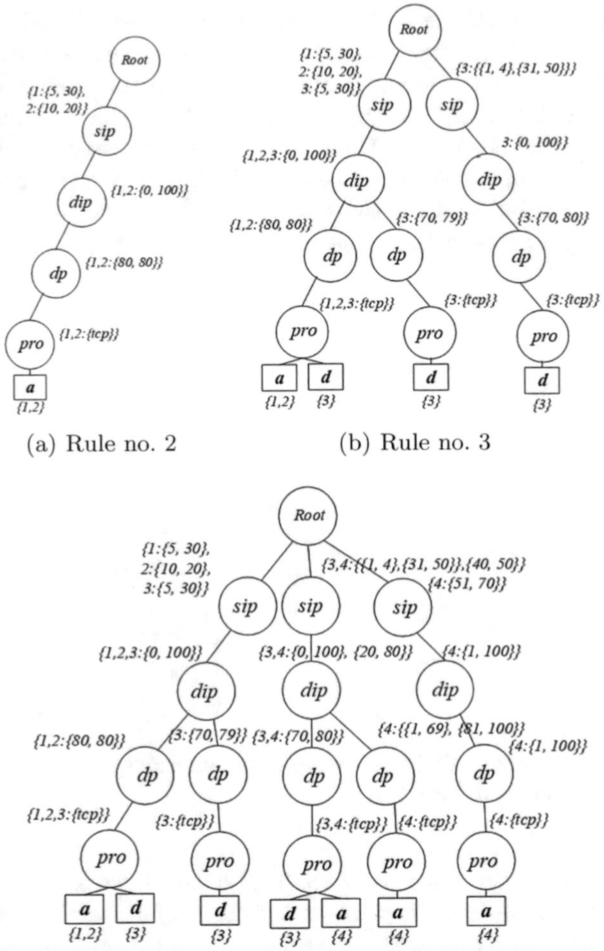

(a) Rule no. 2

(b) Rule no. 3

(c) Rule no. 4

Fig. 6. Building rule no. 2, 3 and 4 on PST

whether to merge or not depends on the administrator itself. If the administrator decides to merge, both rules in PST are merged into one rule immediately, and algorithm will update information for each node from {1, 2} to {1}. In the Table 1, the rules that can be merged are rules no. 1 and 2 only. Other rules appear as other anomalies (no redundancy).

3.5 Detecting and Alerting for Other Anomalies

The PST structure facilitates the detection of other anomalies beyond the redundancy anomaly. Referring to Fig. 6(c), for example, rule no. 1 and 3 are the generalization, rule no. 3 and 4 are the correlation, and rule no. 1 and 4 are not any conflict. All anomalies can be tracked from the PST tree structure efficiently.

4 SELTracker Implementation and Evaluation

SELTracker has developed by Python language version 3.4 (64 bits) on Intel processor Core(TM) i7, the memory size is 8 MB and running on Windows 10. In this paper, SELTracker is tested against Linux firewall rules (IPTables [11]) which are commonly used to configure in the real-world networks. The tested rules are in the Table 2. The number of rules used for testing starts from 1 to 100. Experimental results are displayed in the form of anomaly reports discovered in the firewall rules. Anomalies detected by SELTracker can be recognized by Al-Shaer definition [1]. The reports will be in the form of detected frequency shown in Fig. 7. For the kind of redundancy, the number of the semantics lost (SEL) will be included to the reports.

Table 2. Linux firewall rule examples

IPTables rule examples
iptables -A INPUT -i eth1 -s 0/0 -d 0/0 -j ACCEPT,
iptables -A INPUT -i lo -s 0/0 -d 0/0 -j ACCEPT,
iptables -A INPUT -i eth0 -s x.y.z.s/32 -j DROP,
iptables -A INPUT -i eth0 -s x.y.z.c/32 -j DROP,
iptables -A INPUT -i eth0 -s 192.168.0.0/24 -jDROP,
iptables -A INPUT -i eth0 -s 127.0.0.0/8 -j DROP,
iptables -A INPUT -p tcp -s 0/0 -d x.y.z.m/32–destination-port 25 –syn -j ACCEPT,
iptables -A INPUT -p tcp -s 0/0 -d 0/0–destination-port 80 –syn -j ACCEPT,
iptables -A INPUT -p tcp -s 0/0 -d 0/0–destination-port 443 –syn -j ACCEPT,
iptables -A INPUT -p tcp -s 0/0 -d 0/0–destination-port 22 –syn -j ACCEPT,
iptables -A INPUT -p tcp -s x.y.z.s/32 -d 0/0 –destination-port 20 –syn -j ACCEPT,
iptables -A INPUT -p tcp -s x.y.z.s/32 -d 0/0 –destination-port 21 –syn -j ACCEPT,
....

```
Python 3.4.2 Shell
File Edit Shell Debug Options Windows Help
Python 3.4.2 (v3.4.2:ab2c023a9432, Oct  6 2014, 22:16:31) [MSC v.
D64)] on win32
Type "copyright", "credits" or "license()" for more information.
>>> ================================ RESTART ====================
>>>
***************************
* SELTracker program v1.0 *
***************************
Number of tests  1  Number of tested rules  10
Kinds of anomaly:
Shadowing ->  {R1, R3}, {R1, R5} = 2
Correlation ->  {R2, R3} = 1
Generalization ->   -
Redundancy ->  {R4, R4} = 1, SEL = 1
Irrelevant ->   -
--------------------------------------------------
Number of tests  2  Number of tested rules  50
Kinds of anomaly:
Shadowing ->  {R3, R6} = 1
Correlation ->  {R1, R3}, {R4, R5}, {R5, R10} = 3
Generalization ->  {R12, R13} = 1
Redundancy ->  {R8, R14}, {R15, R16} = 2, SEL = 2
Irrelevant ->   -
--------------------------------------------------
Number of tests  3  Number of tested rules  100
Kinds of anomaly:
Shadowing ->  {R5, R9}, {R15, R22} = 2
Correlation ->  {R2, R12}, {R5, R26}, {R18, R24}, {R45, R68} = 4
Generalization ->  {R45, R62}, {R28, R84} = 2, SEL = 2
Redundancy ->  {R5, R12}, {R35, R77}, SEL = 2
Irrelevant ->   -
>>>
```

Fig. 7. An example of SELTracker program v1.0

5 Conclusion and Future Work

This paper has developed the system for detecting and alerting the semantics loss of firewall rules, and is able to investigate all other anomalies defined by Al-Shaer, called SELTracker. Based on the test, the proposed system is effective in detecting all irregular rules, especially the semantics loss, which is a very rare detecting. In such anomaly, most researchers ignore or do not care, but in practice it has a direct impact on the corporate policy. The future research will be an attempt to resolve the anomalies using artificial intelligence paradigms.

References

1. Al-Shaer, E.S., Hamed, H.H.: Firewall policy advisor for anomaly discovery and rule editing. In: Proceedings of IFIP/IEEE Eighth International Symposium on Integrated Network Management, pp. 17–30 (2003)
2. Liu, A.X., Torng, E., Meiners, C.R.: Firewall compressor: an algorithm for minimizing firewall policies. In: Proceedings of the 27th Conference on Computer Communications. Phoenix, Arizona, USA (2008)
3. Hu, H., Ahn, G.J., Kulkarni, K.: Detecting and resolving firewall policy anomalies. IEEE Trans. Dependable Secure Comput. **9**(3), 318–331 (2012)

4. Zhang, L., Huang, M.: A Firewall rules optimized model based on service-grouping. In: Proceedings of the 12th Web Information System and Application Conference, pp. 142–146 (2015)
5. Yoon, M., Chen, S., Zhang, Z.: Minimizing the maximum Firewall rule set in a network with multiple Firewalls. IEEE Trans. Comput. **59**(2), 218–230 (2010)
6. Katic, T., Pale, P.: Optimization of Firewall rules. In: Proceedings of the 29th International Conference on Information Technology Interfaces, pp. 685–690 (2007)
7. Al-Shaer, E.S., Hamed, H.H.: Modeling and management of Firewall policies. IEEE Trans. Netw. Serv. Manag. **1**(1), 2–10 (2004)
8. Khummanee, S., Khumseela, A., Puangpronpitag, S.: Towards a new design of Firewall: anomaly elimination and fast verifying of firewall rules. In: Proceedings of the 10th International Joint Conference on Computer Science and Software Engineering, pp. 93–98 (2013)
9. Daly, J., Liu, A.X., Torng, E.: A difference resolution approach to compressing access control lists. IEEE/ACM Trans. Netw. **24**(1), 610–623 (2016)
10. Liu, A.X., Gouda, M.G.: Diverse Firewall design. IEEE Trans. Parallel Distrib. Syst. **19**(9), 1237–1251 (2008)
11. The netfilter.org "iptables" project [Internet]. netfilter/iptables (2018). http://www.iptables.org/projects/iptables/index.html. Accessed 15 Dec 2017
12. Yuan, L., Chen, H., Mai, J., Chuah, C., Su, Z., Mohapatra, P.: FIREMAN: a toolkit for firewall modeling and analysis. In: Proceedings of 2006 IEEE Symposium on Security and Privacy, Berkeley/Oakland, CA, 21–24 May 2006

Cryptanalysis of a Multi-factor Biometric-Based Remote Authentication Scheme

Sirapat Boonkrong(✉)

School of Information Technology, Suranaree University of Technology, Nakhon Ratchasima, Thailand
sirapat@sut.ac.th

Abstract. Since the birth of the Internet technology, many remote user authentication protocols have been designed and developed so that secure communication between a user and a server is possible over an insecure channel. One of the first smart card-based biometric authentication schemes is known as the Li-Hwang protocol. Many researchers have found vulnerabilities and attempted to improve its security. Even if that is the case, this research still finds several weaknesses in the latest protocol proposed by Park *et al.* The vulnerabilities found in this research include an attack on the integrity of a protocol message, the possibility of a replay attack and the lack of proof of message authenticity.

Keywords: Authentication · Cryptanalysis Cryptography · Protocols

1 Introduction

The Internet technology has allowed for ease of communication between users and servers, together with user-friendly services such as e-commerce and e-government. Unfortunately, data transmission over insecure channels leads to many security challenges, including confidentiality and integrity of messages. That is, adversaries are usually considered and assumed to have control over communication channels between users and servers. In order to address the issue, a user and a server must prove their identity to one another as well as establish a session key for secure communication. This is where authentication protocols are needed.

Remote user authentication is an approach used to verify the authorisation of a user before accessing a service on a network or the Internet. One commonly used method is one-factor authentication, using password only. However, the password-only schemes have suffered many forms of attacks [1], including brute force attack, password dictionary attack and key logger attack. In order to reduce the risk of these attacks, several biometric-based user authentication protocols have been proposed [2–6]. Instead of using a password to prove the identity of a

H. Unger et al. (Eds.): IC2IT 2018, AISC 769, pp. 232–242, 2019.
https://doi.org/10.1007/978-3-319-93692-5_23

user, biometric information such as fingerprints and irises is considered a unique identifier. Therefore, biometric-based remote user authentication is said to be more secure than the password-only schemes.

Even though biometric-based authentication provides better security than the password-only schemes, biometric on its own is not commonly used and many protocols have adopted what is known as multi-factor authentication instead [7,8]. Multi-factor authentication is a method where users need to provide more than one legitimate credential before being allowed to enter a system. Those credentials may include a password, biometric information and a smart card.

In 2010, Li and Hwang were one of the first to propose an efficient biometric-based remote user authentication scheme using smart cards [2]. The authors claimed that the proposed protocol was secure due to the use of a one-way hash function, biometric verification and a smart card. Since then, many researchers have used this scheme as the basis of their authentication and key agreement protocol [9,10]. A year later, Das [3] pointed out some design flaws of the Li-Hwang scheme. The author proposed an improvement to the Li-Hwang protocol by incorporating a password and random nonce into the protocol. Later, An found several weaknesses in the Das protocol [4]. They included user impersonation attack, server masquerading attack, password guessing attack as well as the lack of mutual authentication between a user and a server. An, therefore, provided some improvements to the protocol to enhance its security.

Although a couple of improvements had been made to the original Li-Hwang protocol, Cao and Ge were still able to demonstrate that the An protocol was vulnerable to replay attacks which led to server masquerading [5]. Cao and Ge then proposed an improved multi-factor biometric-based authentication protocol to address the issues. Most recently, in 2017, Park *et al.* [6] showed that the Cao-Ge protocol still failed to withstand several attacks, including identity guessing attack and server masquerading attack.

Even if many attempts have been made to design a more secure multi-factor biometric-based remote authentication protocol, it is unfortunate that we have also found some vulnerabilities in the protocol proposed by Park *et al.* This paper, therefore, discusses the weaknesses in the Park *et al.*'s scheme. We show them by providing a formal security analysis using the Gong-Needham-Yahalom or the GNY logic [11] on the Park *et al.*'s protocol.

The rest of the paper is organised as follows. Section 2 provides the description of the protocol proposed by Park *et al.* and an overview of the GNY logic. The cryptanalysis and formal security analysis of the protocol are given in Sect. 3. Section 4 concludes the paper.

2 Background Knowledge

This section provides the preliminary knowledge related to the rest of the paper. This includes the description of the Park *et al.*'s remote user authentication protocol and an introduction to the logic of GNY.

2.1 Park *et al.*'s Remote User Authentication Protocol

The multi-factor biometric-based remote user authentication protocol proposed by Park *et al.* [6] consists of three main phases. They are the registration phase, the login phase and the authentication phase. The main aim of the protocol is for a user and server to be mutually authenticated with a shared session key established at the end of the authentication phase. The notations of the scheme [6] are described in Table 1.

Table 1. Notations in the Park *et al.*'s protocol

Notation	Meaning
C_i	User i
R_i	Trusted Registration Centre i
S_i	Remote Server i
ID_i	Actual Identity of C_i
VID_i	Virtual Identity of C_i
DID_i	Dynamic Identity of C_i
PW_i	Password of C_i
B_i	Biometric Template of C_i
SC_i	Smart Card of User C_i
A_i	Adversary i
X_S	Secret Key of S_i
x	Master Key of R_i

Notation	Meaning
K	Random number for Registration of C_i
R_C	Random number generated by C_i
R_S	Random number generated by S_i
$\|\|$	Concatenate Operation
$\oplus$	Bitwise XOR Operation
$h()$	Secure Hash Function
$H()$	Bio-Hashing Function
y_i	A Unique Number of C_i generated by R_i
T_i	Timestamp

Registration Phase

A user C_i must complete the registration at the registration centre R_i before accessing any service on the server S_i. C_i begins by choosing his or her identity ID_i and password PW_i. He or she then imprints biometric information B_i and carries out the following steps.

(R1) A user C_i chooses a random number K, and computes $(PW_i \oplus K)$ and $(H(B_i) \oplus K)$. C_i then submits $ID_i, (PW_i \oplus K), (H(B_i) \oplus K)$ to R_i via a secure channel.

(R2) R_i chooses a unique number y_i of C_i and computes:
$f_i = h(H(B_i) \oplus K)$, $r_i = h(PW_i \oplus K) \oplus f_i$, $e_i = h(ID_i||X_S) \oplus r_i$, $VID_i = h(y_i||X_S) \oplus ID_i \oplus h(PW_i \oplus K||h(H(B_i) \oplus K))$, $Z_i = y_i \oplus h(x)$ and $G_i = h(h(ID_i||XS))$.

(R3) R_i creates an entry of this identity ID_i and the virtual identity VID_i for this record.

(R4) R_i stores $\{VID_i, h(), H(), f_i, e_i, Z_i, G_i\}$ on a smart card SC_i. The smart card is then delivered to C_i via a secure channel.

(R5) Once C_i receives the smart card SC_i, he or she stores the random number K on the smart card.

Login Phase
This is the phase where the user C_i logs in to a remote server S_i. In order to login, C_i carries out the following steps using his or her smart card SC_i.

(L1) C_i imprints his or her biometric information B_i and computes $H(B_i)$. The smart card SC_i then computes $h(H(B_i) \oplus K)$ and compares it with f_i already on the card. If it is valid, the next steps are carried out.
(L2) C_i chooses a random number R_C and inputs (ID_i, PW_i, R_C) into the smart card, SC_i, which then computes:
$r'_i = h(PW_i \oplus K) \oplus f_i$, $M_1 = e_i \oplus r'_i$, $M_2 = M_1 \oplus R_C$, $M_3 = h(M_1||R_C||T_1)$.
C_i generates a dynamic identity DID_i by computing $DID_i = VID_i \oplus h(h(y_i||X_S)||T_1)$, where $h(y_i||X_S) = VID_i \oplus ID_i \oplus h(PW_i \oplus K||h(H(B_i) \oplus K))$.
(L3) The user C_i send the login request to S_i with the message: $\{DID_i, Z_i, M_2, M_3, T_1\}$.

Authentication Phase
The authentication phase is when the remote server S_i and user C_i prove their identity to one another. The followings steps are carried out.

(A1) Having received the message $\{DID_i, Z_i, M_2, M_3, T_1\}$ from C_i, the server S_i verifies whether $T_1 - T \leq \Delta T$. If the verification fails, S_i stops the session. If the verification succeeds, S_i checks the validity of DID_i by comparing VID'_i with VID_i stored in the account database, where $VID'_i = DID_i \oplus h(h(y_i||X_S)||T_1)$ and $y_i = Z_i \oplus h(x)$.
(A2) If the checking fails, S_i stops the session. Otherwise, S_i computes $M_4 = h(ID_i||X_S)$, $M_5 = M_2 \oplus M_4$, and checks whether $M_3 = h(M_4||M_5||T_1)$ or not.
(A3) If the verification fails, S_i stops the session. Otherwise, S_i randomly chooses a number R_S and computes $M_6 = M_4 \oplus R_S$, $M_7 = h(M_4||R_S||T_2)$. S_i then sends the message $\{M_6, M_7, T_2\}$ to C_i.
(A4) C_i verifies $T_2 - T \leq \Delta T$. If the verification fails, C_i terminates the sessions. If the verification succeeds, C_i computes $M_8 = M_1 \oplus M_6$. C_i then checks the validity of $M_7 = h(M_1||M_8||T_2)$.
(A5) If the checking fails, C_i terminates the session. Otherwise, C_i computes $M_9 = h(M_1||R_C||M_8||T_2)$ and sends the message $\{M_9, T_3\}$ to S_i. C_i then computes a session key $SK = h(R_C||M_8||T_2||T_3||M_9)$.
(A6) S_i verifies the timestamp $T_3 - T \leq \Delta T$. If the verification fails, S_i terminates the session. If it succeeds, S_i validates whether $M_9 = h(M_4||M_5||R_S||T_2)$. If not, S_i terminates the session, else S_i computes the session key $SK' = h(M_5||R_S||T_2||T_3||M_9)$. S_i sends $h(SK')$ to C_i.
(A7) C_i compares the received $h(SK')$ with his or her own $h(SK)$. If they match, C_i and S_i share the same session key SK.

2.2 The Logic of GNY

The logic of Gong-Needham-Yahalom or the logic of GNY is a method that will be used to analyse the Park *et al.*'s protocol. The GNY logic was introduced in [11] as an improved method on the BAN logic [12] used in the analyses of many existing multi-factor biometric remote authentication schemes. It was decided in this paper that GNY logic would be applied for the anlaysis because BAN logic was said to have several limitations [11].

The GNY logic provides a way to analyse and examine the components of the protocol messages when they are sent and received. It allows us to analyse cryptographic protocols step-by-step according to the specified postulates or rules. They are the T or being-told postulates, the P or possession postulates, the F or freshness postulates, the R or recognisability postulates, the I or Message Interpretation postulates and the J or Jurisdiction postulates.

Let P and Q be protocol principals. According to the GNY logic, the main notations are presented in Table 2. For more detail of the GNY notations and postulates, see [11].

Table 2. Main notations of the GNY logic

Notation	Meaning
$P \triangleleft X$	P is told formula X.
$P \ni X$	P possesses formula X.
$P \mid\sim X$	P once conveyed formula X.
$P \mid\equiv \sharp X$	P believes that formula X is fresh.
$P \mid\equiv \phi X$	P believes that formula X is recognisable.

Notation	Meaning
$P \mid\equiv P \overset{S}{\leftrightarrow} Q$	P believes that S is a suitable secret for P and Q. Only P and Q know S.
$P \triangleleft * X$	P is told formula X that he did not convey previously.
$P \mid\equiv C$	If C is a statement, P believes that the statement C holds.

3 Cryptanalysis of Park *et al.*'s Authentication Protocol

As stated above, Park *et al.* analysed and found weaknesses in the Cao-Ge authentication scheme [6]. A new and improved protocol was also proposed by Park *et al.* In this section, we cryptanalyse the Park *et al.*'s remote authentication protocol and show that there are still several vulnerabilities to the scheme. It is assumed that an adversary can have the following capabilites. Note that these are the same assumptions as when Park *et al.* analysed the Cao-Ge authentication protocol.

First of all, an adversary A_i has the total control over the communication channel during the registration, login and authentication phases. In other words, A_i can intercept, insert, modify and delete any message transmitted via the communication channel. Secondly, A_i can either steal the user's smart card or obtain the user's password. Thirdly, the adversary A_i can extract the information stored on the smart card.

3.1 Cryptanalysis of the Registration Phase

The registration phase of Park *et al.*'s protocol consists of two messages. We look at each one in turn. The first message is sent by C_i to R_i as follows.

$$C_i \rightarrow R_i : ID_i, (PW_i \oplus K), (H(B_i) \oplus K).$$

The first vulnerability of this message is that an adversary A_i can make a modification to the message, and it will not be possible for the registration centre R_i to realise it. In other words, A_i can make changes to any of the components of the message such as changing ID_i to ID_j. R_i will not be able to detect those changes at all. This is because the message does not contain any mechanism, specifically a cryptographic hash function, which can help detect the message modification. The consequence is that R_i will store wrong information on the smart card as well as in its own database. Therefore, the first vulnerability of the first message of the registration phase is the lack of message integrity detection mechanism.

Let us idealised this message into the format of the logic of GNY so that it can be analysed further. The idealised form of the first message can be written as follows.

$$R_i \triangleleft *ID_i, *(PW_i \oplus K), *(H(B_i) \oplus K)$$

Park *et al.* did not create any assumptions when analysing the registration phase. Therefore, nothing will be assumed in this paper, either. The GNY analysis on this first message of registration is carried out as follows.

Applying the Being-Told postulate T1: $\dfrac{R_i \triangleleft *ID_i, *(PW_i \oplus K), *(H(B_i) \oplus K)}{R_i \triangleleft ID_i, (PW_i \oplus K), (H(B_i) \oplus K)}$

It is learned that the registration centre R_i receives the message and all its components.

Applying the Possession postulate P1: $\dfrac{R_i \triangleleft ID_i, (PW_i \oplus K), (H(B_i) \oplus K)}{R_i \ni ID_i, (PW_i \oplus K), (H(B_i) \oplus K)}$

This means that the registration centre R_i possesses the components ID_i, $(PW_i \oplus K)$ and $(H(B_i) \oplus K)$.

The next step is to apply the freshness rule or the postulate F1. However, the analysis of this message has to terminate here. This is because the registration centre R_i does not believe that any components it has received is fresh or is newly created. The consequence is that A_i can intercept this message and re-send it to R_i without R_i knowing that this is a re-used message. This implies that a replay attack is possible since there is nothing provably fresh here in the message. This is, therefore, the second vulnerability of the first registration message.

The second message and all its components of the registration phase are sent from R_i to C_i, albeit in the smart card SC_i, as follows.

$$R_i \rightarrow C_i : VID_i, h(), H(), f_i, e_i, Z_i, G_i.$$

It can be seen that there are seven separate components on the smart card SC_i. However, only changes in two components f_i and r_i can be detected by C_i if occurred. This is because f_i and r_i are computed with the items known and generated by C_i. The rest of the components on the smart card present a problem similar to the weakness in the first message. In other words, since A_i

can get hold of the smart card SC_i and extract information from it, it is possible for the attacker to also make changes to the information on SC_i, other than f_i and r_i. The problem is that there is no mechanism that can assists the user C_i in checking the integrity of such information at all. This is, therefore, another vulnerability of this part of the protocol.

Again, let us analyse the components sent from R_i to C_i using the logic of GNY. We begin by writing the message in the idealised form as follows.

$$C_i \triangleleft *VID_i, *h(), *H(), *f_i, *e_i, *Z_i, *G_i$$

This idealised form illustrates that the components that C_i received are not originated at C_i. The GNY analysis on this message is carried out as follows.

Applying the Being-Told postulate T1: $\dfrac{C_i \triangleleft *VID_i, *h(), *H(), *f_i, *e_i, *Z_i, *G_i}{C_i \triangleleft VID_i, h(), H(), f_i, e_i, Z_i, G_i}$

It is learned that C_i receives all the components that R_i sent.

Applying the Possession postulate P1: $\dfrac{C_i \triangleleft VID_i, h(), H(), f_i, e_i, Z_i, G_i}{C_i \ni VID_i, h(), H(), f_i, e_i, Z_i, G_i}$

This means that C_i now possesses the components $VID_i, h(), H(), f_i, e_i, Z_i, G_i$. The next step of the analysis is to apply the postulate F1 or the freshness rule. We run into a problem here since there is no provably fresh component in the message or on the smart card at all. This implies that the adversary A_i can get hold of the smart card, extract the information and re-use it for a replay attack. One can argue that y_i is fresh because it is unique to C_i. However, if R_i were to issue a new smart card to C_i, would y_i be changed? This has not been specified by Park *et al.*

For the sake of further analysis, suppose y_i is freshly generated every time a new smart card is issued to C_i.

Applying the Freshness postulate F1: $\dfrac{C_i \mid\equiv \sharp Z_i}{C_i \mid\equiv \sharp(VID_i, h(), H(), f_i, e_i, Z_i, G_i)}$

This states that if y_i is fresh, it is inevitable that C_i will believe that Z_i is fresh. Therefore, C_i believes that the information on the smart card sent by R_i is also fresh.

Applying the Recognisability postulate R1: $\dfrac{C_i \mid\equiv \phi f_i}{C_i \mid\equiv \phi(VID_i, h(), H(), f_i, e_i, Z_i, G_i)}$

The postulate shows that C_i believes that he or she recognises the component f_i. This is because, from the postulate P1 and the first message, $C_i \ni h(), H(B_i \oplus K)$. This implies that C_i believes that the information on the smart card is recognisable.

The next postulate to be applied is the message interpretation postulate. However, no message interpretation rules are satisfied, because the message lacks necessary components such as a one-way function on the message. Therefore, the GNY analysis terminates here. This implies that C_i does not and cannot believe that the information as well as the smart card was really conveyed by the registration centre R_i.

3.2 Cryptanalysis of the Login and Authentication Phases

The login phase of Park *et al.*'s authentication protocol occurs when the user C_i sends a request to log into the remote server S_i. The message is as follows.

$$C_i \rightarrow S_i : DID_i, Z_i, M_2, M_3, T_1$$

Even though Park *et al.* had already analysed the security of the message, a vulnerability can still be found according to the GNY analysis. Before carrying out the analysis, the login message is idealised into the following form.

$$S_i \triangleleft *DID_i, *Z_i, *M_2, *M_3, *T_1$$

In the login and authentication phases, Park *et al.* made the following assumptions before they carried out their analysis using BAN logic.

$$\begin{array}{ll} C_i \mid\equiv C_i \overset{h(y_i||X_S)}{\longleftrightarrow} S_i & S_i \mid\equiv C_i \overset{h(y_i||X_S)}{\longleftrightarrow} S_i \\ C_i \mid\equiv C_i \overset{h(ID_i||X_S)}{\longleftrightarrow} S_i & S_i \mid\equiv C_i \overset{h(ID_i||X_S)}{\longleftrightarrow} S_i \\ S_i \mid\equiv \sharp R_C & C_i \mid\equiv \sharp R_S \\ S_i \ni x & S_i \ni X_S \end{array}$$

As before, no additional assumptions are made here in our analysis of the message. This is so that the analysis is consistent with how it was done by Park *et al.* The following is our analysis on the login message using the logic of GNY.

$$\text{Applying the Being-Told postulate T1:}\frac{S_i \triangleleft *DID_i, *Z_i, *M_2, *M_3, *T_1}{S_i \triangleleft DID_i, Z_i, M_2, M_3, T_1}$$

By applying T1, we obtain that the remote server S_i received the message and all its components.

$$\text{Applying the Possession postulate P1:}\frac{S_i \triangleleft DID_i, Z_i, M_2, M_3, T_1}{S_i \ni DID_i, Z_i, M_2, M_3, T_1}$$

Having received the message, P1 gives us the fact that the remote server S_i now possesses all the components.

$$\text{Applying the Freshness postulate F1:}\frac{S_i \mid\equiv \sharp R_C}{S_i \mid\equiv \sharp(DID_i, Z_i, M_2, M_3, T_1)}$$

By the assumption, the remote server S_i believes that the component R_C is fresh. Therefore, S_i is bound to believe that the whole message is also fresh, i.e., no replay attack is possible here. In this message, we understand that Park *et al.* intended for S_i to check whether M_2 or M_3 had been received prior to this message, because R_C is one of the components used to compute both M_2 and M_3.

$$\text{Applying the Recognisability postulate R1:}\frac{S_i \mid\equiv \phi VID_i}{S_i \mid\equiv \phi(DID_i, Z_i, M_2, M_3, T_1)}$$

At this stage, S_i computes VID'_i from $h(h(y_i||X_S)||T_1) \oplus DID_i$. If it matches with the VID_i stored in the database then S_i believes that the component VID_i is recognisable. Hence, all the components of the message is also believed to be recognisable.

However, the problem we have found is that if the component DID_i was altered during the transmission from C_i to S_i, S_i would not be able to detect that it had been changed. S_i would reject this login message since VID_i would not be found in the database. Another issue is that S_i needs to compare whether or not M_3 is equal to $h(M_4||M_5||T_1)$. They, to us, appear to be mathematically impossible to be equal since S_i is comparing $h(h(ID_i||X_S) \oplus r_i \oplus h(PW_i \oplus K) \oplus f_i)||R_C||T_1)$ with $h(h(ID_i||X_S)||r_i \oplus h(PW_i \oplus K) \oplus f_i \oplus R_C||T_1)$. The consequence is that the protocol will always terminate here.

For the sake of completeness of the protocol analysis, let us assume that the protocol continues with the next message where S_i sends the following to C_i.

$$S_i \rightarrow C_i : M_6, M_7, T_2$$

This message can be idealised into the format of GNY logic as follows.

$$C_i \triangleleft *M_6, *M_7, *T_2$$

The GNY analysis is done on this message using the same assumptions as above.

$$\text{Applying the Being-Told postulate T1:} \frac{C_i \triangleleft *M_6, *M_7, *T_2}{C_i \triangleleft M_6, M_7, T_2}$$

By applying this postulate, we obtain that C_i received the message and all its components from the remote server S_i.

$$\text{Applying the Possession postulate P1:} \frac{C_i \triangleleft M_6, M_7, T_2}{C_i \ni M_6, M_7, T_2}$$

Having received the message, C_i now possesses the components M_6, M_7 and T_2.

$$\text{Applying the Freshness postulate F1:} \frac{C_i \mid\equiv \sharp R_S}{C_i \mid\equiv \sharp(M_6, M_7, T_2)}$$

By the assumption, C_i believes that the component R_S is fresh. That means that C_i also believes that the whole message was freshly generated. No replay attack is possible here. We believe that it is Park *et al.*'s intention to have C_i checks for the freshness of the message by analysing whether or not $R_S = M_6 \oplus h(ID_i||X_S)$ is fresh.

The next step of the analysis is to apply the recognisability postulate R1. The problem is found here. In other words, from the received message and its components, C_i does not recognise anything sent from S_i. There are no components generated and sent by C_i, which S_i used to respond. In this phase of the protocol, no challenge-and-response mechanism exists, which means there is not proof to C_i that the message was really from S_i, the other protocol principal. Further, this implies that man-in-the-middle attack could be possible.

The GNY analysis of the Park *et al.* terminates here since a weakness is found and the continuation of the protocol is not possible.

4 Conclusion

Many smart card-based biometric remote authentication have been designed and proposed with the hope of improving the security of communication. The remote authentication protocol of Park *et al.*'s was the most recent scheme to be proposed in 2017. However, this paper carried out a formal analysis using the logic of GNY and found several weaknesses in the protocol.

The Park *et al.*'s protocol was designed to have three phases - registration, login and authentication. The registration phase contained three main problems. The first was the lack of message modification detection mechanism in both registration messages. The second was the possibility of a replay attack, also in both of the registration messages. The third was the fact that there was component in the second registration message that suggested that the message or the smart card was really conveyed by the registration centre.

The problems found during these stages were the lack of message modification detection mechanism, the impossibility of comparing the components based on the given items, and the potential man-in-the-middle attack due to the lack of challenge-and-response mechanism.

References

1. Lin, C., Sun, H.M., Hwang, T.: Attacks and solutions on strong-password authentication. IEICE Trans. Commun. **84**(9), 2622–2627 (2001)
2. Li, C.T., Hwang, M.S.: An efficient biometrics-based remote user authentication scheme using smart cards. J. Netw. Comput. Appl. **33**(1), 1–5 (2010)
3. Das, A.K.: Analysis and improvement on an efficient biometric-based remote user authentication scheme using smart cards. IET Inf. Secur. **5**(3), 145–151 (2011)
4. An, Y.: Security analysis and enhancements of an effective biometric-based remote user authentication scheme using smart cards. J. Biomed. Biotechnol. **2012**, 1–6 (2012)
5. Cao, L., Ge, W.: Analysis and improvement of a multifactor biometric authentication scheme. Secur. Commun. Netw. **8**(4), 617–625 (2015)
6. Park, Y.H., Park, K.S., Lee, K.K., Song, H., Park, Y.H.: Security analysis and enhancements of an improved multi-factor biometric authentication scheme. Int. J. Distrib. Sens. Netw. **13**(8), 1–12 (2017)
7. Bhargav-Spantzel, A., Squicciarini, A.C.A., Modi, S., Young, M., Bertino, E., Elliott, S.J.B.: Privacy preserving multi-factor authentication with biometrics. J. Comput. Secur. **15**(5), 529–560 (2007)
8. Boonkrong, S.: Internet banking login with multi-factor authentication. KSII Trans. Internet Inf. Syst. **11**(1), 511–535 (2017)
9. Grzonkowski, S., Corcoran, P.M.: Sharing cloud services: user authentication for social enhancement of home networking. IEEE Trans. Consum. Electron. **57**(3), 1424–1432 (2011)
10. Mishra, D., Das, A.K., Mukhopadhyay, S.: A secure user anonymity-preserving biometric-based multi-server authenticated key agreement scheme using smart cards. Expert Syst. Appl. **41**(18), 8129–8143 (2014)

11. Gong, L., Needham, R., Yahalom, R.: Reasoning about belief in cryptographic protocols In: Proceedings of 1990 IEEE Symposium on Research in Security and Privacy, pp. 234–248. IEEE Computer Society Press (1990)
12. Burrows, M., Abadi, M., Needham, R.: A logic of authentication. ACM Trans. Comput. Syst. **8**, 18–36 (1990)

Haptic Alternatives for Mobile Device Authentication by Older Technology Users

Kulwinder Kaur and David M. Cook(✉)

Edith Cowan University, Joondalup, Australia
kulwind0@our.ecu.edu.au, d.cook@ecu.edu.au

Abstract. Turing tests are used to secure the human interaction on the Internet. Tests such as CAPTCHA are based on visual or auditory recognition of symbols and are difficult to distinguish by elderly people. A study examining the consistency of a tactile feedback-based Turing test identified an alternative to mainstream tests. This approach examines the vibration-based sensitivity which is detectable through skin surfaces when used to touch the screen of a mobile device. The study concentrated on a range of rough, smooth, sticky and coarse textures as possible differentiators for swipe-based tactile authentication using mobile devices. This study examined the vibration-based touch screen capabilities of 30 elderly people over the age of 65. The results of this study showed that tactile differentiation can be a viable alternative for device and security authentication for Turing tests such as those used for CAPTCHA and reCAPTCHA verification.

Keywords: Tactile · Authentication · Haptic · Elderly · CAPTCHA reCAPTCHA

1 Introduction

Older people have difficulty in accessing critical services such as money from online bank accounts because they are unable to successfully answer Turing tests such as CAPTCHA authentication checks. CAPTCHA (Completely Automated Public Turing test to tell Computers and Human Apart) is a commercially accepted security mechanism on many websites. It is used as an anti-spam test to deter automated programs and bots' such as "Brute-force" or "Spam-bots" [1–3].

Despite their prominence, CAPTCHA tests are difficult to use online [4]. There are many examples of visual and audio-based CAPTCHAs that are deployed as part of Multi-factor authentication controls. Many of them are problematic for older people with decreased abilities in terms of sight and sound [5]. These security features present humans with a challenge either in text-based, or audio-based formats. CAPTCHAs use distorted characters and numbers with noisy backgrounds. The recognition of these features make it difficult for people with physical and mental disabilities to cope with mainstream online Turing tests [5]. Therefore there is a need for alternative forms of CAPTCHA tests that offer user-friendly (easy for humans) options, yet remain robustly secure for humans (difficult for bots).

H. Unger et al. (Eds.): IC2IT 2018, AISC 769, pp. 243–254, 2019.
https://doi.org/10.1007/978-3-319-93692-5_24

The purpose of this research is to create an alternative haptic and tactile based CAPTCHA as a viable alternative to distorted visual and noisy audio authentication tests for people aged 65 and over. An alternative haptic and tactile-based Turing test is significant because human skin and the sense of touch is acknowledged as a reliable receptor for communicating information [6]. Sensations including vibration and pressure changes are therefore useful for communication, since they allow sensory nerves to convey the tactile response to the brain.

The need for alternative Turing tests is growing because when people reach maturity in adulthood, their ability in terms of vision, hearing, physical and cognitive are known to decline [7]. In order to make the elderly population more comfortable with online authentication, an alternative approach using haptic and tactile feedback may be useful in passing the Turing test to satisfy multi-factor authentication controls.

2 Literature Review

2.1 Variations of CAPTCHAs

Text-Based CAPTCHAs. There are many CAPTCHA versions available in the form of texts and images. These are usually either distorted or incorporate cluttered backgrounds using colour, waves and patterns [7, 8]. Examples of these types are Gimpy, EZ-Gimpy, MSN, and Baffle. Gimpy CAPTCHAs use a list of 10 random words were picked from an "*850-word dictionary based on Ogden's Basic English*" and use different colours to create image noise. For identification, a user must recognize at least three words out of it [9].

EZ-Gimpy images are different images that are predominantly used to prevent bots from infiltrating chat messaging in Yahoo Messenger. These images use one random word from a standard dictionary against a white background using faded swirl-style text [10]. Both Gimpy and EZ-Gimpy CAPTCHAs are not the most reliable method of online security because they include words from a known dictionary source. They have been successfully broken using scripted programs with repetition [11, 12].

A third image format occurs in MSN CAPTCHA used by Microsoft as well as in online services like MSN, Hotmail and Windows Live. It uses warping and swirling backgrounds to add distortion to an image with eight characters. It uses only numbers and uppercase letters. MSN graphics are robust and segmentation-resistant. They provide a challenge for users to identify letters and symbols such as J and arc. These type of CAPTCHA images are considered usable by people with impaired vision. However, their segmented format contributes to their reduced security, and such images have been regarded as vulnerable to attack, with recorded break rates of over 60% [13].

A fourth image format, known as Baffle text, is a modified Gimp-style image where the letters are randomly picked to make a word that is pronounceable but that does not come from a dictionary. Users are often challenged to find the correct word. This version is difficult for users with poor cognitive ability [14, 15].

Audio-Based CAPTCHAs. Audio CAPTCHAs have been created for the visually impaired. Users have to listen and write in a given box. Such CAPTCHA formats are accompanied by background noise and are considered challenging to pass. They use the same interface as text-based CAPTCHA forms [16].

CAPTCHA Challenges. There are a range of limitations found in different vision-based CAPTCHAs. In terms of security, some of these images (such as Gimpy formats) were cracked by simple computer programs and seemingly robust security can easily be compromised. In addition, usability in visual CAPTCHAs becomes difficult for people with low vision, cognitive and impaired hearing. Swirling, wavy, distorted and colorful backgrounds are known to increase the confusion [17]. As physiological abilities decline with age, older people find difficulty in distinguishing between varying colors and shapes that are used in CAPTCHAs [18]. A study on young and elderly responses to distorted text-based CAPTCHAs revealed that older people of increased age find these CAPTCHA tests much more difficult than younger people [19]. These difficulties include longer response times, higher error rates and greater vision stress while performing text-based and audio-based CAPTCHAs [19]. In addition, audio CAPTCHAs show older people having problems in reliably recognizing alphabetical or numerical sounds. The need for robust security has made it difficult for humans to solve online text-based and audio-based schemas. Accessibility is a formidable issue for an emerging population of elderly people worldwide [17, 20].

Usability vs Security and Robustness. The significant features in designing for web-based security using tests such as CAPTCHAs are usability and robustness. Both are inter-related and affect one another when developers place stress on one of them. The difference between usability and the security is a difficult process to balance. On one hand, designers are pushed to make web security robust whereas the usability is declined on the other side [21]. For example, CAPTCHA designs used distorted characters, font type, size, lines, waves, sound, and colours (Table 1). These design issues affect the readability and ability of users. For older people (65 and over) users often complain of vision and hearing difficulties. Many CAPTCHAs use confusing and interchangeable letters and symbols such as ‘W’ and ‘VV’ or ‘1’ and ‘I’. These are difficult to negotiate [21, 22].

Table 1. CAPTCHA Test variations in usability and security [15]

CAPTCHA type	Variation	Disadvantages	Security/usability
Text based	Gimpy	Different colours, and distortions affect poor vision	Not secure - difficult
Text based	Ez-Gimpy	Noisy background affects vision impaired	Not secure
Text based	MSN CAPTCHA	Difficult for vision impaired	Can be compromised
Text based	Baffle Text	Requires good cognitive ability	Not secure
Audio based	Audio CAPTCHA	Noise and distortion affect hearing impaired	Secure and difficult

The primary purpose of CAPTCHAs is to facilitate human access with a high success rate with an accompanying low access rate by scripted solutions using bots [23]. However, in terms of CAPTCHAs the vulnerable section of society are not the deaf and the blind, but rather those who endure varying degrees of vision impairment and hearing loss [24].

Physiological Challenges: Cognitive Skills and the Aging Brain. The growing number of older people across the world makes the challenge of CAPTCHAs an important and significant challenge. According to the Australian Bureau of Statistics, the proportion of older people aged 65 and over is steadily increasing by 15% in 2014 (3.5 million people) and is projected to rise beyond 21% in 2054 (8.4 million Australian people) [25]. Additionally the ABS statistics reveal that 5000 Baby Boomers turn 65 in Australia every day whilst the accompanying challenges to accessibility are also increasing [26]. At the same time CAPTCHA systems are becoming more widespread. In Australia there are thousands of new websites that use reCAPTCHA as a security mechanism for online protection [27].

Physical changes to people over time and with aging is an undeniable and complex process where tissues and various parts in the human body get declined and lower in performing activities [28]. When people become old their vision and hearing abilities decline over time [7, 29–33] (Table 2).

Table 2. Examples of physiological issues related to aging [28]

Hearing Loss	Vision Impairment	Cognitive Ability	Physical Ability
National institute of health reported that ageing caused the problem like Presbycusis A person cannot differentiate the words like S or X [29]	Vision-based problems like cataract, glaucoma and macular degeneration caused when people get old [30]	The studies depict the cognitive ability become slow with the growing age of elderly in recalling or familiarizing new words even it exists in their mind [31]	Ageing decrease the muscular strength and body weight which affect the mobility [32, 33]

A significant percentage of people (35%) suffer from impaired hearing whilst elderly people suffer varying degrees of long and short sightedness. One in ten people are affected by cataracts; eight percent of people are borne with macular degeneration and glaucoma; and approximately one in fifty people are blind [34]. Such global statistical evidence highlights the demand for Turing tests such as CAPTCHAs that place greater emphasis on cohorts who collectively demonstrate challenges such as reduced vision, impaired hearing, and low memory capabilities. There is therefore a growing need for older people to use options that are both more accessible and yet robustly secure in order to protect their information on the Internet [35].

Rise in Problems with Online Banking. Increased online usage in areas such as banking has decreased the number of physical visits to bank branches [36]. However at the same time the rise in Internet banking has augmented the difficulties for older people to authenticate online access [37]. Typically such authentications involve CAPTCHA tests alongside other Multi Factor Authentication (MFA) processes. MFA allows users to validate two or more different security options such as biometrics, passwords, PIN, and links via email or messages [38]. The rise in MFA relies heavily on features such as usable CAPTCHA tests. For older people it is reasonable that a user might forget their password or PIN code. Such occurrences are regularly identified as the physiological decline of cognitive ability within older people [39]. Similarly, one-time passwords (OTPs) are challenging for older people without a mobile phone. Multi Factors Authentication can make account security more robust but also provide reduced accessibility for people with physiological age-related challenges [39, 40].

3 Conceptual Framework

3.1 The Concept of Alternative Tactile-Based Turing Tests

The Concept of Security for Older People. Based on the literature review, security on web-based services is a significant issue damaged by scripted bots and malicious programs that generate users' data automatically [41]. Older adults are especially vulnerable. To make a contribution to as new multifactor authentication or Turing test, a haptic-tactile based concept was tested on a sample of people over the age of 65.

Beyond human sensing using vision and hearing, human skin is considered as the good receptor for human interaction. Haptic sensation felt through skin (in this case using a finger) can assist in the detection of different aspects that can contribute to the recognition of a human over a bot [42]. In this concept, haptic or tactile feedback is termed as "*use of advanced vibration patterns and waveforms*" to transmit information to a user or operator. Haptics use vibrations from an actuator which could be embedded underneath the screen of a mobile device. A microcontroller decides and controls the period of vibration. In this example an actuator returns a sensation to the users' finger by charging a film/surface [43]. Haptic responses with tactile feedback can be achieved by piezo-electric actuator usage on devices [44]. The concept of using actuators to generate touch sensitive differences that can be detected by humans and not scripted bots represents the objective aim of this study.

3.2 Hypothesis and Research Question

Hypothesis. In order to determine the need for further research into the use of an alternate CAPTCHA this research looked to define and measure what a possible alternative CAPTCHA test might look like. The following hypothesis was considered:

H1. That a mobile device screen using a variable tactile interfacing surface can be used as a Turing test that allows reliable interaction by older people.

This study examined the use of a mobile device screen with 12 mm actuators to provide a variable surface that older people could touch, swipe and sense in order to verify a human interaction to the exclusion of scripts, bots that serves as a Turing test. Older people were tested to determine the usability of a haptic swipe test where older people could correctly detect sticky versus smooth surfaces.

4 Research Method

4.1 Apparatus

Apparatus and Tools. In order to test people to ascertain positive haptic interaction a small apparatus consisting of a box with an electronic screen was used. An apparatus was created that emulated a standard smartphone touch screen. The outside measurement was L = 138.176 mm × W = 67.056 mm. The thickness of the glass substrate was 0.2 mm. The piezo elements were affixed onto the edges of the glass screen. The elements were placed over the screen with a substrate that allowed participants to touch the screen and to have a haptic (response-based) interaction.

The substrate used 4 piezo-actuators that give micro vibrations at differing frequencies so that in some instances the screen felt smooth and at other times the screen felt sticky (Fig. 1). A simple box was created to generate a range of variables to as to alter the vibrations across the glass screen substrate. After testing under laboratory conditions two main test variations were chosen for the apparatus. The first created a smooth surface across the screen surface when a "swipe" action using a finger was applied. The second gave a rougher (sticky) surface when the same type of swipe finger action was applied to the surface of the screen (Table 3).

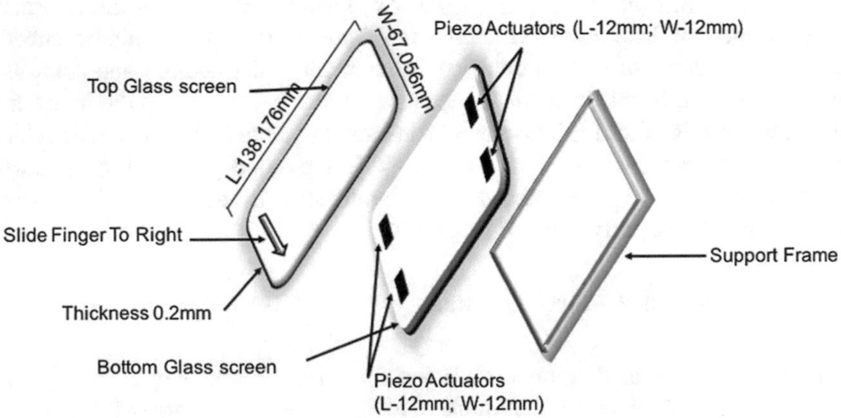

Fig. 1. Haptic interaction test apparatus

Table 3. Variable conditions to distinguish tactile sensation of smooth and sticky surface textures

Tactile sensation	Wave form	Amplitude Vpp.	Frequency Hz
Smooth surface	Sine wave	7.7 Vpp	50 Hz
Sticky surface	Square wave	7.7 Vpp	50 Hz

The frequencies and amplitude will range between 80 Hz to 400 Hz with Volt per peak from 80 and under 150. Differing frequencies allow for a range of tactile sensory interactions such as rough, smooth, sticky or coarse.

4.2 Participants

Participants. A group of randomly selected 35 subjects aged over 65 years of age took part in this Turing testing procedure. These people were in good health. These people were Australian citizens and the test was conducted in Australia. Participants were invited to undertake an experiment where they touched a screen (mounted atop a mobile device) and respond as to whether the screen button feels smooth, sticky, or rough.

Procedure

1. Participants were invited to approach the mounted screen apparatus and asked to a single finger to touch and swipe a specific part of the screen.
2. They were asked to report on whether they sensed the area as rough, sticky, or smooth.
3. The test was repeated twice for each participant and the responses were recorded on a test sheet. Variations were achieved by a combination of different ppV voltages, changes in amplitude, and different wave signals so as to create surfaces that were either sticky or smooth.
4. After each participant had received a selection of smooth and rough (sticky) variations they were asked to describe their experiences in terms of difficulty, confidence, and identification.

5 Results

5.1 Data and Analysis

The results of tests carried out on 35 older participants showed that there was a very high level of agreement with the proposition that older people could identify the difference between rough (sticky) and smooth (silky) textures when touched using an index finger on a sample swipe-screen substrate. Participants demonstrated a range of results whilst swiping their fingers over varying substrates that were designed to elicit responses over either smoothness or stickiness. The initial results of a simple touch-based swipe test revealed the following findings (Table 4).

Table 4. Detection of different textures by swipe/touch across a surface

Ease of tactile sensation	Participants
Very easy	9%
Easy	86%
Neither easy or difficult	0%
Slightly difficult	3%
Difficult	3%

The results of ascertaining whether older people would have difficulty in identifying the difference between the different textured samples showed that the majority of people were able to identify the differences with relative ease. Approximately 86% of participants stated that the identification was easy, whilst a further 9% stated that it was fairly easy. Of the remainder, 3% of participants found the test slightly hard and a further 3% found the test difficult.

Whilst the results showed a high level of success in identifying between rough (sticky) and smooth (silky) textures, there was a wide variety of descriptors from participants in the initial testing. Participants were asked to describe the textures of both smooth (silky) and rough (sticky) surfaces. Their results are shown below (Table 5).

Table 5. Variations in descriptions of texture surfaces

Descriptions of the smooth (silky) sample surface	Agreement with description	Descriptions of the rough (sticky) sample surface	Agreement with description
Smooth	29%	Rough	51%
Very smooth	14%	Extremely rough	17%
Slightly smooth	11%	Coarse	9%
Furry	11%	Sandpaper-like	37%
Slightly soft	9%	Velchro-like	9%
Velvet-like	3%	Wire mesh	3%
Silky	3%	Hard	3%
Like a computer screen	3%	Scratchy	3%
Glass-like	3%	Emery board	3%
Cardboard	3%	Sticky	6%
Not smooth	11%	-	-
Difficult to identify	9%	-	-

The findings show that the result of ascertaining the ease in identifying different types of textures is offset by the number of different descriptions of surfaces. There were 22 different descriptions recorded against two possible surface textures, one being smooth and the other being rough. Whilst participants identified the binary difference with a high level of accuracy, the large number of recorded descriptions suggests that tactile/haptic feedback is determined differently under subjective conditions.

Table 6. Participant comments describing textured surface

Different participant descriptions
"Wasn't exactly smooth and I could identify irregularities (it wasn't like silk)."
"Rough was sticky and smooth was little but less than really silky."
"Smooth was not very smooth I felt some texture there also."
"Mildly textured like the back of the mobile phone."

Additionally, whilst 94% of participants correctly identified rough from smooth, only 35% stated that they were confident in identifying the correct texture. Participant comments suggest that answers were individual and idiosyncratic (Table 6).

6 Discussion

6.1 Discussion of Findings

This research shows that participants (over the age of 65) were able to successfully identify two different textures Rough (sticky) and Smooth (Silky) when presented with a simulated mobile device screen with both rough and smooth textures. The literature for acceptance of accuracy in CAPTCHA testing describes around 90% (not required to be 100% attack resistant) as an appropriate level of successful completion of a CAPTCHA test. This study demonstrates an appropriate number to satisfy the requirements of a secure CAPTCHA Turing-based log-in test [23, 45]. Based on the Nguyen et al. acceptance criteria, the use of a mobile device screen by applying different textures generated through symmetrically positioned actuators on a device screen demonstrates that haptics can be harnessed as an alternate form of Turing test. Both laboratory-based testing to determine appropriate frequency/voltage/and wave form, as well as participant-based testing on a sample of 35 over 65 participants suggest that actuators behind mobile device screens can provide sufficient surface differentiation so that a haptic swipe test can form an additional test for the multifactor authentication of secured entry to online systems.

This study acknowledged the possibility of a scripted bot identifying haptic–driven information by detecting frequencies and signal shapes that are transmitted to a device using this type of CAPTCHA option. Although the researchers did not integrate a specific security measure in this study to repel such detection, further development could include a randomization method along the lines of Bianchi et al. [46], using random alternate vibration patterns to obfuscate the haptic interaction and retain a high level of security.

6.2 Limitations

In order to prove the haptic tactile based alternative for CAPTCHA various conditions were examined. These has included the difficulty in test, texture identification, confidence, certainty and familiarity with the touch devices. The findings from this study, whilst convincing in terms of a binary differentiation between smooth and sticky surface textures, also includes the adverse finding that texture descriptors remain individual and idiosyncratic. The finding of 22 different texture descriptions suggests that further research is required to more accurately determine optimal surface textures that can be more uniformly described by cohorts of older people.

7 Conclusion

This research shows the successful identification of different tactile surfaces using a simulated mobile device screen. The findings support the hypothesis that haptic-driven screen surface variations can be used as a reliable method of testing for human interaction by older people. The findings further indicate that tactile sensing is likely to provide a valid means of differentiation between humans and scripted programs that seek to compromise security.

An online tactile-based Turing test will not only help older people with impaired vision and hearing but also other people with good health. It is anticipated that tactile-haptic response feedback on mobile devices as a security feature will increase usability and accessibility in online systems usage.

References

1. Padave, K.: The Evolution of CAPTCHA – How an Anti-Spam Measure Has Grown? EXEIdeas – Let's Your Mind Rock (2014). http://www.exeideas.com/2014/11/the-evolution-ofCAPTCHA.html. Accessed 4 Aug 2017
2. Gafni, R., Nagar, I.: CAPTCHA: Impact on user experience of users with learning disabilities. Interdiscip. J. e-Skills Life Long Learn. **12**, 207–223 (2016)
3. Pope, C., Kushpreet, K.: Is it human or computer? Defending e-commerce with CAPTCHAs. IT Prof. **7**(2), 43–49 (2005)
4. Baecher, P., Fischlin, M., Gordon, L., Langenberg, R., Lutzow, M., Schroder, D.: CAPTCHAs: the good, the bad and the ugly. In: Frieling, F.C. (ed.) Sicherheit. (2010). LNI, vol. 170, pp. 353–365. GI (2010)
5. Divyashree, N., Kumar, S.T.: A survey on CAPTCHA categories. Int. J. Eng. Comput. Sci. **5**(5), 16458–16462 (2016)
6. Shull, P.B., Damian, D.D.: Haptic wearables as sensory replacement, sensory augmentation and trainer – a review. J. NeuroEng. Rehabil. **12**, 1–13 (2015)
7. Cook, D.M., Kumar, A., Satiah: C.U.: Loyalty cards and the problem of CAPTCHA: 2nd tier security and usability issues for senior citizens. In: 13th Australian Information Security Management Conference, pp. 101–111. SRI Security Research Institute, Edith Cowan University, Perth (2015)
8. Moy, G., Jones, N., Harkless, C., Potter, R.: Distortion estimation techniques in solving visual CAPTCHAs. In: Proceedings of the 2004 IEEE Computer Society Conference on Computer Vision and Pattern Recognition, pp. 23–28. IEEE, Washington, D.C. (2017)
9. TCP: The CAPTCHA Project, Gimpy Dictionary. http://www.captcha.net/captchas/gimpy
10. Mori, G., Malik, J.: Results on Gimpy (2002). https://www.cs.sfu.ca/~mori/research/gimpy/hard/
11. Merriman, J.: Human or Computer? Take This Test. The New York Times. https://www2.eecs.berkeley.edu/Research/Projects/CS/vision/mori-nyt/page2.html

12. Aboufadel, E., Olsen, J., Windle, J.: Breaking the Holiday Inn priority club CAPTCHA. Coll. Math. J. **36**(2), 101–108 (2005)
13. Yan J., El Ahmad, A.: A low-cost attack on a Microsoft CAPTCHA. In: Proceedings of the 15th ACM Conference on Computer and Communications Security, pp. 543–554. ACM Press, Virginia (2008)
14. GITS: Gujarat Institute of Technical Studies CAPTCHA. http://gitsgtu.blogspot.com.au/2013/10/table-of-contents-acknowledgments.html
15. Kaur, K., Behal, S.: CAPTCHA and its techniques: a review. Int. J. Comput. Sci. Inf. Technol. **5**(5), 6341–6344 (2014)
16. Yilmaz, S., Zavrak, S., Bodur, H.: Distinguishing humans from automated programs by a novel audio-based CAPTCHA. Int. J. Comput. Appl. **132**(13), 18–22 (2015)
17. Choudhary, S., Saroha, R., Dahiya, Y., Choudhary, S.: Understanding CAPTCHA: text and audio based CAPTCHA with its applications. Int. J. Adv. Res. Comput. Sci. Softw. Eng. **3**(6), 106–115 (2013)
18. Liu, S.F., Lin, P.Y., Wang, M.H.: The effects of the transparency of the guiding diagrams on the phone interface for the elderly. In: Zhou, J., Salvendy, G. (eds.) Human Aspects of IT for the Aged Population. Applications, Services and Contexts, ITAP 2017. Lecture Notes in Computer Science, vol. 10298. Springer, Cham (2017)
19. Hsu, C.H., Lee, Y.L.: Effects of age groups and distortion types on text-based CAPTCHA tasks. In: Jacko, J.A. (ed.) Human-Computer Interaction. Users and Applications, HCI 2011. Lecture Notes in Computer Science, vol. 6764. Springer, Heidelberg (2011)
20. Roshanbin, N., Miller, J.: ADAMAS: Interweaving unicode and color to enhance CAPTCHA security. Future Gener. Comput. Syst. **55**, 289–310 (2016)
21. Nguyen, D.V.: Contributions to text-based CAPTCHA security. Doctoral dissertation, University of Wollongong database (2014)
22. Beheshti, S., Liatsis, P.: CAPTCHA usability and performance, how to measure the usability level of human interactive applications quantitatively and qualitatively? In: 2015 International Conference on Developments of E-Systems Engineering (DeSE), pp. 131–136. IEEE, Duai (2015)
23. Nguyen, V., Chow, Y., Susilo, W.: A CAPTCHA scheme based on the identification of character locations. In: Huang, X., Zhou, J. (eds.) Information Security Practice and Experience, ISPEC 2014. Lecture Notes in Computer Science, vol. 8434, pp. 60–74. Springer, Cham (2014)
24. Earl, C.: Can CAPTCHAs Be Made Accessible. https://www.afb.org/blog/afb-blog/can-captchas-be-made-accessible/12
25. Australian Institute of Health and Welfare: Australia's Welfare 2015: Growing Older. http://www.aihw.gov.au/australias-welfare/2015/growing-older
26. Australian Bureau of Statistics: Disability, Ageing and Carers, Australia: First Results, Canberra (2015). http://www.abs.gov.au
27. Telarus: What can google-eyed robots and CAPTCHAs tell us about cybersecurity? https://www.telarus.com.au/news/month/feb/can-googly-eyed-robots-captchas-tell-us-cybersecurity
28. Nigam, Y., Knight, J., Bhattacharya, S., Bayer, A.: Physiological changes associated with ageing and immobility. J. Ageing Res. **2012**, 1–2 (2012)
29. NIH, National Institute of Health: Hearing, ear infections, and Deafness: Age related hearing Loss. NIH Pub. No. 97-4235. https://www.nidcd.nih.gov/health/age-related-hearing-loss
30. Kozarsky, A.: Eye problems: what to expect as you age. http://www.webmd.com/eye-health/vision-problems-ageing-adults#1
31. Howieson, B.D.: Cognitive skills and the Ageing Brain: What to Expect. The Dana Foundation (2015)

32. Martin, J.A., Ramsay, J., Hughes, C., Peters, D.M., Edwards, M.G.: Age and grip strength predict hand dexterity in adults. PLoS ONE **10**(2), e0117598 (2015)
33. Milanovic, Z., Jorgić, B., Trajković, N., Sporis, G., Pantelić, S., James, N.: Age-related decrease in physical activity and functional fitness among elderly men and women. Clin. Interv. Aging **8**, 549–556 (2013)
34. AIHW, Australian Institute of Health and Welfare: Ageing and the health system: challenges, opportunities and adaptations. Australia's Health Series no. 14 Cat no. Aus. 178. AIHW, Canberra (2014)
35. Neustatl, G.: Accessibility for All: Business Benefits. Blog accessibility and assistive technology. https://www.visionaustralia.org/business-and-professionals/digital-access-consulting/resources/blog—accessibility-and-assistive-technology-blog/blog
36. Jones, R.: Mobile banking on the rise as payment via apps soars by 54% in 2015. The Guardian. https://www.theguardian.com/business/2016/jul/22/mobile-banking-on-the-rise-as-payment-via-apps-soars-by-54-in-2015
37. Haxton, N.: Internet banking proving difficult for some elderly and disabled people. ABC News. http://www.abc.net.au/pm/content/2014/s4149952.htm
38. Rouse, M.: Multifactor Authentication (MFA). TechTarget. http://searchsecurity.techtarget.com/definition/multifactor-authentication-MFA
39. Ihalainen, P.: What is Multi-Factor Authentication (MFA)? GlobalSign Blog. https://www.globalsign.com/en/blog/what-is-multi-factor-authentication-mfa
40. Barbir, A.: Multi-factor Authentication Methods Taxonomy. Presentation, Oasis (2017)
41. Al-Mukhtar, M.M.A., Al-Taie, R.R.K.: A more robust text based CAPTCHA for security in web applications. Int. J. Emerg. Trends Technol. Comput. Sci. (IJETTCS) **3**(2), 7–12 (2014)
42. Tan, H.Z., Gray, R., Young, J.J., Taylor, R.: A haptic back display for attentional and directional cueing. Haptics-e Electron. J. Haptics Res. **3**(1), 1–20 (2003)
43. Bau, O., Poupyrev, I., Israr, A., Harrison, C.: TeslaTouch: electrovibration for touch surfaces. In: Proceedings of the 23rd Annual ACM Symposium on User Interface Software and Technology, pp. 283–292. ACM, New York (2010)
44. Lylykangas, J., Surakka, V., Salminen, K., Raisamo, J., Laitinen, P., Rönning, K., Raisamo, R.: Designing tactile feedback for piezo buttons. In: Proceedings of the SIGCHI Conference on Human Factors in Computing Systems, Vancouver, BC, Canada, pp. 3281–3284 (2011)
45. Chew, M., Tygar, D.J.: Image recognition CAPTCHAs. In: Zhang, K., Zheng, Y. (eds.) Information Security, ISC 2004. Lecture Notes in Computer Science, vol. 3225, pp. 268–279. Springer, Heidelberg (2004)
46. Bianchi, A., Oakley, I., Kwon, D.S.: The secure haptic keypad: a tactile password system. In: Proceedings of the 10th SIGCHI Conference on Human Factors in Computing Systems, pp. 1089–1092. ACM, New York (2010)

The New Modified Methodology to Solve ECDLP Based on Brute Force Attack

Kritsanapong Somsuk[1] and Chalida Sanemueang[2(✉)]

[1] Department of Computer and Communication Engineering, Faculty of Technology, Udon Thani Rajabhat University, UDRU, Udon Thani, Thailand
kritsanapong@udru.ac.th

[2] Office of Academic Resources and Information Technology, Udon Thani Rajabhat University, UDRU, Udon Thani, Thailand
Chalida.sanemueng@gmail.com

Abstract. Elliptic curve cryptography (ECC) is one of public key cryptography suitable for the limited storages and low power devices. The reason is that ECC has the same security level with other public key cryptographies, although bits length is very small. However, ECC is based on Elliptic Curve Discrete Logarithm Problem (ECDLP) that is very difficult to be solved. At present, many algorithms were introduced to solve the problem. Nevertheless, the efficiency of each algorithm is based on the characteristic of k, $Q = kP$, when Q and P are known points on the curve, and type of curve. Deeply, brute force attack is one of techniques to solve ECDLP. This algorithm has very high performance when k is small. However, to find k, $2P$, $3P$, $4P$, $\cdots$, $(k-1)P$ and kP must be computed. Thus, numbers of inversion process are $k - 1$. Moreover, for traditional brute force attack, y's points must be computed all loops computation. In this paper, the new method based on brute force attack, is called Resolving Elliptic Curve Discrete Logarithm Problem by Decreasing Inversion Processes and Finding only x's points (RIX-ECDLP), is proposed. The key is to remove some inversion processes and y's points out of the computation. In fact, every two point additions can be done with only one inversion process. The experimental results show that RIX-ECDLP can reduce time about 10–20% based on size of k and prime number.

Keywords: ECC · ECDLP · Point doubling · Point addition
Inversion process

1 Introduction

The communication using electronic device via network is enormous at present. However, the network is unsecured channel. Hence, many algorithms to protect information sent over this channel were introduced. Cryptography is the technique to hide data from intruders by encryption and decryption process. There are two types for cryptography, symmetric key cryptography and asymmetric key cryptography (or public key cryptography). In general, symmetric key cryptography is always faster than public key cryptography. However, the secured channel must be chosen for key

H. Unger et al. (Eds.): IC2IT 2018, AISC 769, pp. 255–264, 2019.
https://doi.org/10.1007/978-3-319-93692-5_25

exchanging process because it uses the same key, is called secret key, for both of encryption and decryption process. Nevertheless, public key cryptography can be chosen to exchange secret key because two different keys are used for the different tasks. One is called public key, disclosed to everyone in group. The other is called private key and it is kept secretly. RSA [1] is the well-known public key cryptography using a pair of keys for encryption and decryption. However, key size of modulus is at least 1024 bits taking computation cost. That means, it takes very high time consuming in both of encryption and decryption. Later, V.C. Miller and N. Koblitz proposed Elliptic curve cryptography (ECC) [2–4] which is one of public key cryptography for hiding information from the intruders. In deep, ECC has the same security level with other public key cryptography, like RSA but smaller key size is required. For example, the security of 1024 bits length of RSA has the same security level with about 160–163 bits length of ECC [5, 6]. Therefore, ECC suits for small computing power device such as smart phone. Nevertheless, ECC is based on the Elliptic Curve Discrete Logarithm Problem (ECDLP) [7] that the process to find point $Q = kP$, when point, P, and integer, k, are published, over Elliptic curve is very easy and fast. On the other hand, it is very difficult to recover k from P and Q. At present, many different algorithms were proposed to solve ECDLP. The efficiency of each algorithm is different. Brute force attack is the efficient algorithm when k is small. However, if k is large, it becomes the inefficient method.

In this paper, the new modified algorithm, called Resolving Elliptic Curve Discrete Logarithm Problem by Decreasing Inversion Processes and Finding only x's points (RIX-ECDLP), is proposed to solve ECDLP. In fact, this algorithm is the improvement of brute force attack. The key is that y will be calculated only the result of x equal to the solution is found. On the other hand, the others not in this case will be leaving out from the computation. Furthermore, the inversion processes can be also decreased because one improved inversion equation can be applied with two point additions.

2 Related Works

2.1 Elliptic Curve Cryptography (ECC)

ECC is one of public key cryptography suitable for the low power devices such as smart phone because of smaller bits length. In fact, the security level for ECC is similar to others public key cryptography such as RSA. At present, there are many types for ECC. However, only ECC over the sets $\#E(F_p)$ will be mentioned in this paper. Assuming $P = (x_1, y_1)$ and $Q = (x_2, y_2)$ are represented as the points on the curve $y^2 = x^3 + ax + b \bmod p$, $4a^3 + 27b^2 \neq 0$. Addition law which is the process to find $R = (x_3, y_3) = P + Q$ is the key for ECC as follows:

Case 1: $P = Q$,

$$m = ((3x_1^2 + a) * (2y_1)^{-1}) \bmod p \quad (1)$$

Case 2: $P \neq Q$,

$$m = ((y_2 - y_1) * (x_2 - x_1)^{-1}) \bmod p \tag{2}$$

Then, x_3 and y_3 can be found as follows:

$$x_3 = \left(m^2 - x_1 - x_2\right) \bmod p \tag{3}$$

$$y_3 = (m * (x_1 - x_2) - y_1) \bmod p \tag{4}$$

In fact, if m is from (1), then $R = P + Q$ is called point addition. On the other hand, it is called point doubling whenever m is from (2).

2.2 The Method to Compute 2*P* + *Q* by Leaving *y*'s Computing

In 2003, K. Eisentrager et al. [8] proposed the algorithm to speed up ECC. This algorithm is focused on $2P + Q$ leaving the computing point of y from $P + Q$. The process begins with $P + Q$ $((x_1, y_1) + (x_2, y_2))$. Assuming $R = (x_3, y_3)$ is represented as the result of $P + Q$, then only x_3 will be computed. Therefore, after finishing this process, x_3 and $m = (y_2 - y_1) * ((x_2 - x_1)^{-1})$ are known. The last process, is to compute the result of $S = (x_4, y_4) = 2P + Q$, $(P + Q) + P$. Because, y_3 is unknown value, the process to find $m' = (y_3 - y_1) * ((x_3 - x_1)^{-1})$ must be changed as $-m - 2y_1 * (x_3 - x_1)^{-1}$. In addition, the performance costs for computing traditionally $2P + Q$ are 2 multiplications, 3 squarings and 2 divisions. However, only 1 multiplication, 2 squarings and 2 divisions are included for the improved method.

2.3 The Method to Compute 3*P* + *Q* by Decreasing Inversion Processes

Later, the technique to reduce inversion processes to find $3P + Q$ was proposed [9]. In fact, only one inversion process which can be chosen to find $R = 2P$ and $T = P+Q$ is performed. Then, the addition point is selected to compute $3P + Q = T + R$. Therefore, the process to find $3P + Q$ requires only two inversion process. Furthermore, the improvement $3P + Q$ requiring only one inversion process was also presented again in that paper.

In Sects. 2.4, 2.5, 2.6 and 2.7, the methods for finding k will be introduced. Generally, assuming P is represented a generator point; the key concept behind ECDLP is that it is very easy to compute point $Q = kP$. On the other hand, it becomes very hard to find k whenever Q is disclosed. In addition, the efficiency of each algorithm is different. For example, the performance for brute force attack is based on size of k. However, Pohlig-Hellman attack is different from brute force attack because this algorithm is depended on small factors of order for ECC.

2.4 Brute Force Attack

Brute force attack is the technique suitable for the small value of k because the process begins with the smallest integer and it continues increasing until $Q = kP$ is found. With the reason, it becomes inefficient algorithm when k is large.

2.5 Pohlig-Hellman Attack

Pohlig-Hellman attack [11] is the method that can be used for solving both of Discrete Logarithm Problem (DLP) which is the key behind Elgamal cryptography and ECDLP. In fact, this technique has very high performance when the factors of the order of ECC over the sets $\#E(F_p)$ are small. Thus, to avoid this attack, $\#E(F_p)$ should be prime or almost prime.

2.6 Baby-Step-Giant-Step

Baby-Step-Giant-Step is the attacking method proposed by D. Shanks [7]. This method is another algorithm for solving ECDLP. However, the disadvantage of this method is that it wastes time to find all iP, $0 < i < m$, where m is a constant which is larger than $\sqrt{\#E(F_p)}$. Therefore, Baby-Step-Giant-Step becomes inefficient when m is large.

2.7 Pollard-Rho Algorithm

In 1978, J. Pollard [12] proposed the method which is called Monte-Carlo or also Pollard –Rho to solve DLP. Later, his method was improved to be applied with ECDLP. At present, this method is known as the best algorithm for recovering k. However, the performance of the method is based on integers chosen randomly.

3 The Proposed Method

In this paper, the improvement of brute force attacked for solving ECDLP is proposed by removing both inversion processes and y_i from point (x_i, y_i) for some loops iteration. In fact, the method proposed in [8–10] are applied with the proposed method. In addition, this algorithm can be modified to find the integer, k, from two published points P and $Q = kP$. Therefore, almost values of y_i are rejected from the computation except when x_i is equal to x_q, $i \geq 2$. Furthermore, inversion processes are also avoided by a half. Assigning, $P = (x_1, y_1)$, $2P = (x_2, y_2)$, $3P = (x_3, y_3)$, $kP = (x_k, y_k)$ are represented as points on Elliptic Curve over $\#E(F_p)$, $p > 3$, E: $y^2 = (x^2 + ax + b) \bmod p$. Every two point additions (or point doublings) are computed with only one inversion process; usually one inversion is required for a point addition or point doubling.

The information in Table 1 is shown the proposed technique to find k. There are three columns in the table. For each row, 2nd column and 3rd column will be processed with a joined inversion process. That means, only one inversion process will be included for each row. In addition, only point addition is chosen for both of 2nd column and 3rd column during 5th row to row k. In fact, there are only points doubling in 3rd

Table 1. The method to find *iP* and (*i* + 1)*P* for reducing inversion process.

i	*iP*	(*i* + 1)*P*
1	$P(x_1, y_1)$	$2P = P + P$ $(x_2, y_2) = (x_1, y_1) + (x_1, y_1)$
3	$3P = P + 2P$ $(x_3, y_3) = (x_1, y_1) + (x_2, y_2)$	$4P = 2P + 2P$ $(x_4, y_4) = (x_2, y_2) + (x_2, y_2)$
5	$5P = 3P + 2P$ $(x_5, y_5) = (x_3, y_3) + (x_2, y_2)$	$6P = 4P + 2P$ $(x_6, y_6) = (x_4, y_4) + (x_2, y_2)$
7	$7P = 5P + 2P$ $(x_7, y_7) = (x_5, y_5) + (x_2, y_2)$	$8P = 6P + 2P$ $(x_8, y_8) = (x_6, y_6) + (x_2, y_2)$
k is even number		
$k - 3$	$(k-3)P = (k-5)P + 2P$ $(x_{k-3}, y_{k-3}) = (x_{k-5}, y_{k-5}) + (x_2, y_2)$	$(k-2)P = (k-4)P + 2P$ $(x_{k-2}, y_{k-2}) = (x_{k-4}, y_{k-4}) + (x_2, y_2)$
$k - 1$	$(k-1)P = (k-3)P + 2P$ $(x_{k-1}, y_{k-1}) = (x_{k-3}, y_{k-3}) + (x_2, y_2)$	$kP = (k-2)P + 2P$ $(x_k, y_k) = (x_{k-2}, y_{k-2}) + (x_2, y_2)$
k is odd number		
$k - 2$	$(k-2)P = (k-4)P + 2P$ $(x_{k-2}, y_{k-2}) = (x_{k-4}, y_{k-4}) + (x_2, y_2)$	$(k-1)P = (k-3)P + 2P$ $(x_{k-1}, y_{k-1}) = (x_{k-1}, y_{k-1}) + (x_2, y_2)$
k	$kP = (k-2)P + 2P$ $(x_k, y_k) = (x_{k-2}, y_{k-2}) + (x_2, y_2)$	$(k+1)P = (k-1)P + 2P$ $(x_{k+1}, y_{k+1}) = (x_{k-1}, y_{k-1}) + (x_2, y_2)$

column of the 1st and 2nd rows. In deep, the key of the proposed method is begun at the 3rd row, $i = 5$. However, 2*P* will be implemented by using point doubling and 3*P* and 4*P* is applied from the method in [8]. Moreover, y_i and y_{i+1}, $i \geq 5$, may not be performed whenever x_i and x_{i+1} are not equal to x_k. Therefore, the method is distinguished as two parts; part 1 is the process to find 3*P* and 4*P*, second is the process to find 5*P*, 6*P*, 7*P*, ⋯, $k - 1P$, *kP*.

Part 1: Process to find 3*P* and 4*P*.

Assuming $P = (x_1, y_1)$, $2P = (x_2, y_2)$, then $R = (x_3, y_3) = 3P = 2P + P$ and $T = (x_4, y_4) = 4P = 2P + 2P$ and be computed as following:

$$m_c = ((2y_2) * (x_2 - x_1))^{-1}$$

$$m_3 = (y_2 - y_1) * (2y_2) * m_c$$

$$m_4 = (x_2 - x_1) * (3x_2^2 + a) * m_c$$

$$x_3 = m_3^2 - x_2 - x_1$$

$$x_4 = m_4^2 - 2x_2$$

Part 2: Process to find $5P$ to kP.

Assuming $(x_j, y_j) = (x_r, y_r) + (x_s, y_s)$, where $m_j = (y_s - y_r) * (x_s - x_r)^{-1}$, $x_j = m_j^2 - x_s - x_r, y_j = m_j * (x_s - x_j) - y_s$ and k is represented as odd number which is larger than 3. The process to find results of 2nd column and 3rd column, iP and $(i + 1)P$ is as following:

Calculating $iP = (i - 2)P + 2P$

$$\begin{array}{ll} \text{From} & m_i = (y_{i-2} - y_2) * (x_{i-2} - x_2)^{-1} \\ \text{Because,} & y_{i-2} = m_{i-2} * (x_2 - x_{i-2}) - y_2 \\ \text{Therefore,} & m_i = (m_{i-2} * (x_2 - x_{i-2}) - 2y_2) * (x_{i-2} - x_2)^{-1} \end{array} \quad (5)$$

Calculating $(i + 1)P = (i - 1)P + 2P$

$$\begin{array}{ll} \text{From} & m_{i+1} = (y_{i-1} - y_2) * (x_{i-1} - x_2)^{-1} \\ \text{Because,} & y_{i-1} = m_{i-1} * (x_2 - x_{i-1}) - y_2 \\ \text{Therefore,} & m_{i+1} = (m_{i-1} * (x_2 - x_{i-1}) - 2y_2) * (x_{i-1} - x_2)^{-1} \end{array} \quad (6)$$

Assigning $I = ((x_{(i-1)} - x_2) * (x_{(i-2)} - x_2)^{-1}$, then Eqs. (5) and (6) can be improved without inversion computing as follows:

$$m_i = (m_{i-2} * (x_2 - x_{i-2}) - 2y_2) * I * (x_{i-1} - x_2) \quad (7)$$

$$m_{i+1} = (m_{i-1} * (x_2 - x_{i-1}) - 2y_2) * I * (x_{i-2} - x_2) \quad (8)$$

Moreover, from Table 1, it implies that loops computation can be reduced by a half in comparison to brute force attack for computing every values of kP, $k = 1, 2, 3, \cdots, k$.

From the proposed algorithm below, it implies that only one inversion process is included whenever iP and $(i + 1)P$, $i > 3$, are computed. Furthermore, if both of x_i and x_{i+1} are not equal to x_k, then y_i and y_{i+1} are left out from the process. In addition, if k is odd number, the result of kP is in row k.

On the other hand, the result is in the row $k - 1$ whenever k is even number. Therefore, $(i - 1)P$ and iP are computed together with only one inversion process to find kP the last process.

Algorithm: RIX-ECDLP

Input: $P = (x_1, y_1)$, $y^2 = x^3 + ax + b$, $Q = kP = (x_k, y_k)$

```
Computing 2P = (x_2, y_2)
m_c = ((2y_2)(x_2-x_1))^-1
m_3 = (y_2 - y_1)(2y_2)m_c, m_4 = (x_2 - x_1)(3x_2^2+a)m_c
x_3 = m_3^2 - x_2 - x_1, x4 = m_4^2 - 2x_2
IF x_3 == x_k then
   y_3 = m_3(x_2 - x_3) - y_2
   IF y_3 == y_k then
       k = 3
       End Process
   EndIF
End IF
IF x_4 == x_k then
   y_4 = m_4(x_2 - x_3) - y_2
   IF y_4 == y_k then
       k = 4
       End Process
   End IF
End IF
i=5, m_{i-2}=m_3, x_{i-2}=x_3, m_{i-1}=m_4, x_{i-1}=x_4
while true do
   I = ((x_{i-1}-x_2)(x_{i-2}-x_2)^-1
   m_i = (m_{i-2}(x_2 - x_{i-2})-2y_2)*I*(x_{i-1}-x_2)
   m_{i+1} = (m_{i-1}(x_2 - x_{i-1})-2y_2)*I*(x_{i-2}-x_2)
   x_i = m_i^2 - x_{i-2} - x_2
   IF x_i == x_k then
       y = m_i(x_2 - x_{i-2}) - y_2
        IF y == y_k then
            k = i
            End Process
       EndIF
   EndIF
   x_{i+1} = m_{i1}^2 - x_{i-1} - x_2
   IF x_{i+1} == x_k then
       y = m_{i1}(x_2 - x_{i-1}) - y_2
        IF y == y_k then
            k = i+1
            End Process
       End IF
   End IF
   i += 2,m_{i-2} = m_i, m_{i-1} = m_{i+1}, x_{i-2} = x_i, x_{i-1} = x_{i+1}
End While
```

Output: k

4 Experimental Results

This section will show the comparison about computation time to find k between brute force attack and RIX-ECDLP. In addition, 16–56 bits length of p chosen randomly are selected for the experiment. All values of k in the experiments are close to $\sqrt{p}$. Moreover, Java Language and BigInteger Class which can be represented as unlimited data type are chosen. In order to control the same settings, all experiments were conducted on 2.20 GHz an Intel® Core i7 with 4 GB memory.

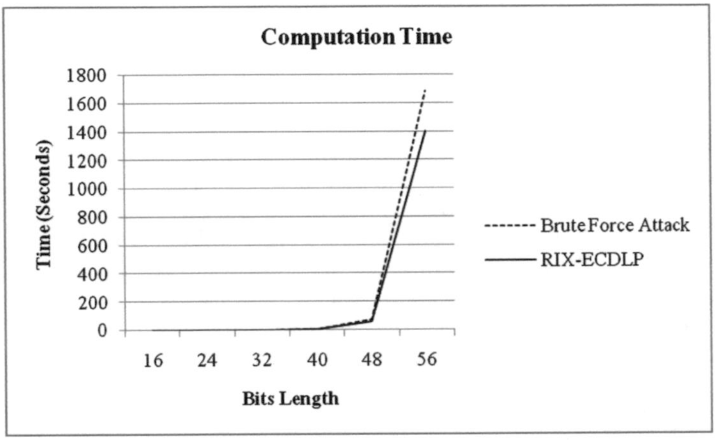

Fig. 1. Comparison about computation time to find k between Brute Force Attack and RIX-ECLDP

The information in Fig. 1 shows that RIX-ECLDP can find k faster than brute force attack. In addition, RIX-ECDLP can decrease time about 10–20% by average. Moreover, from the experimental results, it is predicted that the computation time can be reduced more when larger bits length of p is selected.

5 System Analysis

Usually, 1 Inversion, 1 Multiplication and 1 Squaring are required for a point addition process. Thus, if two point additions are requested, then costs for these processes are 2 Inversions, 2 Multiplications and 2 Squarings. However, Table 2 shows that costs for computing two point additions by using the proposed method are 1 Inversion, 6 Multiplications and 2 Squarings, when both of x_i and x_{i+1} are not equal to x_k. Although, multiplication processes are larger than brute force attack, only one inversion which is the most expensive cost process is required for RIX-ECDLP.

However, if either x_i or x_{i+1} is equal to x_k, multiplication processes are increased as 7. On the other hand, 8 multiplication processes are required when both of x_i and x_{i+1} are equal to x_k.

Table 2. Computation costs to find iP and $(i+1)P$ using RIX-ECDLP

Process	I	M	S
$I = \left(\left(x_{(i-1)} - x_2\right)\left(x_{(i-2)} - x_2\right)\right)^{-1}$	1	-	-
$m_i = \left(m_{i-2}\left(x_2 - x_{(i-2)}\right) - 2y_2\right) * I * \left(x_{(i-1)} - x_2\right)$	-	3	-
$m_{i+1} = \left(m_{i-1}\left(x_2 - x_{(i-1)}\right) - 2y_2\right) * I * \left(x_{(i-2)} - x_2\right)$	-	3	-
$x_i = m_i^2 - x_{i-2} - x_2$	-	-	1
$x_{i+1} = m_{i+1}^2 - x_{i-1} - x_2$	-	-	1
Totals	**1**	**6**	**2**

6 Conclusion

RIX-ECDLP which is the new improvement algorithm for attacking ECC is proposed. This algorithm based on brute force attack and suitable for small value of k requires only one inversion process to compute two point additions, kP and $(k + 1)P$ where k is odd number or $(k - 1)P$ and kP where k is even number. Although, numbers of multiplication process are increased in comparison to computing traditional two point additions, the computation costs are very smaller than costs of inversion process. Therefore, it implies than RIX-ECDLP can reduce time to find k. Furthermore, point additions to find y_j can be left out whenever, x_j is not equal to x_k, j is an integer and x_k is represented as result of point x. The experimental result shows that RIX-ECDLP can reduce time when it is compared with traditional brute force attack. In fact, $2P$ is the main point to be performed all loops.

References

1. Rivest, R.L., Shamir, A., Adleman, L.: A method for obtaining digital signatures and public key cryptosystems. J. Commun. ACM **21**(2), 120–126 (1978)
2. Koblitz, N.: Elliptic Curve Cryptosystems. Math. Comput. **48**, 203–209 (1987)
3. Elbirt, A.J.: Under Standing and Applying Cryptography and Data Security. Auerbach Publications, USA (2009)
4. Miller, V.S.: Uses of elliptic curves in cryptography. In: Williams, H.C. (ed.) Lecture Notes in Computer Science, vol. 218, pp. 417–428. Springer, Heidelberg (1986)
5. Amara, M., Siad, A.: Elliptic curve cryptography and its application. In: 7th International Workshop on Systems, Signal Processing and their Applications, Tipaza, Algeria, pp. 247–250 (2011)
6. Subhranil, S., Rana, M., Sandip, D.: Elliptic curve cryptography: a dynamic paradigm. In: International Conference on Infocom Technologies and Unmanned Systems, Dubai, UAE, pp. 427–431 (2017)
7. Singh, L.D., Debbrama, T.: A new approach to Elliptic curve cryptography. In: International Conference on Advanced Communication Control and Computing Technologies, Ramanathapuram, India, pp. 78–82 (2014)
8. Eisentrager, K., Lauter, K., Montgomery, P.L.: Fast Elliptic curve arithmetic and improved Weil pairing evaluation. In: Joye, M. (ed.) Lecture Notes in Computer Science, vol. 2612, pp. 343–354. Springer, Heidelberg (2003)

9. Ciet, M., Joye, M., Lauter, K., Montgomery, P.L.: Trading inversions for multiplications in Elliptic curve cryptography. J. Des. Codes and Cryptogr. **39**(2), 189–206 (2005)
10. Li, Y., Feng, L.: Overview of scalar multiplication in Elliptic curve cryptography. In: International Conference on Computer Science and Network Technology, Harbin, China, pp. 2670–2673 (2011)
11. Attacks on the curve-based discrete logarithm problem. http://ecc2011.loria.fr/slides/summerschool-vitse.pdf
12. Neamah, A.A.: New Collisions to Improve Pollard's Rho Method of Solving the Discrete Logarithm Problem on Elliptic Curves. https://arxiv.org/ftp/arxiv/papers/1607/1607.05901.pdf

Information Technology

Generations and Level of Information on Mobile Devices Usage: An Entropy-Based Study

Charnsak Srisawatsakul(✉) and Waransanang Boontarig

Faculty of Computer Science, Ubon Ratchathani Rajabhat University, Ubon Ratchathani 34000, Thailand
charnsak@researcher.in.th, waransanang.b@ubru.ac.th

Abstract. Recently, mobile devices are now widely adopted in daily life. Several studies have examined the adoption of mobile devices in different context. However, previous research on the relationship between aging cohort and level of information in mobile devices is not consistent. Therefore, this study aims to investigate the usage behavior of mobile devices in users with different generations using information theory. Mobile applications were represented as a set of possible outcomes. The 95 datasets were collected using an online questionnaire and used to calculate the entropy values. The entropy values were compared within 4 generations including baby boomer, generation X, Y, and Z. The results indicated that baby boomer has the lowest level of information (1.58 bits) following by generation Z (2.48 bits), Y (2.69 bits), and X (3.11 bits), accordingly. The present study proved that self-evaluated questionnaire could be used for measuring the level of information in mobile devices.

Keywords: Generation · Information theory · Entropy
Mobile application · Mobile devices · Cohort

1 Introduction

In recent years, there has been an increasing development of mobile technologies. Mobile devices became an important part of people's life from teenager to older adult. Smartphone becomes one of the most popular mobile devices. It is now equipped with more advanced technologies [1] such as powerful octa-cores CPU (Central Processing Unit) with 64 bits 10-nanometer design, high performance GPU (Graphic Processing Unit), high definition multiple cameras, 4 K UltraHD display, high speed Wi-Fi and cellular connection, GPS (Global Positioning System), gyroscope and etc. Furthermore, the average price of smartphone become affordable for most of the users. On the second quarter of 2017, there were 341.6 million of smartphone units ship worldwide [2]. Moreover, there are 7,740 million global mobile cellular subscribers at the end of the year 2017 [3]. Moreover, the number of smartphone users are predicted to reach 2.87 billion in 2020 [4].

The mobile application is an important success factor that allows users ubiquitously take advantage of their mobile devices for a wide range of applications. Therefore,

H. Unger et al. (Eds.): IC2IT 2018, AISC 769, pp. 267–276, 2019.
https://doi.org/10.1007/978-3-319-93692-5_26

studies over the past decade have provided important information on the adoption of mobile applications [5–7] or the technical aspect of mobile application development. However, previous studies of mobile application have not dealt with how to objectively find the relation between the aging cohort (generation) and the level of information on mobile applications. Hence, the aim of this study is to explore the relationship between the generation and level of information on mobile devices usage by implemented Shannon's information theory (Entropy) [8].

The study was conducted in the form of a quantitative research, with data being gathered via an online questionnaire. The methodological approach taken in this study is based on previous literature [9] that proposed to use entropy for measuring the level of information on mobile devices. However, in the previous study, the researcher uses a mobile application that running in the background to collect the usage data. In the opposite, this study will let the participants estimate their usage behavior by themselves using the questionnaire. The results of this study will prove the concept of using entropy for measuring the level of information in mobile devices and enhance our understanding of how people in different generation use their mobile applications.

The remaining parts of the paper take form of 5 sections. The Sect. 2 begins with the literature review of previous studies. The Sect. 3 is concerned with the methodology and data collection used for this study. The Sect. 4 presents the findings of the research. The Sect. 5 is the interpretation and discussion of the research outcome.

2 Literature Review

2.1 Information Theory (Entropy)

The original literature on information theory (entropy) was proposed by Shannon [8] and Cover [10]. They defined entropy as "a measure of the average uncertainty in the random variable, it is the number of bits on the average required to describe the random variable." Entropy was used to assess the uncertainty in a random variable using a quantitative method. The basic formula of Shannon's entropy is shown below:

$$H(X) = -\sum_{i=1}^{k} P_i \, log_2 \, P_i \tag{1}$$

Where:

X =	a discrete random variable.
$H(X) =$	represents the entropy value in bits. It indicates the uncertainty level of variable X. The value of entropy will always positive.
$k =$	the total number of possible outcomes of variable X.
$P_i =$	the probability that a possible outcome of variable X_i will occur.
$log_2 P_i =$	logarithm base 2 of Pi

To date, several studies used information theory to examine the level of uncertainty in different contexts. For example, Akaike [11] used information theory to match and compare the difference between variables in data mining. It is now a foundation of the maximum likelihood technique that is widely used. Stock [12] applied the entropy with the prediction of the uncertainty in the stock market. The entropy was used to analyze the relationship between the growth rate and wealth of stocks. The objective is to find the optimal competitive portfolio. Sinatra and Szell [13] used entropy to measure the uncertainty of mobile virtual massive multiplayer online game. They found that the movements that users move in the virtual world have nearly the same high levels of predictability in the real world. The data that collected from the online game could be used to predict the behavior. It also requires less of information that needs to gather when compared with other methods.

Spiekermann and Korunovska [14] demonstrated that they could use entropy to measure the level of information on users' display when they are using online social networks. The result suggested that the entropy value will be high in complication screen. Therefore, the high entropy screen will discourage users to share their thought and reduce their creativity. Oruç [15] published a paper in which they applied entropy with the survey scales. They found that the entropy value will be low if the survey has fewer questions. Moreover, participants will have more concentration on the low entropy survey.

Together, these studies indicate that entropy could be used to measure the level of information in different context as long as there is a random variable that has possible outcomes. Therefore, this study used entropy as a measurement of information in the mobile device based on mobile applications and find the relationship between different generations.

2.2 Aging Cohort

Over the past decade, most research on the attitudes and behaviors effect of a different group of people has been characterized by an individual's age. However, the results of different age group may be alteration easily. For example, the older adults in the next 10 years would change because they were the active adults at the time that research carried out. Therefore, this study characterized a group of people using the age cohort as generations to find the effect on the level of information in mobile devices.

Generations are one way to group age cohorts. It was originally elaborated by Mannheim (1970) [16] as the theory of generations. It was used to group the people in the late 1950 s to early 1960. Pilcher (1994) [17] has suggested the revised definition of the generations as "people within a delineated population who experience the same significant events within a given period of time". Thus far, a number of studies have attempted to evaluate the age cohort to make it up to date [18–21]. Hence, the generations now can be grouped as 4 generations, Baby Boomers, Generation X, Generation Y and Generation Z.

1. **Baby Boomer** refer to individuals who were born from 1943–1960 [18, 19]. This group of individuals shares the same experience on the incident of World War II [21]. They are noticeable by their leadership, hardworking, conservative, confidence and optimism. The reason behind these traits was the growing economy of the U.S. after the World War II in the late 1940 s, 1950 s and the 1960 s.
2. **Generation X** is individuals who were born from 1961–1981 [18, 19]. Generation X has been distinct by skepticism. They are an extremely resourceful, celebrate

change, flexibility and independent generation. They also have some high level of technology related because they were born within the birth of personal computer (PC), Internet and mobile phones [21].

3. **Generation Y** is also known as the millennial and echo boomer. They were born from 1982–1994 [18, 19]. They also have the confident and optimistic like the baby boomers. However, they are smart, like challenges, ambitions, adapt to change and flexible. They don't like to follow the rule.
4. **Generation Z** is the latest generation who were born since 1995 [18, 19]. They also are known as the digital native because they were born in the world full of digital technologies. They like to live in the comfort zone and addicted to technologies. They are absolutely living in the digital world.

3 Approach

3.1 Measuring Information on Mobile Devices Using Entropy

In this paper, the information theory will be used in the context of mobile devices usage. End users use mobile devices almost anywhere and anytime. However, when they turn on and use their mobile devices, they did not directly access the inside hardware itself. On the other hand, they have access hardware features through mobile applications. Therefore, when they use mobile device, they interact with several mobile applications that serve them with different proposes.

The usage behavior of mobile applications will be defined as a random variable. The possible outcomes of this variable are the probability that end users randomly open different mobile applications. They may be more predictable if they have fewer number of mobile applications installed. Besides, they also more predictable if they have the higher number of mobile applications installed but they rarely use most of it. This occurrence appears comparable to the information theory; that is to say, the entropy increases as much as numbers of the mobile applications have been used. Thus, the frequency of mobile application that has been launched over a period of time will be used to calculate the probability of the entropy. The equation used for this study are:

$$H(A) = -\sum_{i=1}^{k} P_i \, log_2 P_i \tag{2}$$

Where:

A =	a set of mobile applications of a participant.
$H(A)$ =	the entropy value of a participant. It shows the level of information on the uncertainty of that participant. (Bits)
k =	the total number of mobile applications that installed in the participant's system.
P_i =	the probability that a mobile application number *i* will be launch over a period of time.
$log_2 P_i$ =	the self-information measured of probability *i* using logarithm base 2 of *Pi*.

This paper applied this equation with the collected data and compare it between generations. The generation was divided into 4 groups as defined in the previous section. According to Eq. 2, the probability of possible outcomes (P_i) will be calculated using the frequency of the launch of mobile applications divide by the total number of applications that have been launch. Table 1 shows the example of collected data from a participant.

Table 1. Data collected from a participant.

No	Apps (A)	*Frequency* (f_i)	$P(i)$	$log_2 P_i$	$P_i * log_2 P_i(f_i)$
1	Phone	15	0.16	−2.66	−0.42
2	Facebook	30	0.32	−1.66	−0.53
3	Line	50	0.53	−0.93	−0.49
Total		95	1.00		

According to Table 1 An example participants used 3 mobile applications (K = 3). Hence, there are a total of 3 possible outcomes. The total frequency of launched mobile applications is 95. The entropy value of this participant can be calculated as below:

$$
\begin{aligned}
H(A) &= -\left(\left(P_{phone} * log_2 P_{phone}\right) + \left(P_{Facebook} * log_2 P_{Facebook}\right) + \left(P_{Line} * log_2 P_{Line}\right)\right) \\
H(A) &= -((0.16 * (-2.66)) + (0.32 * (-1.66)) + (0.53 * (-0.93))) \\
H(A) &= -(-0.42) + (-0.53) + (-0.49) \\
H(A) &= -(-1.43) \\
H(A) &= 1.43
\end{aligned}
$$

The entropy value is 1.43 bits of information. This method also applied to entirely collected data. After that, the entropy values were examined using the statistics approach and were grouped and compared by each generation.

3.2 Data Collection

Srisawatsakul [9] proposed a method for collecting the possible outcomes of each mobile application by using a mobile application called "QualityTime" [22]. This mobile application is running in the background and count every time that user launch any mobile applications. However, there are certain drawbacks associated with the use of "QualityTime" mobile application. Firstly, the QualityTime can only run on Android operating system. Secondly, participants need to install the QualityTime on their mobile device and keep it open in the background all the time. Thus, due to the privacy issue, it is very hard to collect data from a large number of participants. Lastly, the QualityTime may consume a lot of memory and battery power, which affected the daily lives of participants. Therefore, qualitative methods can be more useful for collecting the usage data of mobile applications from participants. Therefore, this study uses a self-evaluated online questionnaire to collect the data.

The participants are Thai mobile devices users. The design of the questionnaires was constructed with 3 simple questions in 2 parts. The first part asked about the demographic information including age, gender and operating system of mobile devices. The second part consisted of 2 questions. Firstly, it requested participants to enter the name of mobile applications that they used in daily life with the maximum of 44 mobile applications. However, the first 19 questions were pre-filled with the normal and well-known mobile applications that most of the participants already used it. That are, phone (cellular), SMS, Camera, Facebook, Line, Instagram, Twitter, WeChat, Whatsapp, Google Maps, Google Photos, Safari, Chrome, Gmail, App Store, App Store, Google Classroom, Youtube, and Joox. Participants did not require to answer all the questions. Secondly, the questionnaire will generate questions based on the mobile applications' name from previous inputs and ask the participants on how often they have open each mobile applications on a daily basis in average.

The initial sample consisted of 200 participants. However, after data screening processes. There were only 95 participants that can be used for analysis. The participants were divided into 4 groups based on their age. The demographic variables of participants are shown in Tables 2 and 3.

Table 2. Demographic variables of participants.

Generations	Male	Female	Other	Total
Baby boomer	6	2	0	8
Generation X	7	6	0	13
Generation Y	20	14	1	35
Generation Z	22	16	1	39
Total	56	38	2	95

Table 3. Operating system of participants' mobile devices.

Generations	iOS	Android	Windows phone	Total
Baby boomer	7	2	0	9
Generation X	7	6	0	13
Generation Y	10	30	1	41
Generation Z	20	23	4	47
Total	44	61	5	110

* Each participant can answer more than one operating systems.

4 Results

4.1 Number of Mobile Applications of Different Generations

In the questionnaire, the maximum number of mobile applications that participants could answer is 44. The baby boomer used less number of mobile applications with the average of 14.7 mobile applications (SD = 9.4). However, there was a big gap between the highest (32 apps) and lowest number (6 apps). There are a similar average number for Generation X ($M = 22.42$, $SD = 8.54$) and Generation Y ($M = 21.08$, $SD = 8.27$). The average install of mobile applications in generation Z was 17.7 ($SD = 7.04$). Figure 1 presents the highest, average and lowest number of mobile application installed of each generation.

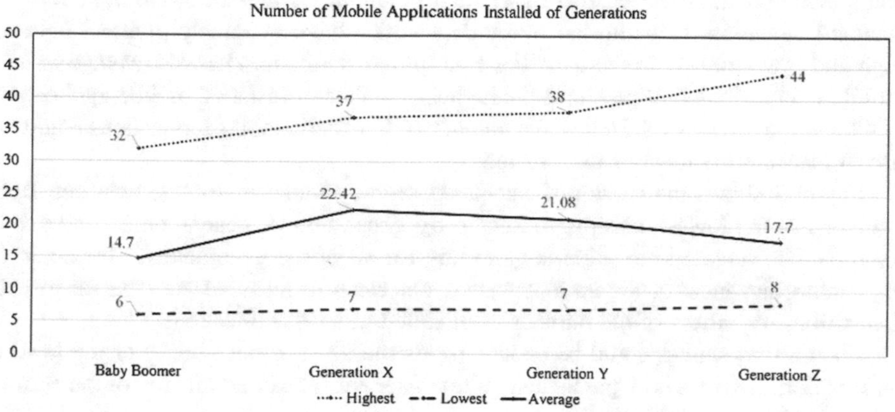

Fig. 1. Number of mobile applications that installed on mobile devices of each generation.

The results obtained from the calculation of entropy using Eq. 2 for each generation are presented in Fig. 2. It provides the highest, average and lowest entropy values of each generation. From the chart, it can be seen that by far the highest average entropy values is in the generation X ($M = 3.11$, $SD = 0.57$) following by generation Y ($M = 2.69$, $SD = 0.66$) and generation Z ($M = 2.48$, $SD = 0.35$). However, baby boomer has lowest average entropy values ($M = 1.58$, $SD = 1.58$).

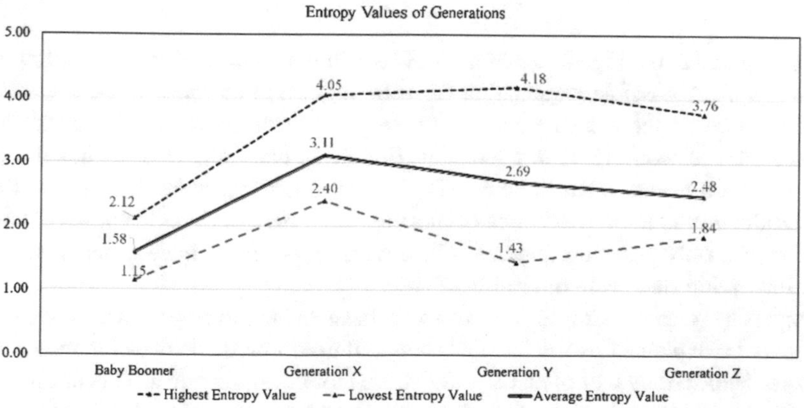

Fig. 2. The entropy values of each generation.

5 Interpretation and Discussion of Results

The present study was designed to determine the relationship between generations and the level of information on mobile devices. The results from Fig. 1. shows the trends that generation X installed less mobile applications on their mobile devices. Generation X installed the highest number of mobile applications following by generation Y and generation Z. However, there may be a potential for bias from the high value of standard deviation. It is affected by outliers which have extremely different between high and low numbers in the data. For example, generation Z has the average of 17.7 mobile applications but they have the highest number of installed mobile applications of 44 and the lowest of 8. Hence, the number of installed mobile applications could not use to measure the level of information.

Figure 2 shows the results of calculated entropy value of each generation. Baby boomer has the lowest entropy, following by generation Z, generation Y, and generation X. The standard deviation is quite low for all of the generations. This could be explaining that most of the entropy values are close to mean of the data set in each generation, regardless of the number of installed mobile applications.

The entropy values could be explaining as the level of uncertainty or the level of information. It shows how predictable of that user are. The interpretation of the entropy values is shown in Table 4.

Table 4. The interpretation of entropy values.

Entropy	Uncertainty	Predictability	Apps usage behavior
High	High	Low	• Oftenly use more of mobile applications that installed over a period of time
Low	Low	High	• Oftenly use less of mobile applications that installed over a period of time
0	No uncertainty	Predictable	• Use only one mobile application • Not use at all

As explained by Fig. 2, generation X has higher entropy values, following by generation Y. A possible explanation for this might be that most of the generation X and Y are now in the working age, which requires using more mobile applications in the processes of working. One interesting finding is that baby boomer has the lowest entropy. These results seem to be consistent with other research which found that the older adults will use fewer features of mobile social network applications [23] as they prefer to use only essential features. Moreover, older adults have a negative attitude toward adopting new technology [24, 25].

Surprisingly, generation Z was found to have lowest average entropy value. This result may be explained by the fact that they still used mobile devices for entertainment proposes. Hence, they will often use only some mobile applications. This finding, while preliminary, confirms that the entropy approach was capable of measuring the level of information on the mobile device. Furthermore, it also shows that self-evaluated questionnaire also can be used to get the usage data instead of installing background mobile applications in previous work [9]. However, the survey also has a potential for unconscious bias.

The number of mobile applications installed on the system may depend on many factors such as the capacity of the mobile phone, the user's education, income, and lifestyle, etc. which are also different in each generation. Therefore, future studies should pay more attention to the detailed factor of the different generation. Moreover, the entropy approach could be applied to other situation for measuring the level of information as long as there is a random variable with possible outcomes.

Acknowledgements. This research was partially supported by the faculty of Computer Science, Ubon Ratchathani Rajabhat University. The authors would like to show our gratitude to Assoc. Prof. Dr. Borworn Papasratorn from the school of information technology, King Mongkut's University of Technology Thonburi, Bangkok, Thailand for the assistance of this research. We also thank the reviewers for their suggestions that greatly improved the manuscript.

References

1. Snapdragon 835 Mobile Platform. https://www.qualcomm.com/products/snapdragon/processors/835
2. Smartphone Volumes Decline Slightly in the Second Quarter of 2017 Amid Anticipation of Strong Second Half Product Launches, According to IDC. https://www.idc.com/getdoc.jsp?containerId=prUS42935817
3. International Telecommunication Union STATISTICS. https://www.itu.int/en/ITU-D/Statistics/Pages/stat/default.aspx
4. Statista: Number of smartphone users worldwide from 2014 to 2020 (in billions). https://www.statista.com/statistics/330695/number-of-smartphone-users-worldwide/
5. Srisawatsakul, C., Quirchmayr, G., Papasratorn, B.: A pilot study on the effects of personality traits on the usage of mobile applications: a case study on office workers and tertiary students in the Bangkok area. In: Boonkrong, S., Unger, H., Meesad, P. (eds.) Recent Advances in Information and Communication Technology, pp. 145–155. Springer, Switzerland (2014)
6. Srisawatsakul, C., Papasratorn, B.: Factors affecting consumer acceptance mobile broadband services with add-on advertising: Thailand case study. Wirel. Pers. Commun. **69**(3), 1055–1065 (2013)
7. Lane, W., Manner, C.: The influence of personality traits on mobile phone application preferences. J. Econ. Behav. Stud. **4**(5), 252–260 (2012)
8. Shannon, C.E.: A mathematical theory of communication. Bell Syst. Tech. J. **27**, 379–423 (1948)
9. Srisawatsakul, C.: Measuring information on mobile devices usage: an entropy-based approach. In: 2016 International Computer Science and Engineering Conference (ICSEC), pp. 1–6. IEEE, Chiang Mai (2016)
10. Cover, T.M., Thomas, J.A.: Entropy, relative entropy and mutual information. In: Elements of Information Theory. Wiley Inc., New York (1991). Chap. 2
11. Akaike, H.: Information theory and an extension of the maximum likelihood principle. In: Parzen, E., Tanabe, K., Kitagawa, G. (eds.) Selected Papers of Hirotugu Akaike, pp. 199–213. Springer, New York (1998)

12. Cover, T.M., Thomas, J.A.: Information theory and the stock market. In: Elements of Information Theory. Wiley Inc., New York (1991). Chap. 5
13. Sinatra, R., Szell, M.: Entropy and the predictability of online life. Entropy **16**(1), 543–556 (2014)
14. Spiekermann, S., Korunovska, J.: The importance of interface complexity and entropy for online information sharing. Behav. Inf. Technol. **33**(6), 636–645 (2014)
15. Oruç, Ö.E., Kuruoglu, E., Gündüz, A.: Entropy applications for customer satisfaction survey in information theory. Front. Sci. **1**(1), 1–4 (2012)
16. Mannheim, K.: The problem of generations. Psychoanal. Rev. **57**, 378 (1970)
17. Pilcher, J.: Mannheim's sociology of generations: an undervalued legacy. Br. J. Sociol. **45**(3), 481–495 (1994)
18. McDonald, K.: How to Market to People Not Like You: "Know It or Blow It" Rules for Reaching Diverse Customers. Wiley Inc., Hoboken (2011)
19. Schroer, W.J., Klein, K.: Defining, Managing, and Marketing to Generations X, Y, and Z. The Portal XL, 9 (2008)
20. Lachman M.L., Brett D.L.: Generation Y: Shopping and entertainment in the digital age. Urban Land Institute (2013)
21. Kim, D.J.: Generation gaps in Engineering? Massachusetts Institute of Technology (2008)
22. ZeroDesktop I: Qualitytime App. http://www.qualitytimeapp.com
23. Srisawatsakul, C., Papasratorn, B.: Effects of age on end users' usage of mobile social network applications. In: The 10th International Conference on e-Business (iNCEB2015), Bangkok, pp. 1–7 (2015)
24. Lee, Y.S.: Older adult's user experiences with mobile phones: identification of user clusters and user requirements. Blacksburg, Virginia (2007)
25. Selwyn, N., Gorard, S., Furlong, J., Madden, L.: Older adults' use of information and communications technology in everyday life. Ageing Soc. **23**(5), 561–582 (2003)

Requirement Patterns Analysis and Design of Online Social Media Marketing for Promoting Eco-Tourism in Thailand

Orasa Tetiwat[1(✉)], Vatcharaporn Esichaikul[2], and Ranee Esichaikul[3]

[1] Faculty of Agricultural Technology and Industrial Technology, Nakhon Sawan Rajabhat University, Nakhon Sawan, Thailand
orasa.t@nsru.ac.th

[2] School of Engineering and Technology, Asian Institute of Technology, Pathumthani, Thailand
vatchara@ait.ac.th

[3] School of Management Science, Sukhothai Thammathirat Open University, Nonthaburi, Thailand
ranee.esi@stou.ac.th

Abstract. This research aimed to identify the requirements of online social media marketing system for promoting eco-tourism in Thailand. It also aimed to design the system. The research methodology used a combination of qualitative interviews with 60 people and quantitative sample survey of 844 people, in order to confirm the findings. The results of the analysis showed that most of the findings were consistent. For the content needed, it included the featured attractions, travel itinerary, pictures, sights atmosphere, and costs/expenses. For the functionality needed, it included a search facility for eco-tourism destinations, a search to identify travel routes, and a capability of a system to share files, photos, and video. For the content structure of website Home Page, it should be organized by type of activity, by region, and by month. Finally, the system analysis and design were based on object-oriented principles in order to develop the system in the future.

Keywords: Requirement patterns · Analysis and design · Online social media Social marketing · Eco-tourism · Thailand

1 Introduction

The tourism industry is very important to Thai economy because it is an industry that draws a large amount of revenues into Thailand. It is the main source of income, creating employment and businesses such as accommodation services, restaurants, tour operators, travel agents, and souvenir stores. One tourism form that is popular today is eco-tourism, a tourism that focused on environmental conservation as a habitat of species of plants, animals and humans. An eco-tourism is supported and promoted by the Thai government. This is because the eco-tourism can help to encourage the local community to concern and protect natural resources and local wisdom in their own communities [1]. In addition, there is a Thai Ecotourism and Adventure Travel

H. Unger et al. (Eds.): IC2IT 2018, AISC 769, pp. 277–285, 2019.
https://doi.org/10.1007/978-3-319-93692-5_27

Association (TEATA), which aims to develop and promote eco and adventure tourism as well as set up the standard of tourism related activities and services. Moreover, TEATA works and collaborates closely with the government, private sectors and local communities to enhance environmental conservation, and to ensure that tourism development goes well with nature, leading to increased awareness of the importance to conserve natural attractions [2]. Thus, eco-tourism becomes popular among Thai and foreign tourists.

Currently, social media is a form of communication that connects people at all levels. Social media is likely to become the major media for people even more in the future. It is widely used for sending and sharing messages in various forms to the audience through interactive online network between senders and receivers. It is also a tool that corporate travel agencies are required to adopt and develop the system used in their organizations. Many government and eco-tourism business agencies are interested in using various advertising media including online computer media, print media, TV, radio, and the Internet. The advertisement actually is a main selling point of each business. Apparently, marketing through Internet nowadays should focus more on mobile users, which will reach customers the most.

The use of social media is a part of today's life. In order to promote the marketing of eco-tourism in Thailand, it is necessary to study the behavior and needs of eco-tourists in using social media to find information on making their travel decisions. This study aims to collect and analyze the requirements to use social media marketing for eco-tourism, as well as to design online social media marketing system to promote eco-tourism in Thailand. The focus is on the main purpose for using social media, and the requirements of eco-tourists in term of contents, functionality, and services.

2 Literature Review

As mentioned, eco-tourism becomes popular and it is considered to be the fastest growing sector of the tourism industry in this decade. Also, social media plays an important role in tourism. The consumers can use social media to research trips, make decisions about their travel plans, and share their travel experiences. This section presents several studies related to the eco-tourism and the use of social media in the tourism industry, from many aspects.

According to William (2016) [3], the ecotourism industry has seen massive growth since the early 1990s. It grew 20%–34% per year. The fast growth of this segment suggests that now is the time to enter this market and gain share as a hotel, resort, or as a country. Eco-tourism falls under a larger category of tourism labeled as "experimental" tourism, which is projected to grow over the next twenty years while mass tourism (sun and sand) is expected to remain flat and mature.

Popesku [4] conducted a research on the use of social media as a marketing tool to support corporate business travel destinations. The objective was to study how the use of social media can support the market in travel business. The research found that social media marketing can help to guide tourists planning trips at the beginning of the trip to the travel destination by using the Internet to communicate with tourists. They also can access the information that they need and help them to plan trip as well as to help in the decision to select destinations.

In addition, Hays et al. [5] conducted a study to examine the use of social media as a marketing tool for tourism organizations, national tourism. Social media is a part of the marketing policy of the tourist market, especially in cases where the government has cut the budget. Social media is a tool to reach people all over the world with limited resources. This research explored the use of social media marketing organization of 10 national tourist attractions for the trip. The study results suggested the best practices of social media marketing to promote tourism.

For a broad picture, Zeng and Gerritsen [6] reviewed and analyzed the research publications focusing on social media in tourism. Through a comprehensive literature review, they identified what is known about social media in tourism, and recommended a future research. They suggested comprehensive investigation into the influence and impact of social media as a part of marketing strategy on all aspects of the tourism industry.

3 Research Methodology

This research was divided into two phases: (1) Requirements analysis of using social media marketing and (2) System analysis and design of online social media marketing system. The first phase was to study and analyze the requirements of using social media marketing to promote eco-tourism in Thailand. In Phase1, this study used a combination of a qualitative method (in-depth interviews for seeking to better understand opinions, motivations, and preferences) and a quantitative survey method to confirm the findings. Therefore, there were two sample groups used for investigating the requirements of using social media to promote eco-tourism. The first purposeful sample group participated in the in-depth interviews consisted of 60 key informants, including 22 eco-tourism operators and 4 members of the Thai Eco-tourism and Adventure Travel Association (TEATA), 10 academicians, and 24 Thai and foreign eco-tourists during August–September 2015. The second sample group used in the survey was derived from a random sampling method. A random sample survey was conducted with 844 respondents, including eco-tourists, eco-tourism operators and managers who work in tour business, at Doi Inthanon National Park and Khao Sok National Park during October–December 2015. The survey samples included 533 Thai respondents (63.15%) and 311 foreign respondents (36.85%). As for data collection instruments, the semi-structured interview questions and questionnaires were developed in order to ask the requirements of using social media marketing to promote eco-tourism. Moreover, semi-structured interview questions and questionnaires were sent to 5 experts including tourism academic staff, entrepreneurs, tour operators and experienced eco-tourists, to verify the content validity and the appropriateness of the language used in order to reduce ambiguity, leading questions, emotive questions, stressful questions. Questionnaire instrument consisted of four parts. The first part was designed to identify general information such as gender, age, hometown, marriage status, education background, occupation, income, status related to ecotourism, information related to eco-tourism and behavior of using information technology and social media of the respondents. The second part focused on the requirements of content and functions of using social media to promote ecotourism in terms of the level of satisfactory and the level of service requirements. All constructs were measured on five-point Likert-type scales, from 1 = the least to 5 = the most. The third part was factors affecting the use of social media for ecotourism. The last part was additional recommendations.

Phase 2 of this study was to analyze and design an online social media marketing system to promote eco-tourism as a responsive web application which supports any devices, including tablets and mobile phones. Users can access information anywhere and anytime. Data from the requirements of using online social media marketing system to promote eco-tourism obtained from the first part was analyzed and designed using the principles of object-oriented analysis and design. Two diagrams of unified modeling language (UML) were used for modeling the system in order to make it easy to understand: a use case diagram (for designing system from the perspective of external users), and (2) an activity diagram for describing the work within the system. This design will be used for developing the online social media marketing system in the future.

4 Research Results

4.1 Findings of the Requirements of Using Online Social Media

4.1.1 In-Depth Interview Data Analysis

The data analysis of interview with 60 key informants regarding the requirements of using online social media to promote eco-tourism can be summarized as follows:

(a) In terms of social media used, the top four widely used social media were Facebook, Line, YouTube, and Google+. The main purpose of using social media were (the top five answers): (1) to receive information and knowledge, (2) to communicate with friends and acquaintances, (3) to search for information, (4) to entertain, and (5) to do business.

(b) In terms of the structure or classification of information, the system should be categorized by (the top five answers): (1) types of activities/ceremonies, (2) month/season, (3) popular places, (4) region/province, and (5) major festivals.

(c) In terms of the contents, the system should include (the top five answers): (1) information of interesting eco-tourism places and their photos, (2) highlights of tourist attractions, (3) location information of tourist attractions, (4) travel itinerary, and (5) events' activities of eco-tourism places.

(d) In terms of the requirements for functional services, the system should provide the functional services of (the top five answers): (1) searching for the eco-tourism route, (2) searching for the eco-tourism tourist attractions, (3) providing shared files/videos, (4) recommending service for popular attractions, and (5) providing service of discussion and sharing files.

(e) In terms of the requirements of the service used, the system should be (the top five answers): (1) easy to use and no hassle, (2) able to use with the application effectively, (3) quickly responding to the use, (4) manually installed by his/himself, and (5) having a variety of functions in the application.

4.1.2 Survey Data Analysis

The survey with 533 Thai eco-tourists and 311 foreign eco-tourists (a total of 844 tourists) regarding the requirements of using online social media to promote eco-tourism can be summarized as follows:

(a) In terms of general information of eco-tourists who use online social media, a majority are women, single, aged around 20–29 years old, and educational background is bachelor degree. The information source affecting decision for eco-trips were their friends and family. In terms of the travel, a majority of Thai eco-tourists liked to travel with their family while most foreign eco-tourists liked to travel with their friends. The primary purpose of using online social media was to get the most information and knowledge. For mobile phone operating system, most Thai eco-tourists used Android while foreign eco-tourists use IOS (iPhone).

(b) In terms of the behavior of tourists using social media, the most popular social media that Thai eco-tourist used was Facebook, followed by LINE. The frequency of use of social media was every day and at least 1–2 h a day. However, most foreign eco-tourists used Facebook, but not many of them used LINE and they used social media less than one hour a day.

(c) In terms of information search, the study found that most Thai eco-tourists used social media to search for information but foreign tourists did not use social media that much for eco-tourism information search. In term of functions, most Thai eco-tourists were satisfied at the high/highest level but most foreign eco-tourists were satisfied in a moderate level.

(d) In terms of the requirements of the service used, the study found that most Thai eco-tourists would like to use the system that: (1) was user friendly and no hassle, (2) had ability to application effectively, (3) was very responsive, (4) can be installed manually, and (5) had a variety of functions in the application. However, most foreign eco-tourists preferred to use the system that: (1) was very responsive, (2) was user friendly and no hassle, (3) had the ability to use the program effectively, (4) had ability to exchange information and interact with others, and (5) can display in real time.

(e) In terms of the content requirements of using online social media, most Thai eco-tourists preferred to have contents related to: (1) information of interesting eco-tourism destinations, (2) highlights of tourist attractions, (3) location information of tourist attractions, (4) travel route/itinerary, and (5) activities/events and details of eco-tourism destinations. This result was similar to the result from the interview. As for most foreign eco-tourists, they would like to have the contents related to: (1) trip review, (2) activities/events and details of eco-tourism destinations, (3) photos of scenery and the places, (4) activities related to nature conservation/ecotourism, and (5) cost of the trip. Thus there are common content requirements that both Thai and foreign eco-tourists preferred. However, some content requirements were somewhat different. Thai eco-tourists would like to know the highlights of tourist attractions, location information of tourist attractions, and travel route/itinerary while foreign eco-tourists would like to have trip review, photos of scenery and the places, and cost of the trip (as shown in Table 1).

Table 1. A comparison of the content requirements of using online social media of Thai tourists and foreign tourists.

the content requirements of using online social media	Thai Tourists			Foreign Tourists		
	Percent of requirement level			Percent of requirement level		
	Less/Least	Moderate	More/Most	Less/Least	Moderate	More/Most
Information about definition of	5.7	22.1	72.2 (18)	15.6	28.7	55.8 (14)
information of interesting eco-tourism	1.9	11.9	86.2 (1)	7.4	25.6	67.0 (7)
activities related to nature conservation	4.3	19	76.7 (12)	6	23.7	70.3 (4)
History information of tourist attraction	2.9	23.8	73.3 (17)	7	25.8	67.2 (6)
Tourism Features	1.8	18.4	79.8 (7)	10.9	34.4	54.8 (15)
activities/events and details of eco-	2.8	16.7	80.6 (5)	5.1	23.8	71.2 (2)
Information about restaurant	4.3	21.4	74.2 (15)	21	30.7	48.3 (19)
Information about hotels/accomodations	3.1	16.8	80.1 (6)	13.7	22.7	63.6 (10)
Information on transportation service	6.3	19.8	73.9 (16)	11.7	26	62.4 (12)
travel route/ itinerary	2.8	11.9	85.3 (4)	12	27	61.0 (13)
location information of tourist attractions	1.6	12.8	85.6 (3)	10.2	24.2	65.5 (9)
Service Center	3.1	19.5	77.4 (11)	22.1	33.9	44.1 (21)
Information about OTOP distribution	7.7	30.5	61.7 (23)	31.1	35.5	33.4 (24)
Information on services infrastructure	3.9	26.2	69.9 (20)	25.3	39.2	35.5 (23)
Information about tourism facilities	2.2	19.7	78.2 (9)	17.9	31.2	50.8 (16)
Promotion Information / trip journey	5.5	22.8	71.8 (19)	21.4	30.3	48.3 (19)
cost of the trip	3.7	18.6	77.8 (10)	9.4	23.2	67.3 (5)
Rules / restrictions in ecotourism	4.6	19.6	75.8 (14)	17.6	33.7	48.7 (18)
security measure	3.4	17.1	79.7 (8)	22.3	32.3	45.3 (20)
trip review	4.3	19.2	76.4 (13)	8	19.3	72.7 (1)
photos of scenery and the places	1.6	12.3	86.2 (1)	7.4	21.6	71.0 (3)
Video explain tourist attractions	6.9	24.9	68.3 (21)	19.8	30.6	49.5 (17)
highlights of tourist attractions	3.3	11	85.7 (2)	8.7	25.2	66.2 (8)
Calendar / time to visit tourist attractions	5.7	16.9	77.4 (11)	13.1	23.6	63.2 (11)
Other	9	22.7	68.2 (22)	19	39.7	41.3 (22)
Overview opinion	4.1	19	76.9	14.5	28.5	57

(f) In terms of the requirements for functional services, the study found that most Thai eco-tourists would like to have a system to: (1) search and mange travel routes functions, (2) recommend new eco-tourism places, (3) search for eco-tourism places, (4) allow users to review the eco-tourism place by giving comments/suggestions or ranking, and (5) share files/image/video. As for foreign eco- tourists, they would like to have a system: (1) search for eco-tourism places, (2) recommend new eco-tourism places, (3) search and mange travel routes functions, (4) allow users to review the ecotourism place by giving comments/suggestions or ranking, and (5) have a wish list to explore the eco-tourism. Thus there are common function requirements that both Thai and foreign eco-tourists preferred. However, some functional requirements were somewhat different. Thai eco-tourists would like to have share files/image/video services while foreign eco-tourists would like to have a wish list to explore the eco-tourism.

(g) In terms of structure or the classification of information, the study found that most Thai eco-tourists agreed that the system should be categorized by: (1) types of activities, (2) region/province, (3) travel program trip, (4) major festivals/ceremonies, and (5) the promotion of tourism. As for most foreign eco-tourists, they agreed that the system should be categorized by: (1) region/province, (2) types of activities, (3) popular tourist places, (4) month/season, and (5) information being searched for the most. The common information classification requirements that both Thai and foreign tourists preferred were types of activities, region/province, and month/season. However, some information classification requirements were somewhat different. Thai eco-tourists would like to categorize information by travel program trip and major festivals/ceremonies while foreign eco-tourists would like to categorize information by popular tourist places, and information being searched for the most.

(h) In terms of factors influencing the use of online social media for ecotourism promotion, the study found that most Thai eco-tourists agreed that key factors were: (1) ease of use, (2) high speed of communication, (3) in-trends/up-to-date content, (4) fondness/favorite, and (5) convenience. As for most foreign eco-tourists, they agreed that the key factors to use social media for ecotourism promotion were: (1) ease of use, (2) friend's influence, (3) information services, (4) individual characteristics, and (5) high speed of communication. The common factors that both Thai and foreign tourists preferred were ease of use and high speed of communication. However, some important factors were somewhat different. Most Thai eco-tourists thought that important factors were in-trends/up-to-date content, fondness/favorite, and convenience while foreign eco-tourists thought that important factors were friend's influence, information services, and individual characteristics.

4.1.3 Confirmation of Findings from In-Depth Interview and Survey

The results from interview (qualitative method) and survey (quantitative method) showed that most of the findings were consistent. Some of the priorities were somewhat different. The contents that were consistent were highlights of tourist attractions, travel route/itinerary, photos of scenery and the places, and cost of the trip. For the functions, the ones that were consistent included search functions for eco-tourism places, search and mange travel routes functions, and share files/image/video. For information classification, the ones that were consistent included types of activities, region/province, month/season, popular tourist places and information being searched for the most.

4.2 Analysis and Design an Online Social Media Marketing System to Promote Eco-Tourism

Data from the requirements of using online social media marketing system to promote eco-tourism obtained from the first part was analyzed and designed using the principles of object-oriented analysis and design. Two diagrams of unified modeling language (UML) were used for modeling the system in order to make it easy to understand. These two diagrams include: (1) a use case diagram for designing system from the perspective of external users as shown in Fig. 1, and (2) an activity diagram for

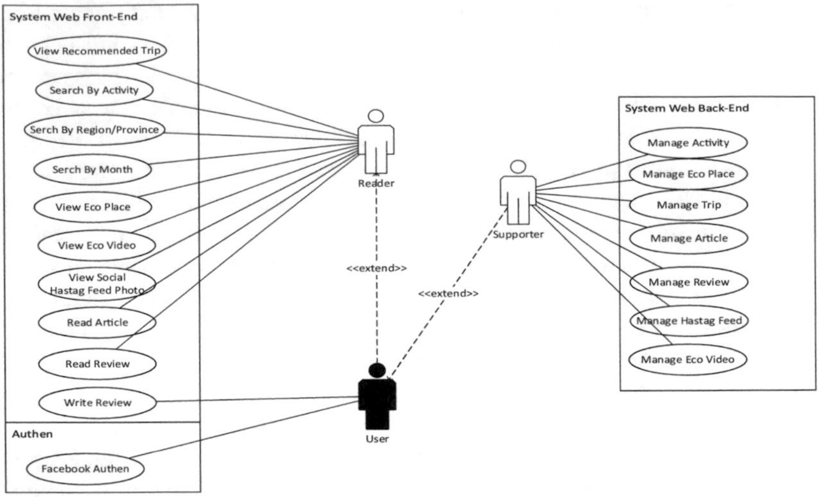

Fig. 1. Use case diagram of online social media marketing system to promote eco-tourism

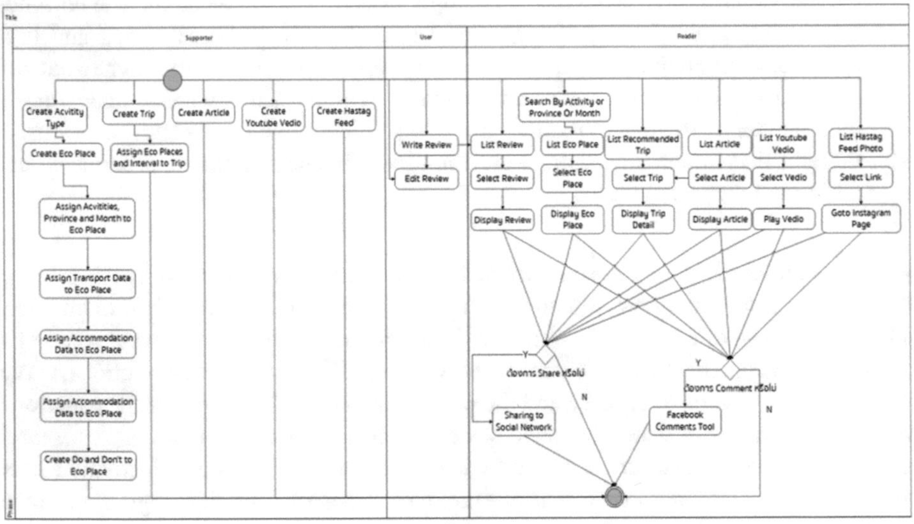

Fig. 2. Activity diagram of online social media marketing system to promote eco-tourism

describing the work within the system as shown in Fig. 2. This design is used for developing the online social media marketing system in the future.

5 Conclusion

The tourism industry is very important to Thai economy and one tourism form that is popular today is eco-tourism. In order to promote eco-tourism, there is a need to use popular social media to attract more eco-tourists to visit Thailand. The purposes of study aims are to determine the requirements of using social media marketing for eco-tourism, and to design an online social media marketing system to promote eco-tourism in Thailand in order to develop the system in the future.

Using a combination of interview and survey to collect the requirements of using social media for eco-tourism promotion, the results show that the contents that eco-tourists need are the featured attractions, travel itinerary, pictures, sightseeing atmosphere, and costs. As for the functionality needed, it included a search facility for ecotourism destinations, a search to identify travel routes, and a capability of a system to share, photos/video. Based on the requirements, an online social media marketing system is designed as a responsive web application which supports any devices, including tablets and mobile phones. Users can access information anywhere and anytime. The system designed was used the principles of object-oriented analysis and design. Two diagrams of unified modeling language (UML) were used for modeling the system in order to make it easy to understand. Two diagrams include a use case diagram (for designing system from the perspective of external users) and an activity diagram for describing the work within the system.

For practical implications, a system design of social media marketing system will be used for developing the full system later. Then, this system can be adopted and used by Thai Eco-tourism and Adventure Travel Association (TEATA) and related organizations in order to promote eco-tourism in Thailand. In addition, a model using social media for tourism promotion and a prototype of social media marketing system developed can also be applied to other tourism promotion.

Acknowledgement. The authors gratefully acknowledge financial support from the Thailand Research Fund (TRF) for this study under Research Project of Guidelines for Eco-Tourism Standards Development and Promotion in Thailand.

References

1. National Research Council. National Tourism Strategic Research B.E. 2555–2559 (2013)
2. TEATA. Mission and Vision of TEATA. http://www.teata.or.th/about.php
3. William, S.: The growth of the global ecotourism industry. http://ecowanderlust.com/ecotourism-2/growth-global-ecotourism-industry/1487
4. Popesku, J.: Social media as a tool of destination marketing organizations. Singidunum J. Appl. Sci. **2**, 715–721 (2014)
5. Hays, S., Page, S.J., Buhalis, D.: Social media as a destination marketing tool: its use by national tourism organisations. Current Issues Tour. **16**(3), 211–239 (2013)
6. Zeng, B., Gerritsen, R.: What do we know about social media in tourism? A review. Tour. Manag. Perspect. **10**, 27–36 (2014)

The Empirical Discovery of Near-Optimal Offline Cache Replacement Policies for Nonuniform Objects

Thepparit Banditwattanawong(✉)

Computer Science Department, Faculty of Science,
Kasetsart University, Bangkok, Thailand
thepparit.b@ku.th

Abstract. Offline cache replacement policies are a solid foundation for developing online schemes. This paper presents the determination of near-optimal offline policies from suboptimal ones through the trace-driven simulation of six existing offline policies, C_0, C_0^*, $Mattson's\ OPT$, *LFD-A*, $SMFD$, and $SMFD^*$, in conjunction with three well-known online ones, *LRU*, $GDSF$, and $LFUDA$. The results indicated that $SMFD$ and $SMFD^*$ were near-optimal over prior policies across byte-hit and delay-saving as well as hit performances in overall. A compelling finding is the achievement of an optimal policy's performance upper bounds by the near-optimal policies.

Keywords: Offline cache · Replacement policy · Nonuniform object

1 Introduction

In efficient information sharing, a network optimization technique, web cache, has been being employed worldwide. Whenever the web cache lacks space for a newly requested object found missing from the cache, a cache replacement policy is enlisted to either make a room or discard the requested object. These mutually exclusive decisions are referred to as mandatory eviction and optional eviction, respectively.

Offline cache replacement policies are the theoretical foundation of online ones. Unlike the online policies, the offline ones require the provision of the complete sequence of object requests in advance to be searched for an efficient caching state sequence. Our previous work [1] has theoretically established a pair of novel offline cache replacement policies for nonuniform object costs named $SMFD$ and $SMFD^*$ that were proved to belong to the suboptimal class of replacement policies just as the other offline ones, C_0 [2], C_0^* [2], $Mattson's\ OPT$ [3], and *LFD-A* [4]. The same paper [1] also provided analysis results in which they all belonged to the same class of quadratic time complexity, $O(mn\log_2 n + m^2n)$. A known optimal offline cache replacement policy, which is referred to throughout this paper, is $Hosseini - Khayat's\ OPT$ [5], which has $O(n^{3m})$ [1].

H. Unger et al. (Eds.): IC2IT 2018, AISC 769, pp. 286–296, 2019.
https://doi.org/10.1007/978-3-319-93692-5_28

Even though [1] has identified a set of suboptimal policies along with their time complexities, we have had no knowledge yet about which of the suboptimal policies perform best among the others or even near optimal. Intuitively, if we successfully uncover near-optimal offline policies among the suboptimal ones, we will have the development framework of near-optimal online cache replacement policies for real deployment in nonuniform object environments such as WWW, big data, and cloud computing. Therefore, this paper presents the study of how to gain such knowledge by means of the trace-driven simulation of not only the offline policies but also well-known online ones supported by Squid caching proxy [6], LRU, $GDSF$, and $LFUDA$ as performance benchmarks to verify the authenticity of the near-optimal attribute.

The following subsections concisely revisit the notions of the suboptimal offline policies and performance metrics that are the major components of this study.

1.1 Suboptimal Offline Policies

The algorithmic operations of the suboptimal offline policies are formally defined below, which is excerpted from [1] where the executing demonstration of these policies has been provided therein.

Definition 1: Let $\mathcal{U}$ be an object universe, $O = \{o_j \in \mathcal{U} | o_j is\, a\, unique\; cacheable\, object$ for $j = 1, 2, ..., N \wedge N\ is\ the\ total\ number\ of\ o_j\}$, p_j be the probability of o_j to be requested, c_j be the cost of fetching o_j from an original server to a cache, and $V_t \subset O$ be the set of objects in the cache at time t. Suppose that cache replacement takes place at t, C_0 evicts o_j iff $(j = argmin(p_j c_j)) \wedge (o_j \in V_t)$.

Definition 2: Let o_t be a missing object requested at time t. Suppose that cache replacement takes place at t, C_0^* evicts o_j iff $(j = argmin(p_j c_j)) \wedge (o_j \in V_t \cup \{o_t\})$.

Definition 3: $Mattson's\ OPT$ evicts $o_j \in V_t$ if o_j *has no more request* else if o_j *has next request furthest in time.*

Definition 4: $LFD\text{-}A$ evicts $o_j \in V_t \cup \{o_t\}$ if o_j *has no more request* else if o_j *has next request furthest in time.*

Definition 5: $SMFD$ evicts $o_j \in V_t$ if o_j *has no more request* else if o_j *has the final request the soonest.*

Definition 6: $SMFD^*$ evicts $o_j \in V_t \cup \{o_t\}$ if o_j *has no more request* else if o_j *has the final request the soonest.*

1.2 Performance Metrics

We formerly justified in [1] that cost-saving ratio (CSR) is more suitable metric than hit rate in the realm of nonuniform object costs.

Definition 7: CSR is defined as follows where n is the number of objects in a cache, c_i is the cost of fetching object i from an original server to the cache, h_i is how many times the valid copy of i is fetched from the cache, and r_i is the total number of requests to i.

$$CSR = \frac{\sum_{i=1}^{n} c_i h_i}{\sum_{i=1}^{n} c_i r_i} \tag{1}$$

Since c_i is generic, CSR allows two concrete variants, byte-hit ratio (BHR) and delay-saving ratio (DSR), both of which are conventional performance metrics for web caching. They simply specialize c_i as follows.

Definition 8: Let s_i be the size of i, BHR is defined as follows.

$$BHR = \frac{\sum_{i=1}^{n} s_i h_i}{\sum_{i=1}^{n} s_i r_i} \tag{2}$$

BHR aims to reflect the degree of WAN's downstream bandwidth reduction.

Definition 9: Let l_i is the downloading latency of i from a remote server, DSR is defined as follows.

$$DSR = \frac{\sum_{i=1}^{n} l_i h_i}{\sum_{i=1}^{n} l_i r_i} \tag{3}$$

DSR aims to reflect the degree of downstream delay reduction.

The other employed metrics are object removal rate (ORR) [7] and hit rate (HR). ORR aims to reflect the total number of cache storage's I/O operations and total CPU time for cache replacement policy processing. HR is a classical metric suitable for uniform caching (utilized in this paper as benchmarks in the uniform cost world).

Definition 10: Let e_i be the total number of evictions of object i from a cache, ORR is defined as follows.

$$ORR = \frac{\sum_{i=1}^{n} e_i}{\sum_{i=1}^{n} r_i} \tag{4}$$

Unlike the other metrics, the lower ORR, the better scalability on a caching proxy.

2 Experiment on Near-Optimality

2.1 Data Sets

We utilized a pair of real HTTP traces (i.e., cache logs) [8] that differed in the time periods of collections, collecting durations, end-user groups, and the population of each group. Requests to dynamic objects were not precluded from the traces to reflect real-world performances and because some of the dynamic objects may be cacheable [9], thus affecting the performances. The traces were preprocessed to provide necessary data for all simulated policies by retaining only necessary fields (*request timestamp*, URL, *object size*, and *downloading latency*) and adding newly generated ones (*invalid-request timestamp*, *future request frequency*, and *next-request timestamp*) associated with each object request record. The URL represented an object's identity. The preprocessing engaged the heuristics of conjectures 1 to 4 that extended those in [10]. The conjectures 1 to 3 are used to figure out the *invalid-request timestamp*. The last conjecture calculates the *next-request timestamp* for every final hit. The qualitative effects of these heuristics have been verified further in the following $SMFD$'s and $SMFD^*$'s performance measurement in Sect. 2.3.

Conjecture 1: An object's *invalid-request timestamp* is set to the timestamp of the first request to the same object's URL whose *object size* was found changed.

Conjecture 2: As long as the size of an object remained unchanged throughout its entirely succeeding requests, its *invalid-request timestamp* was set to its succeeding *request timestamp* furthest toward to the end of a trace plus small time elapse (0.0001 s) so that the furthest request itself would not be regarded invalid.

Conjecture 3: An object apparent only once throughout the rest of a trace (e.g., dynamic one) has its *invalid-request timestamp* set to its own *request timestamp* plus the small time elapse so that the only request was not made on the invalid object.

Conjecture 4: An object appearing more than once in a trace with no longer request simply had a *next request timestamp* set to zero.

The resulting preprocessed traces have possessed various characteristics showed in Table 1.

2.2 Simulation Configuration

To identify near-optimal policies among the suboptimal ones by performance measurement, optimum benchmarks must be defined first: Since a limitless cache size always produces performance upper bounds for the optimal replacement policy, the simulated policies that deliver performances closest or even identical

Table 1. Characteristics of simulation traces.

Attribute	7-day trace	15-day trace
Total requests	849 556	1 613 616
Total requested bytes	35 258 412 511	63 244 272 792
Total unique objects	488 030	984 574
Maximum bytes of total unique objects	31 039 605 385	39 888 711 208

to the bounds were considered near-optimal of a kind. We defined the limitless cache size for each trace in a finite manner with the exact value of the maximum bytes of total unique objects denoted in Table 1.

To observe the policies' behaviors given varied available cache spaces, each pair of policy and trace has been simulated against four deviated cache sizes, 30% down to 5%. Each of them is proportional to the maximum bytes of the total unique objects of each trace. A 100% cache size is effectively equivalent to the limitless one, which completely avoids cache replacement.

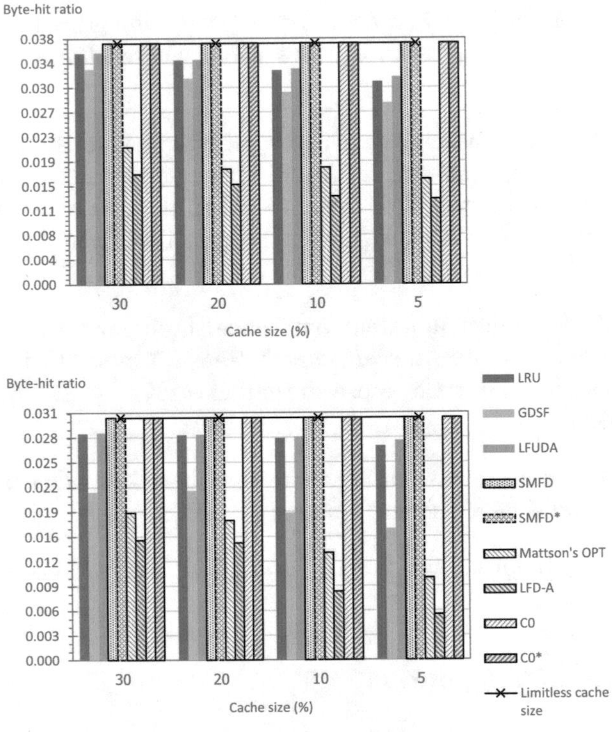

Fig. 1. *BHR*s based on the 15-day trace (top) and the 7-day trace (bottom)

2.3 Performance Results

Simulation performance results have been analyzed and discussed metric by metric based on both distinct traces through Figs. 1, 2, 3 and 4. The numerical performance data is provided in Appendix A.

Figure 1 illustrates comparative BHR results with the following findings of our focus:

- $SMFD$, $SMFD^*$, C_0 and C_0^* produced exactly the same BHRs as those of the limitless cache size (0.0373 and 0.0304 for the 15-day and 7-day traces, respectively).
- $SMFD$, $SMFD^*$, C_0 and C_0^* were so tolerate to the increasing cache size constraints that they maintained the stable BHRs across all simulated cache sizes based on both of the traces.
- All of the online policies have even outperformed some offline ones (*Mattson's OPT*, and *LFD-A*) at all cache sizes for both traces especially *Mattson's OPT*, which is optimum in the world of uniform object costs.

The first two findings of supreme capabilities are our first radical finding. Since the optimal policy requires the a priori knowledge of an entire request

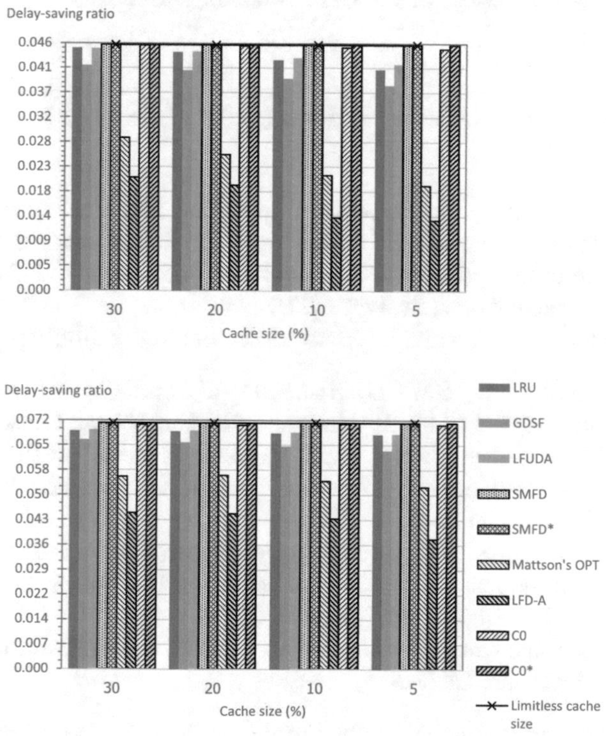

Fig. 2. DSRs based on the 15-day trace (top) and the 7-day trace (bottom)

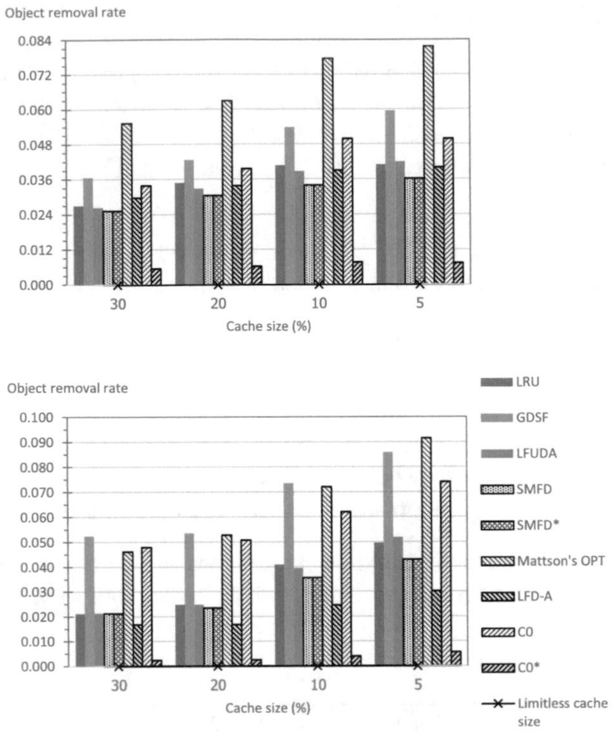

Fig. 3. *ORR*s based on the 15-day trace (top) and the 7-day trace (bottom)

stream to produce an optimum performance, the last finding implied that a historical request stream was able to estimate the future request stream and in turn enabled the online policies to approach the optimal one. This phenomenon is also known as certainty equivalence [11].

Figure 2 shows *DSR* results in comparison leading to findings as follows:

- With the 7-day trace, *SMFD*, *SMFD**, C_0 and C_0^* delivered the limitless cache size's *DSR* (0.0712) whereas with the 15-day trace only *SMFD* and *SMFD** produced exactly *DSR* (0.0458) identical to that of the limitless cache size. Both performance values were also constant for both traces regardless of the cache sizes.
- The 7-day trace showed an interestingly opposite observation in which the *DSR* increases against decreasing cache sizes: *Mattson's OPT* using 30% and 20%, and C_0 based on 20% down to 5%.
- Independent of the cache sizes and traces, all of the online policies defeated *Mattson's OPT* and *LFD-A*.

This first finding is accounted the second radical finding. Similar to that of *BHR*, the property that past statistic approximated the future requests also held in a *DSR* context based on the last finding.

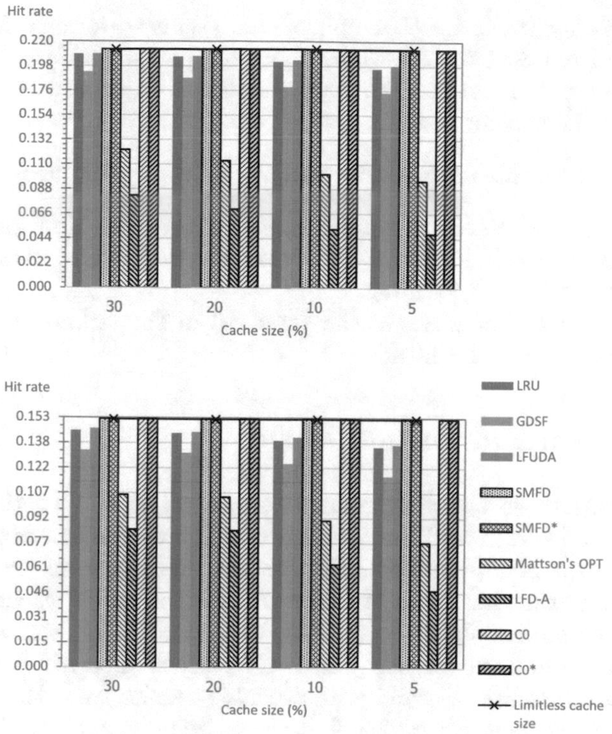

Fig. 4. *HR*s based on the 15-day trace (top) and the 7-day trace (bottom)

Since in the context of nonuniform object cost, CSR is a key performance indicator and has been substantiated more specifically by the BHR and DSR radical findings. $SMFD$ and $SMFD^*$ were obviously the near-optimal policy of our ultimate discovery.

Ones might question how significant $SMFD$ and $SMFD^*$ outperformed the best of the online policies (i.e., $LFUDA$ in terms of CSR). We translate the performance gaps between both policies based on a representative scenario from [10] where an organization consumes data from an off-premise private cloud through 10 Gbps Metro Ethernet with 50% average downstream bandwidth utilization for 8 work hours a day and 260 workdays/year requires the total amount of cloud data-out transfer 4570.313 TB/year. As the results, that $SMFD$ or $SMFD^*$ surpassed $LFUDA$ by 0.521% BHR and the 0.362% DSR based on the 5% cache size of the 15-day trace is equivalent to saving 23.811 TB/year (or 2023.77 USD/year based on the AWS's average data-out charge rate of 0.083 USD/GB) and 7.53 h/year, respectively.

Figure 3 portrays ORR results, which include the following findings.

- Based on both of the traces, both $SMFD$ and $SMFD^*$ produced the same ORR for each cache size while only C_0^* attained the minimum ORR values that were inversely proportional to the cache sizes.

- As of the 7-day trace, $LFD\text{-}A$ unexpectedly caused lower ORR at smaller cache sizes from 30% to 20%. This could happen if the larger cache before the policy executed had been filled up with more objects to be evicted by the policy than the smaller cache.

Figure 4 demonstrates HR results and the following findings.

- Only $SMFD$ and $SMFD^*$ consistently delivered the limitless cache size's HRs, 0.213 or 21.31% for the 15-day trace and 0.152 or 15.21% for the 7-day trace, as the last radical finding.
- Regardless of the traces and cache sizes, all of the online policies defeated $Mattson's\ OPT$ and $LFD\text{-}A$.

3 Conclusion and Future Work

This paper presents work in progress in all empirical aspects continued from our theoretical one [1] and toward the seamless development of a near-optimal online cache replacement policy. The contributions of this paper are twofold. Firstly, major novelty is the determination of near-optimal offline cache replacement policies for nonuniform object costs: $SMFD$ and $SMFD^*$ was revealed to be a near-optimal policy consistently across all performance metrics, cache sizes, and simulation data sets. Secondly, remarkable results were that $SMFD$ and $SMFD^*$ produced the three radical findings as being capable of duplicating the stable efficiency of the limitless cache size in all performance metrics. C_0 and C_0^* also manifested the near-optimum performances in limited cases.

We plan to conduct more comprehensive simulation to distinguish between $SMFD$ and $SMFD^*$ and to verify whether the radical findings are true-positive groundbreaking that lead to a breakthrough in the field. Yet, either $SMFD$ or $SMFD^*$ will be exploited to serve as our algorithmic framework for the systematic investigation of near-global optimum online solutions rather than conducting a series of trial-and-error experiments that may lead to local optimum results.

A Appendix

Figures 5 and 6 provide the numerical performance data of all of the charts in Figs. 1, 2, 3 and 4. Notice that the highlighted numbers denote equality to the limitless cache sizes' performances.

BHR

Cache replacement	Cache size (%)			
	30	20	10	5
Limitless cache size	0.037259390297			
LRU	0.035721117490	0.034661362638	0.033052583969	0.031333061644
GDSF	0.033251532371	0.031818341016	0.029725808125	0.028060068583
LFUDA	0.035764789824	0.034672678002	0.033258160022	0.032046694673
SMFD	0.037259390297	0.037259390297	0.037259390297	0.037259390297
SMFD*	0.037259390297	0.037259390297	0.037259390297	0.037259390297
Mattson's OPT	0.021215578767	0.017854105426	0.018067458215	0.016245622183
LFD-A	0.017016749035	0.015457688275	0.013591632349	0.013175167683
C_0	0.037259390297	0.037259390297	0.037259390297	0.037259390297
C_0^*	0.037259390297	0.037259390297	0.037259390297	0.037259390297

DSR

Cache replacement	Cache size (%)			
	30	20	10	5
Limitless cache size	0.045755513286			
LRU	0.045069220733	0.044355687534	0.042934422098	0.041144794053
GDSF	0.041899259504	0.040947219555	0.039450573081	0.038219124918
LFUDA	0.045006668520	0.044464820797	0.043291565648	0.042131395683
SMFD	0.045755513286	0.045755513286	0.045755513286	0.045755513286
SMFD*	0.045755513286	0.045755513286	0.045755513286	0.045755513286
Mattson's OPT	0.028412838880	0.025338845634	0.021448882688	0.019551353192
LFD-A	0.021074499350	0.019627314104	0.013605379823	0.013116266856
C_0	0.045591783159	0.045500583568	0.045223089002	0.044966965401
C_0^*	0.045755512840	0.045755512840	0.045749413798	0.045749413798

ORR

Cache replacement	Cache size (%)			
	30	20	10	5
Limitless cache size	0.000000000000			
LRU	0.026983495454	0.034738748252	0.040598878544	0.040937249011
GDSF	0.036649363913	0.042655129845	0.053781073068	0.059402608799
LFUDA	0.026360670692	0.032749427373	0.038664713290	0.041868697385
SMFD	0.025248882014	0.030384552459	0.033780651654	0.036156061913
SMFD*	0.025248882014	0.030384552459	0.033780651654	0.036156061913
Mattson's OPT	0.055267176329	0.062885469653	0.077483118660	0.081656354424
LFD-A	0.029671247682	0.033701326710	0.038985731426	0.040056618179
C_0	0.033866173860	0.039618471805	0.049822882272	0.049726824722
C_0^*	0.005483956530	0.006165655274	0.007393332738	0.007838915826

HR

Cache replacement	Cache size (%)			
	30	20	10	5
Limitless cache size	0.213085393303			
LRU	0.208664886813	0.206154995984	0.201643389753	0.195671089032
GDSF	0.192697643058	0.187188897482	0.179267558081	0.174237241078
LFUDA	0.208778916421	0.206812525409	0.202989434909	0.198324756324
SMFD	0.213085393303	0.213085393303	0.213085393303	0.213085393303
SMFD*	0.213085393303	0.213085393303	0.213085393303	0.213085393303
Mattson's OPT	0.122806169498	0.113297091749	0.101209953297	0.095183116677
LFD-A	0.082149036698	0.070108377706	0.052190236091	0.048031873754
C_0	0.213068660697	0.213063083162	0.213050068914	0.213031477130
C_0^*	0.213084773577	0.213084773577	0.213084153851	0.213084153851

Fig. 5. 15-day trace-based performance.

BHR

Cache replacement	Cache size (%)			
	30	20	10	5
Limitless cache size	0.030368220426			
LRU	0.028392471717	0.028213991106	0.027817511032	0.026854533644
GDSF	0.021095189432	0.021270092145	0.018440554685	0.016465699663
LFUDA	0.028495701435	0.028263562935	0.027977664896	0.027491522788
SMFD	0.030368220426	0.030368220426	0.030368220426	0.030368220426
SMFD*	0.030368220426	0.030368220426	0.030368220426	0.030368220426
Mattson's OPT	0.018489249305	0.017523299973	0.013484683573	0.010353826988
LFD-A	0.015074238122	0.014729680550	0.008598868026	0.005715765704
C_0	0.030368220426	0.030368220426	0.030368220426	0.030368220426
C_0^*	0.030368220426	0.030368220426	0.030368220426	0.030368220426

DSR

Cache replacement	Cache size (%)			
	30	20	10	5
Limitless cache size	0.071229884867			
LRU	0.068919507717	0.068724209118	0.068229719780	0.067790209756
GDSF	0.066389286822	0.065531413349	0.064450596060	0.063180035055
LFUDA	0.069210666484	0.068992345673	0.068550688989	0.068023502987
SMFD	0.071229884867	0.071229884867	0.071229884867	0.071229884867
SMFD*	0.071229884867	0.071229884867	0.071229884867	0.071229884867
Mattson's OPT	0.055785478845	0.056145838646	0.054482923611	0.052762484605
LFD-A	0.045191121076	0.045076334504	0.043674868779	0.037716861457
C_0	0.070810924422	0.070614985018	0.070970565881	0.070697072698
C_0^*	0.071229884867	0.071229884867	0.071229884867	0.071229884867

ORR

Cache replacement	Cache size (%)			
	30	20	10	5
Limitless cache size	0.000000000000			
LRU	0.021095725297	0.024788242329	0.040924906657	0.049684776519
GDSF	0.052423854343	0.053449095763	0.073491329589	0.085904872663
LFUDA	0.021422955049	0.024723502630	0.039508872870	0.051824717853
SMFD	0.021298183993	0.023392218994	0.035671574328	0.042880045577
SMFD*	0.021298183993	0.023392218994	0.035671574328	0.042880045577
Mattson's OPT	0.046179416071	0.052735781985	0.071975243539	0.091533695248
LFD-A	0.016748748758	0.016713436195	0.024450418807	0.030116908126
C_0	0.047901492074	0.050735913819	0.062012392356	0.073979820047
C_0^*	0.002380066764	0.002461285660	0.004055059349	0.005465207708

HR

Cache replacement	Cache size (%)			
	30	20	10	5
Limitless cache size	0.152108866278			
LRU	0.145091082872	0.143580882249	0.139165634755	0.135069377416
GDSF	0.133020071661	0.131445131339	0.125053557388	0.117280085127
LFUDA	0.146442376959	0.144313029394	0.141149023725	0.136427734016
SMFD	0.152108866278	0.152108866278	0.152108866278	0.152108866278
SMFD*	0.152108866278	0.152108866278	0.152108866278	0.152108866278
Mattson's OPT	0.105954169001	0.104478103857	0.090280099252	0.076857793954
LFD-A	0.084765453955	0.084020358870	0.063471978304	0.047389459906
C_0	0.152098272509	0.152094741253	0.152098272509	0.152093564168
C_0^*	0.152108866278	0.152108866278	0.152108866278	0.152108866278

Fig. 6. 7-day trace-based performance.

References

1. Banditwattanawong, T.: On optimal replacement of nonuniform cache objects. Int. J. Future Comput. Commun. (2018, in press)
2. Bahat, O., Makowski, A.M.: Optimal replacement policies for nonuniform cache objects with optional eviction. In: 22nd Annual Joint Conference of the IEEE Computer and Communications Societies, pp. 427–437 (2003). https://doi.org/10.1109/INFCOM.2003.1208694
3. Mattson, R.L., Gecsei, J., Slutz, D.R., Traiger, I.L.: Evaluation techniques for storage hierarchies. IBM Syst. J. **9**(2), 78–117 (1970). https://doi.org/10.1147/sj.92.0078
4. Seda, Č: Replacement problem in web caching. Master thesis, Bilkent University (2002)
5. Hosseini-Khayat, S.: On optimal replacement of nonuniform cache objects. IEEE Trans. Comput. **49**(8), 769–778 (2000). https://doi.org/10.1109/12.868024
6. Squid Configuration Directive Cache Replacement Policy. http://www.squid-cache.org/Doc/config/cache_replacement_policy/

7. Romano, S., ElAarag, H.: A quantitative study of web cache replacement strategies using simulation. In: Web Proxy Cache Replacement Strategies. SpringerBriefs in Computer Science, pp. 17–60. Springer, London (2013). https://doi.org/10.1007/978-1-4471-4893-7_4
8. Weekly Squid HTTP Access Logs. http://www.ircache.net/
9. Chemolli, F.: Caching dynamic content using adaptation (2018). https://wiki.squid-cache.org/ConfigExamples/DynamicContent/Coordinator
10. Banditwattanawong, T., Masdisornchote, M., Uthayopas, P.: Multi-provider cloud computing network infrastructure optimization. Future Gener. Comput. Syst. **55**, 116–128 (2016). http://www.sciencedirect.com/science/article/pii/S0167739X15002800
11. Water, V.H., Willems, J.: The certainty equivalence property in stochastic control theory. IEEE Trans. Autom. Control **26**(5), 1080–1087 (1981). https://doi.org/10.1109/TAC.1981.1102781

Organic Farming Policy Effects in Northern of Thailand: Spatial Lag Analysis

Ke Nunthasen and Waraporn Nunthasen(✉)

Faculty of Economics, Maejo University, Chiang Mai, Thailand
{ke_nunt, garn007}@hotmail.com

Abstract. The spatial exploratory and spatial lag are applied to analyze the spatial analysis of policy effects on organic farming in northern of Thailand. The empirical result explores that organic farming on one positive district affects to organic farming in neighbouring districts. Moreover, the spatial analysis indicates that relationship between farmers becomes an important factor that affect to grow organic farming. Therefore, to increase organic farming, the policy-maker should lay emphasis on implementing the proper organic farming policy in each area and creating the connection between farmers and organic communities.

Keywords: Organic farming policy · Spatial lag regression
Spatial exploratory

1 Introduction

Thailand is an agricultural country. The sustainable agriculture including organic farming is clearly mentioned firstly in the national framework in the 8th National Economic and Social Development Plan (1997–2001). After that, the sufficiency economy philosophy (SEP), Thai development approach attributed to the King Rama IV, has included in the agricultural management since the 9th National Economic and Social Development Plan (2002–2006). Moreover, agricultural knowledge management of local farmers is accepted as an important agricultural learning process as well as many successful farming enterprises which applying the local innovations have been used as the demonstration or learning centers to stimulate new organic farming.

However, Thailand organic farming is still at an early development stage although it has been promoted for a long period of time. The record found that there were only 0.11% of organic farming areas which missed the target of 20% at the end of the 9th plan and there had only 0.17% at the end of 2010. This might imply that government failed to translate the plan and to apply some concrete measures in organic farming policy. Similarly, to Koesling [1] mentioned that Norway's goal of organic areas could not be achieved because there was only 4% of conventional farmers converted to organic farming while the nation goal was 10%. Moreover, government of Thailand goals to achieve 600,000 Rai of organic farming in 2021 [2]. However, the record showed that total areas of organic farming equal 187,981 Rai and 273,881 Rai in 2009

H. Unger et al. (Eds.): IC2IT 2018, AISC 769, pp. 297–305, 2019.
https://doi.org/10.1007/978-3-319-93692-5_29

and 2015, respectively. This goal becomes the big challenge of policymaker to translate the plan for any concrete activities and measures for achieving the target.

Many researches presented the obstacles to achieve the organic farming goal. The main obstacle of organic farming development in the developing country were the infrastructure problems such as the appropriate market, convenient place to support organic farmers to produce organic crops [3]. Stolze and Lampkin [4] also suggested that organic farming development policy should focus on three main instruments; (1) legal or regulation instruments (organic control and standard management), (2) financial instruments (payment to support conversion to organic production), and (3) communicative instruments (information, communication, research, training and advice). Moreover, Thapa and Rattanasuteerakul [5] mentioned that the government organization, non-government organization, community members and farmers' groups, training programs attendance, price satisfaction, and pest hazard intensity were able to motivate farmers to produce organic vegetable farming. However, policy implementation should be suitable in each area [6].

This study, therefore, aims to examine the effect of government policy on organic farming in northern of Thailand for implementing the appropriate policy.

2 Methodology

This research uses organic farming and household data from the agricultural census of northern (district level) and geographical data from database of Global Administrative Areas (GADM) [7]. The agricultural census data consisted of 4 categories; (1) socio-economic factors, (2) government training programs, (3) agricultural infrastructure and (4) marketing options.

According to, nature of spatial relationship between observations in each area causing the classic linear regression problem from non-constant error variances (spatial heteroscedasticity) and non-zero error covariance (spatial autocorrelation) so the classic linear regression is not Best Linear Unbiased Estimators (BLUE) and need non-conventional model to correct these problems. This study first explores the spatial effect using Moran's I statistic for the whole data set and the local indicators of spatial association "LISA" for investigating the spatial effect on observations.

2.1 Exploratory Spatial Autocorrelation

The spatial autocorrelation and spatial heteroscedasticity in spatial data violates the BLUE properties in the classic linear regression. Therefore, this study uses Moran's I statistic for the whole data set (global spatial autocorrelation) to explore the spatial autocorrelation. A set of spatial data consists of measurements (data values) taken at specific locations (data sites) in geographic space (study region). Many measures of spatial association have been developed to examine the nature and extent of spatial dependence on data. Global measures of spatial association are applied to derive a single value of entire study region by using the complete data set. These measures emphasize the average or typical characteristics of the complete data set. An approach is the explicit assumption that the single value holds throughout the study region. The measurement of global spatial autocorrelation is based on Moran's I statistic, which gives a formal indication on the degree of linear association between vector of observed values and the vector of spatial weighted averages of neighboring values, called spatial lagged vector.

The local statistics for each observation give an indication of the extent of significant spatial clustering of similar values around that observation [8]. Local indicator of spatial association (LISA) is the statistic that satisfies the following two requirements. Firstly, the LISA for each observation gives an indicator of the extent of significant spatial clustering of similar values around that observation and secondly, the sum of LISAs for all observations is proportional to a global indicator of spatial association. The exploratory of the spatial autocorrelation between observation (local spatial auto-correlation) using Moran's I scatter plot and local indicators of spatial association "LISA". Inspection of local spatial instability is carried out by the means of the Moran scatter plot, which plot the spatial lag against the original values. The four different quadrants of the scatter plot present four types of local spatial association between a district and its neighbours. Quadrant I represents a district with high value surrounded by district with high values (High-High), Quadrant II represents a district with low value surrounded by district with high values (Low-High), Quadrant III represents a district with low value surrounded with low values (Low-Low) and Quadrant IV represents a district with a high value surrounded by district with low values (High-Low). Consequently, Quadrant I and III express a positive spatial autocorrelation and quadrant II and IV represent a negative spatial autocorrelation of dissimilar values. Bright colours highlight spatial clusters in LISA. Hot spots (high-high regions) are coloured by bright red. A positive spatial association arises from their own and neighbouring high values of the attribute variable. Cold spots (low-low regions) are coloured by bright blue. A Positive spatial autocorrelation emerges from their own and neighbouring low values of the variable. In term of LISA cluster map, the colour code presents the significant locations of spatial autocorrelation. There are six colour codes in the legend; dark red for high-high, dark blue for low-low, pink for high-low, light blue for low-high, white for not significant, and grey for neighbouring less.

2.2 Model Specification Testing

Six tests for spatial dependence model are applied to identify the appropriate model because its ability to test spatial lag and spatial error specifications [7]. The appropriate model is examined from five statistics tests for spatial dependence in linear regression models. Simple LM test for a missing spatially lagged dependent variable (Lagrange Multiplier (lag)), the simple LM test for error dependence (Lagrange Multiplier (error)), variants of these robust to present of the other (Robust LM (lag)) and Robust (LM (error)) which tests for error dependence in the possible presence of missing lagged dependent variable, and LM (SARMA). The possible model is considered from the significant of the lag and error tests. We also use the robust tests of the lag and error to specify which spatial model is suitable.

2.3 Spatial Lag Regression

The spatial lag regression is used to examine spatial effects of the model. The spatial lag specification is added a new variable on the right-hand side of the equation, a spatially lagged dependent variable (Wy), to capture the spatial interaction effect as a weighted average of neighbouring observations [9]. Organic farming households in this study are assumed as affected by the variables from four categories: socio-economic, government training programs, agricultural infrastructure and marketing options. The empirical model specification is expressed as follow:

$$ORG = \alpha + \rho W_{ORG} + \sum\nolimits_{i=1}^{n} \beta_i X_i + \varepsilon \tag{1}$$

where ORG is the number of organic farming, α is a constant term, W_{ORG} is spatial lag dependent variable and ρ is the spatial lag coefficient, β_i is the coefficients for explanatory variables X_i, and ε is an error term. The explanatory variables of the model are described in Table 1. All explanatory variables are assumed as positive effects to the number of organic farming, which means that if the government puts more supports on policies, the number of organic farming will increase. The high or low effects are depending on the coefficient and significant of the explanatory variables.

Table 1. Variable definitions of organic farming model

Variables	Definitions	Expected sign
ORG	Number of organic farming household	
Socio-economic factors		
HH	Number of household	+
POP	Number of population	+
Education		
MID	Middle school graduated	+
HIGH	High school graduated	+
DIP	Diploma or equal graduated	+
BA	Bachelor or higher graduated	+
INCOME	Household income (1,000 Thai Baht)	+
AREA	Total farm area (square mile)	+
Government training programs		
TAP	Number of farmers attending the agricultural production training program	+
TAPTMS	Number of farmers attending the trade and marketing training program	+
TIE	Number of farmers attending the integrated agriculture education program	+
TNRENV	Number of farmers attending the natural resources and environment conservation training program	+

(*continued*)

Table 1. (*continued*)

Variables	Definitions	Expected sign
TEKT	Number of farmers attending the organic farming technology training	+
Agricultural infrastructure		
AINF	Agricultural infrastructure	+
Marketing options		
CMAP	Number of agricultural central market	+
MAP	Number of markets for agricultural products	+
CS	Number of cooperative shops	+
FGS	Number of farmer group's shops	+
FCM	Number of field crop markets	+

3 Experimental Results

3.1 Global Spatial Autocorrelation

Figure 1 shows that the organic farming variables occur the autocorrelation problem from the spatial relationship between organic farming in each area. The Moran's I statistic test can reject H0 at 90% confidence.

3.2 Local Spatial Autocorrelation

GeoDa creates a LISA Cluster Map (Fig. 2) for identifying local clusters and spatial outliers of a geo-referenced variable. In each region, the Local Moran statistic is tested for significance by randomization (Fig. 1). Significant local clusters and spatial outliers are portrayed in different colours.

The result of LISA test (Fig. 2) indicates the spatial autocorrelation between organic farming in each district. The four-colour codes present four quadrants in Moran scatter plot. This study presents the border based weight matrix. The LISA cluster map shows that 34 of 182 districts occur the spatial autocorrelation problem.

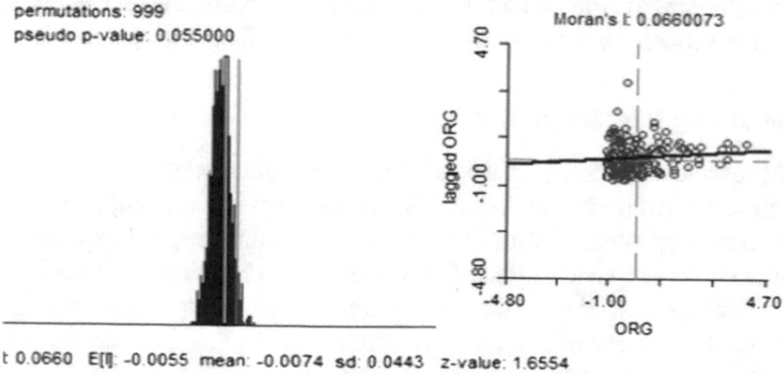

Fig. 1. Moran's I statistic and Moran's I scatter plot

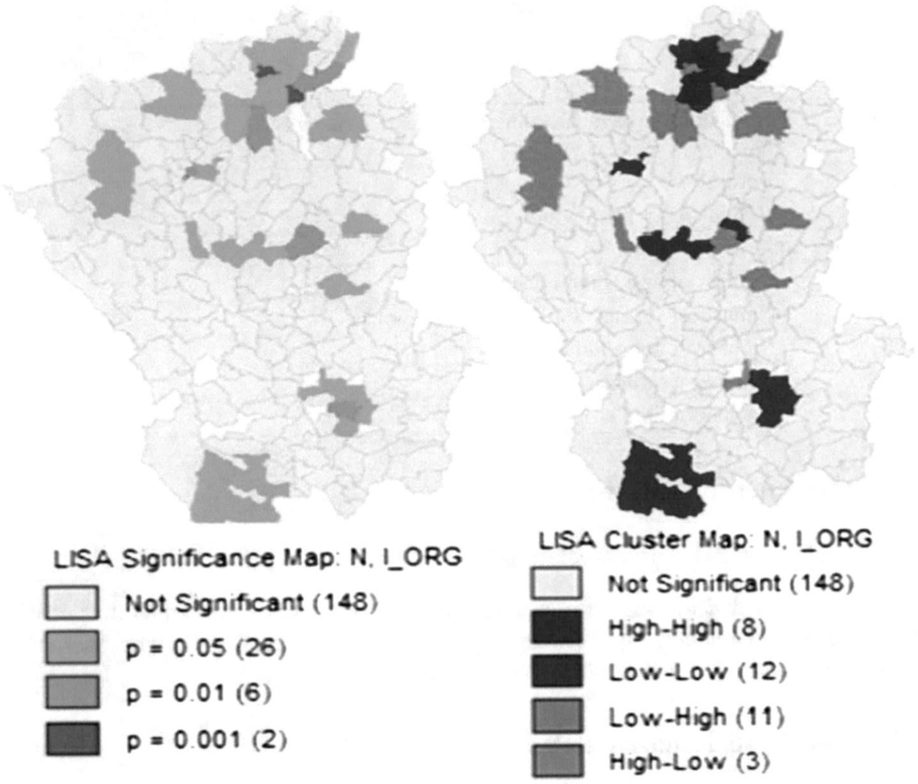

Fig. 2. Local Indicators of Spatial Association (LISA)

3.3 Model Specification Results

There are six tests for spatial dependence model (Table 2). Moran's I score shows highly significant which indicates that there is a strong spatial autocorrelation of residual in the classic linear regression model and the classic linear regression analysis is not suitable estimator for this data set. In term of the model selection, the results show the significant of Robust LM (lag) indicates the spatial dependence problem and the spatial lag model are better or may be appropriate than the spatial error model.

3.4 Spatial Lag Regression Result

This part presents the results of spatial lag regression model to identify the factors affecting to organic farming including four categories; socio-economics, government support factors, agricultural infrastructure, and marketing options. Spatial lag regression of organic farming and organic farming areas are tested and identified.

To examine the best fit of spatial lag model, the spatial weight matrix using Queen weight matrix (first order of share border district) is applies. For the coefficients estimated, the maximum likelihood technique with GeoDa software is used.

Table 2. Diagnostics for spatial dependence test results (n = 182)

Test	MI/DF	Prob.
Moran's I (error)	0.1042	0.00887
Lagrange Multiplier (lag)	1	0.00240
Robust LM (lag)	1	0.03436
Lagrange Multiplier (error)	1	0.02864
Robust LM (error)	1	0.81823
Lagrange Multiplier (SARMA)	2	0.00973

Table 3. Maximum Likelihood estimation, spatial lag model (ORGW1) (n = 182)

Variables	Coefficient	Prob.
Socio-economic factors		
HH	**0.132**	**0.019**
POP_L	**297.138**	**0.036**
Education		
MIDSCH	**0.031**	**0.017**
HIGHSCH	0.018	0.836
DIP	−0.153	0.337
BA	0.039	0.665
LABOUR	−0.050	0.150
INCOME_L	**−179.425**	**0.052**
AREA	0.000	0.924
Government training programs		
TAP	**0.291**	**0.005**
TAPTMS	−0.540	0.174
TIE	**0.594**	**0.038**
TNRENV	**0.289**	**0.001**
TEKT	**2.300**	**0.000**
Agricultural infrastructure		
AINF	−0.293	0.741
Marketing options		
CMAP	−13.984	0.264
MAP	21.584	0.282
CS	**28.609**	**0.088**
FGS	**67.580**	**0.017**
FCM	−5.970	0.392
W_ORG	**0.219**	**0.007**
CONSTANT	**−576.594**	**0.003**
Pseudo R-squared	**0.608**	
Likelihood Ratio Test	**7.680**	
PROB	**0.006**	
Akaike info criterion	**2858.440**	
Schwarz criterion	**2928.930**	
Breusch-Pagan test	**75.616**	
Prob.	**0.000**	

In term of spatial lag regression model results (Table 3), the spatial dependence seems to have a relevant influence on the spatial distribution of organic farming and it is highly significant. The positive sign of the spatial lag coefficient implies that organic farming on one district positive affects to organic farming in neighbouring districts. The meaning of coefficient, organic farming in neighbouring district increase by one percentage point affecting to organic farming in considered district will increase by percentage points equal the coefficient of spatial lag variable.

From the results, the significant of socio-economic factors including household, population, and labor indicate that organic farming is labor incentive sector. According to organic farming, need more labors and more management than conventional agriculture. As well, organic farming has a limited income compared to other sectors (negative sign of coefficient) and it implies that organic farming still has lower income than other sector. In term of government training programs, the agricultural production training program, integrated agriculture education program, and natural resources, environment conservation training program and organic farming technology training were significant which show that the area which getting more government support measures has more organic farming. Moreover, the marketing options; the cooperative shops, and farmer groups' shops are significant. These two significant marketing options are used for selling local organic products and the other marketing options are used for selling conventional agriculture products.

4 Conclusion

Although the first national framework that clearly mentioned the organic agriculture of Thailand since 1997, the organic agriculture is still at an early development stage. This might implies that government failed to translate the plan into the concrete activities therefore this study analyzes the government policy effect to organic farming. The spatial exploratory and spatial lag are used to analyze the policy effects on organic farming. The results show that the government training programs are highly significant to the organic farming. As well, organic farming in each area are affected to their neighbours' organic farming. Therefore, government should lay emphasis on applying organic farming policies in each area.

5 Suggestions

To increase the organic agriculture, the government needs to understand the most influent factor which affecting to farmers to stay in organic farming for example farm income. If their income is lower than other sectors, the organic sector still becomes unsustain. The empirical spatial analysis also examines that relationship of each farmer affects to the organic farming hence the connection between organic communities or farmers' networks may increase organic farming areas.

The empirical results also show that government training and extension projects including agricultural production, integrated agriculture education, natural resources and environment conservation and organic farming technology are affected to the increasing of organic farming areas especially organic farming technology. Therefore, government sector should pay more attention to these issues.

References

1. Koesling, M.: Factors influencing the conversion to organic farming in Norway. Int. J. Agric. Res. Gov. Ecol. **7**(1/2), 78–95 (2008)
2. Strategies for Developing Thailand's Organic Agriculture. http://www.oae.go.th/download/download_journal/2560/OrganicAgricultureStrategy.pdf
3. Veisi, A., Gholami, M., Shiri, N.: What are the barriers to the development of organic farming? Sci. Papers Ser. Manag. Econ. Eng. Agric. Rural Dev. **13**(3), 321–326 (2013)
4. Stolze, M., Lampkin, N.: Policy for organic farming: rationale and concepts. J. Food Policy **34**, 237–244 (2009)
5. Thapa, B., Rattnansuteerakul, K.: Adoption and extent of organic vegetable farming in Mahasarakham province, Thailand. Appl. Geogr. **31**, 201–209 (2011)
6. Sukallaya, K., Gopal, B.T.: Crop diversification in Thailand: status, determinants, and effects on income and use of inputs. J. Land Use Policy **28**, 618–628 (2011)
7. Thailand geographical data. http://gadm.org
8. Anselin, L.: Local Indicators of Spatial Association—LISA. Geogr. Anal. **27**, 93–115 (1995)
9. Anselin, L.: Spatial Econometrics: Methods and Models. Kluwer Academic Publishers, Dordrecht (1988)

Approaching Mobile Constraint-Based Recommendation to Car Parking System

Khin Mar Win(✉) and Benjawan Srisura(✉)

Information Technology Laboratory, Vincent Mary School of Science and Technology, Assumption University, Bangkok, Thailand
Khinmar26win@gmail.com, benjawan@scitech.au.edu

Abstract. Although the car parking system has been emerged since last decade, it seems that helping car owners to find their suitable parking lots has been required, continuously. This can been seen from an effort in developing infrastructure as well as adopting technology of many parking areas, continuously. In the era of Internet of Thing (IOT), mobile application allows individual to access to the system easily. Moreover, the system itself adopts techniques to facilitate users as much as possible. One of those popular technique is the recommendation technique - not limited to E-Commence system. This paper proposes the constraint based recommendation techniques that can support car owners to find their parking lots via mobile application, efficiently. An experiment was set up to test the performance of recommendation and found that it was acceptable.

Keywords: Recommender system · Parking system · Mobile parking
Constraint-based recommendation · NFC mobile parking

1 Introduction

Car parking system has been implemented for long time and found in every parking area especially in metropolitan. This system was initially implements in form client-server until now widely used in form of web base system and mostly support operational function of the car park owner. When the technology has been rapidly changed, this system has been developed in various dimension such as network infrastructure and communication, functionality can facilitate both the car park owner and car owner.

In recent years, the car parking system has been adopted Internet of Thing (IOT) to support a car owner to a suitable parking slot comfortably. To monitor whether there is available parking slot in the parking area or not, pi-camera has been firstly initiated [1] however it requires high cost budget. In the later on, many researchers proposed alternative optimized cost network such as wireless sensor network (WSN) [2], infrared sensor [3, 4], Wireless Fidelity (WiFi) [5, 6], Zigbee wireless network [7, 8] and Visible Light Communication (VLC) or Li-Fi or optical wireless technology [9] – such as the infrared sensor. Due to the reasonable performance and cost, the infrared sensor (IR sensor) has become popular to use in industrial and use as an infrastructure component in our propose system as well. Another component must be used with IR

H. Unger et al. (Eds.): IC2IT 2018, AISC 769, pp. 306–313, 2019.
https://doi.org/10.1007/978-3-319-93692-5_30

sensor is microcontroller [2, 9, 10] which we will use to collect and send IR sensor status in the parking area to the server.

In the part of designing how to communicate with the car owner, sending Short Message Service (SMS) via Global System for Mobile communication (GSM) has been used [7, 10, 11] to allow the system to send to the recommended lots to the car owner. In a recent year, building a mobile application to communicate with a cloud-base server has been increasingly concerned [2, 3, 12]. The mobile application can provide several kind of functionalities such as displaying information about total number of vacant parking lots, computing parking fee and time however the care owner is requested to log in to the application before.

To authenticate the car owner at the entrance and exit, Radio Frequency Identification (RFID) has been early presented [2, 3]. By using RFID, a car owner does not waste a long time to stop at the KIOS. This can reduce traffic jam problem in the parking area. Nowadays, Near Field Communication (NFC) can be found in every mobile phone [13]. Therefore, we proposes a mobile application which facilitate a car owner to log in to the parking system at the entrance via tapping NFC on their mobile device as well.

Moreover, one of the pivot functionality that has required by the car owner and has been continuous studied by researcher is finding an available parking lot. This function requires a supporting of a suitable recommendation technique. The first generation of recommendation methods were developed with Dijkstra algorithm [17] and Breadth-first search algorithm [11]. These required computation time consume and complex data structure representation like graph.

Recently, the recommendation techniques in literature [14] – such as Content-based recommendation (CR), Collaborative Filtering (CF) and Knowledge-based recommendation (KR) have been applied in E-Commerce system such as Amazon, NetFlix. There are a little attention to apply in the parking system context. A goal to be achieved by this research is to find an efficient recommendation technique for the mobile parking system.

Finally, the constraint-based recommendation (PR) which is a subset of KR technique is proposed to be a most suitable recommendation. In the literature, representation of knowledge is defined as a framework, this paper explain how the constraint-based recommendation can be implement in the mobile parking system works with acceptable performance.

2 Proposed Mobile Recommendation for Car Parking System

2.1 Knowledge Base

Extracting data used in recommending is concerned to be a prior stage that must be prepared carefully. Every recommender system must design and construct their own knowledge. In the constraint-based recommendation, all interested items (parking zones) and three main knowledge gathered from user - including users' privilege, preference in term of floor and visited departments or stores are required to be collected in a database. These information is summarized in Fig. 1.

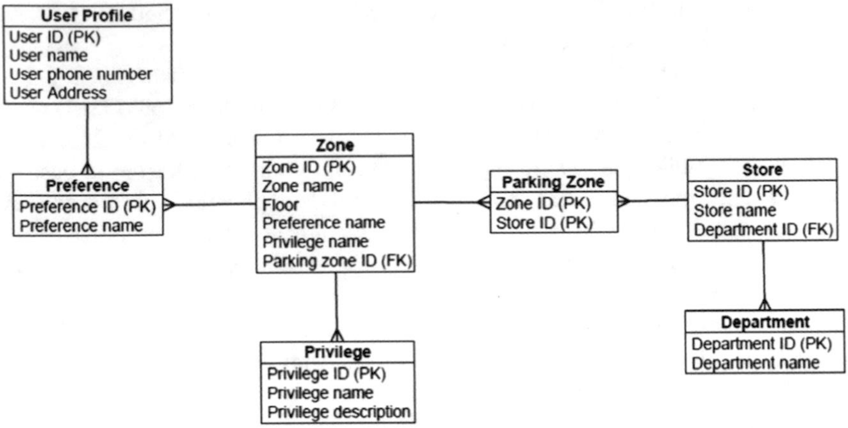

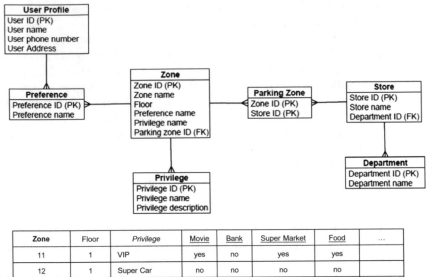

Zone	Floor	*Privilege*	Movie	Bank	Super Market	Food	...
11	1	VIP	yes	no	yes	yes	
12	1	Super Car	no	no	no	no	
24	2	Privilege Card1	no	no	no	no	
31	3	User	no	yes	no	yes	
41	4	User	yes	no	no	yes	

Fig. 1. Knowledge base representation

2.2 Requirements and Constraints

There are several patterns of representing the user requirement and constraint [15]. This paper uses a classical pattern [16] that can be described by a-tuple (V, C, R) as follows:

Table 1. Requirement and constraint representation

	Definition	Representation
V	Set of attributes prepared in the knowledge base shown in Fig. 1	{Floor(0..4), Privilege(Special, VIP, Super Car, Privilege Card 1), Movie(0,1), Bank(0,1), …, Food(0,1), …}
C	Constraints that is generated from the system or limitation. In the context of parking system, it is interpreted as reserved zone	$\{c_1$: Privilege = VIP -» Zone = 11$\}$ $\wedge$ $\{c_2$: Privilege = Super Car -» Zone = 12$\}$ $\wedge$ $\{c_3$: Privilege = Privilege Card 1 -» Zone = 24$\}$
Z	Data set (total record zones) in the table "Parking Zone"	$\{z_1$: (Zone = 11 $\wedge$ Floor = 1 $\wedge$ Privilege = VIP $\wedge$ … $\wedge$ Food = 1) $\vee$ z_2 $\vee$ z_3 $\vee$ z_4 $\vee$ z_5: (Zone = 12 $\wedge$ Floor = 1 …)
R	User requirement or preference based on particular attributes in the knowledge base	$\{r_1$: Floor = 1, 3$\}$ $\wedge$ $\{r_2$: Privilege = VIP$\}$ $\wedge$ $\{r_3$: Movie = 1$\}$ $\wedge$$\{r_4$: Food = 1$\}$

2.3 Recommendation Task

Task of identifying a set of parking zones ranked by the maximum similarity score between user's requirements (R) with entire corresponding zones in the parking zone (Z) and must not violate with the constraint set (C). Since there are three concerned knowledge – including the user's preference in term of floor, the user's privilege and the visited stores in term of its department, the proposed model initialed the average weights for three components as 0.4, 0.3 and 0.3 in sequence. In the third component use in recommendation, it requires to prior load all dimensions (D) in department table to be a set of dimension of the main table and used in computing zone that is nearest. Therefore, the similarity measure is defined as:

$$Similarity(z, R) = \frac{\sum_{r=1}^{3} w_r * sim(r, D)}{\sum_{r \in R} w_r} \tag{1}$$

$$sim(r, D) = \frac{\sum_{r=1}^{R} x}{\sum_{d=1}^{D} 1}; x = \begin{cases} 1, r\, is\, true\, on\, dimension\, d \\ 0,\ r\, is\, false\, on\, dimension\, d \end{cases} \tag{2}$$

Refer to the given user's requirement set (R) – including r_1, r_2, r_3 and r_4, and the constraint set (C) consists of c1, c2 and c3 in Table 1 and the parking zone table (Z) in Fig. 1. The similarity computation returned as follows:

Zone	Floor	*Privilege*	Movie	Bank	Super Market	Food	Z - C	Similarity(z, R)	Rank
11	1	VIP	1	0	1	1	0	0.4(1)+0.3(1)+0.3((1+1)/4) = 0.85	1
12	1	Super Car	0	1	0	0	1	-	-
21	2	Privilege Card1	0	0	1	0	1	-	-
31	3	User	0	0	0	1	0	0.4(1)+0.3(0)+0.3(1/4) = 0.48	2
41	4	User	1	0	0	1	0	0.4(0)+0.3(0)+ 0.3((1+1)/4) = 0.15	3

Fig. 2. Recommendation list

Considering a recommendation shown in Fig. 2, since zone 12 and zone 21 were reserved and identified by the constraint set, only privilege user can park this zone. Therefore, in order to save the computational time, all zone that are not in the constraint set (Z - C) will be evaluated by computing their similarity. Next, the first item in the list – consists of three components that are matched to the requirements and are not violet to the constraints, will be chosen as a recommendation's zone.

Finally, the proposed recommendation will find an available parking lot within the recommended zone whose similarity score is maximum or rank is minimum to the car owner. If there is no available parking lot found in the first recommended zone the next zone in the list will be chosen, sequentially.

Unfortunately, the recommended parking lot is stolen or a parking conflict problem occurs, the system can find the next available parking lot in that zone automatically. If that zone is full, the system have enough zones in the recommendation list to be recommended in sequence. List of recommendation in term of zone level has been designed for enlarging the related parking lots that can be recommended to the user when any ad-hoc problems or requests come up (Fig. 3).

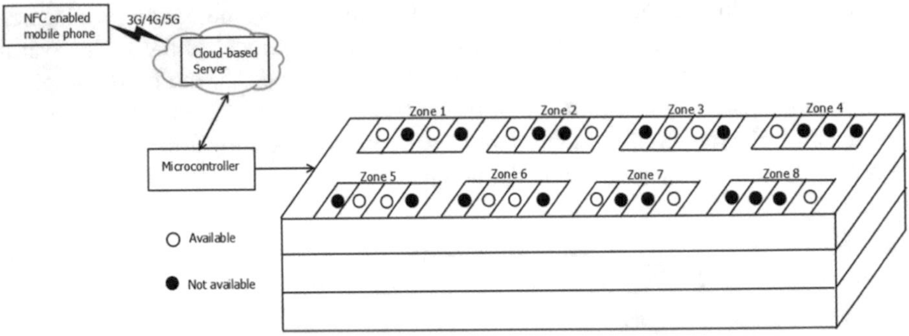

Fig. 3. Architecture of the proposed system

2.4 Mobile Recommender Interaction

The infrastructure of car parking system is also designed for facilitating the car owner to interact with the system easily. Firstly, Infrared sensor (IR sensor) will be installed in every parking spaces to monitor a vacant parking space. The microcontroller will then collect all information from the IR sensor and send the updated status of parking lots in the server. Moreover, the cloud-based server, it is not only used to serve all of parking requests from the car owner's mobile but also update the status of available parking slot in the parking area. Thirdly, Kiosk with NFC reader must be installed in every entry gate for logging in to the system. Finally, mobile phone application that allow the car owner to register their profile, privilege and requirement; receive the recommended parking lot from the system and communicate or ask the next recommendation when the recommended parking lot is not available or parking conflict occur.

3 Experimental Design and Result

Due to lack of the standard data set related to the interested domain – parking system, an experiment is needed to set up by simulating 400 test cases (user requirements) with a parking area of a department store which consist of 4 floors, 8 zones per floor and 3 parking lots per zone. Therefore, there are totally 96 parking lots in the given parking area. The dimensions of zone are collected based on three concerned components - including user profile, preference, privilege, visited dimensions and parking slots data, to compare the performance of different algorithms against each other (Table 2).

The acceptable performance of this study can be evaluated in term of the hit rate and F-Measure. When the correct recommendation is true positive (TP), the correct omission is true negative (TN), the recommendation's rejection is false positive (FP) and Non-recommendation's acceptance is false negative (FN) (Table 3).

Table 2. Comparative recomendation

	Recommendation technique	Description
1	Content-based recommendation	CR [18] uses tf-idf keyword matching between the user's requirement and the zone in knowledge base
2	Collaborative Filtering	CF cannot be done since there is not rating information
3	Knowledge-based recommendation	KR [19]
4	Constraint-based recommendation	PR is the proposed recommendation technique

Table 3. Performance matrix

Predicted condition	True condition	
	Correct	Not correct
Recommended	True Positive (TP)	False Positive (FP)
Not recommended	False Negative (FN)	True Negative (TN)

Monitoring the outcome of 400 test cases, the result showed that the proposed model constraint-based recommendation, returned the maximum hit rate (165). This is meant that PR recommendation returns the most correction comparing with CR and KR (Table 4).

Table 4. Hit rate and false positive performance

Recommendation	FN False Negative	TN True Negative	TP True Positive (Hit rate)	FP False Positive
CR	160	90	85	65
KR	115	85	124	76
PR	83	80	165*	72*

However, considering the false positive (FP), PR – (72) returned higher rate than CR – (65), smaller than KR – (76). Since the false positive of PR is higher than CR, in order to ensure that PR performance is acceptable, F-Measure is required to monitor further (Table 5).

Table 5. F-Measure performance

Model	Accuracy (TP + TN)/N	Recall (R) TP/TP + FN	Precision (P) TP/TP + FP	F-Measure (2PR/(R + P))
CR	0.44	0.57	0.35	0.43
KR	0.52	0.62	0.52	0.56
PR	0.61	0.70	0.67	0.68*

Although the false positive of PR is higher than CR, F-Measure of PR is highest – (0.68). Therefore, the performance of proposed recommendation technique – PR, is the acceptable.

4 Conclusion and Recommendation

This study proposes the constraint-based recommendation which is suitable to be used in the mobile parking system. The experiment confirmed that the recommendation's performance of proposed technique is acceptable. However, it is required to re-test when the infrastructure of proposed parking system is fully installed. When considering the high false positive score – the high the recommendation's rejection rate, it may occur from several causes. One of them is facing with the parking conflict - recommended parking lots has already been stolen and force the car owner to reject the recommendation. In actually, the proposed system has already design to solve this problem by re-recommending the next suitable parking lot to the car owner who send a request to the system again. This may require us to research for more improvement and implementation further.

References

1. Basavaraju, S.R.: Automatic smart parking system using Internet of Thing (IoT). Int. J. Sci. Res. Publ. **5**(12), 629–632 (2015)
2. Radha, P.R., Pooja, B.L., Atul, N., Atul, S., Shingade, P.: An intelligent parking system based on Cloud using IoT technologies. Int. Adv. Res. J. Sci. Eng. Technol. **4**, 120–122 (2017)
3. Buchade, A.C.: Sophisticated parking availability prediction system in IoT network. Int. J. Comput. Sci. Eng. **5**, 132–136 (2017)
4. Ilhan, A., Karakose, M., Ebru, K.: A navigation and reservation based smart parking platform using genetic optimization for smart cities. In: 5th International Istanbul on Smart Grid and Cities Congress and Fair (ICSG), pp. 120–124, Turkey (2017)
5. Raghavender, R.Y.: Automatic smart parking system using Internet of Things (IOT). Int. J. Eng. Technol. Sci. Res. **4**, 225–228 (2017)
6. Vishwanath, Y., Aishwarya, D.K., Debarupa, R.: Survey paper on smart parking system based on Internet of Thing (IoT). Int. J. Recent Trends Eng. Res. **2**(3), 156–160 (2016)
7. Srikurinji, S., Prema, U., Sathya, S., Manivannan, P.: Smart parking system architecture using infrared detector. Int. J. Adv. Inf. Commun. Technol. **2**(11), 1113–1116 (2016)
8. Abu, A., Kishore, K.C., Muhammad, F.M.: A time and energy efficient parking system using ZigBee communication protocol. In: Proceedings of the IEEE SoutheastCon, pp. 1–5. IEEE Press, USA (2015)
9. Abinayaa, B., Arun, R.A.: Visible light communication based smart parking system using MSP430. ARPN J. Eng. Appl. Sci. **11**(2), 1337–1342 (2016)
10. Vaibhav, I.V., Ramya, A.: A review on smart parking management system using vehicle authentication. Int. J. Adv. Res. Electr. Electron. Instrum. Eng. **5**(3), 1752–1755 (2016)
11. Khang, S.C., Hong, T.J., Chin, T.S., Wang, S.: Wireless mobile-based shopping mall car parking system (WMCPS). In: 2010 IEEE Asia-Pacific Services Computing Conference, pp. 573–577 (2010)

12. Shipra, S., Vaibhav, K.: Smart-parking system using Internet-of-Things technologies and RFID. Int. J. Res. Appl. Sci. Eng. Technol. **5**, 1363–1367 (2017)
13. Aishwarya, I., Sagar, M., Priyanka, S., Prashant, D.: Optimized solution for parking automation implement as an application of Internet of Things-IoT using near field communication. Intern. J. Eng. Sci. Comput. **6**, 635–6360 (2016)
14. Dietmar, J., Markus, Z., Alexander, F., Gerhard, F.: Recommender Systems, 1st edn. Cambridge University Press, Cambridge (2011)
15. Tsang, E.: Foundations of Constraint Satisfaction. Academic Press, New York (1993)
16. McSherry, D.: Increasing dialogue efficiency in case-based reasoning without loss of solution quality. In: Proceedings of the Eighteenth International Joint Conference on Artificial Intelligence, pp. 121–126. Morgan Kaufmann, San Francisco (2003)
17. Farida, P., Vishnuvardhan, D.: Design of NFC based vehicle parking system using smartphone. Int. J. Sci. Res. **3**, 819–821 (2015)
18. Shinde, S., Bhagwat, S., Pharate, S., Paymode, V.: Prediction of parking availability in car parks using sensors and IoT:SPS. Int. J. Eng. Sci. Comput. **6**, 4751–4755 (2016)
19. Srisura, B., Wan, C., Sae-lim, D., Meechoosup, P., Mar Win, K.: User preference recommendation on mobile car parking application. In: 2018 6th IEEE International Conference on Mobile Cloud Computing, Services, and Engineering (MobileCloud), pp. 59–64. Bamberg, Germany (2018)

Software Engineering

A Comparative Study on Scrum Simulations

Sakgasit Ramingwong(✉), Lachana Ramingwong, Kenneth Cosh, and Narissara Eiamkanitchat

Department of Computer Engineering, Faculty of Engineering, Chiang Mai University, Chiang Mai, Thailand
{sakgasit,narisara}@eng.cmu.ac.th,
lachana@gmail.com, drkencosh@gmail.com

Abstract. Learning through simulation is an efficient method for illustrating the essentials of Scrum software development to those who are interested in further understanding modern software engineering techniques. Several models of simulations have been proposed in attempts to match needs and simulation environments. This paper compares the efficiency of two variations of Scrum simulation; the original "Scrum simulation with LEGO bricks" and the economic alternative "Plasticine Scrum". Similarities, differences, advantages and disadvantages of both models are discussed in the findings.

Keywords: Software engineering · Software development
Scrum · Simulation · Education

1 Introduction

While software developers strive for perfection in their products, many factors play important roles in either supporting or hindering their success. Realistic objectives, competent team, support from executives and users, efficient requirement management and adequate quality assurance are some of factors that can have a high impact [1, 2]. The chosen development model is also one of the most influential factors. Appropriate implementation and adaptation of a suitable model can greatly contribute to the success of a software product.

Due to many factors in the modern software industry, traditional product development methods, such as the classic Waterfall or Spiral models, can be undeniably bulky and inefficient for certain scenarios. More recently the concepts of agile and lightweight development have emerged in order to tackle such challenges. Due to their high flexibility, change orientation, increased communication and short time-to-market, they have become increasingly popular approaches [3]. Scrum is one of the more popular agile development models which has been largely adopted internationally [4]. Similar to other processes, extensive practice is inevitably needed in order to master the concept of Scrum and mastering its philosophy can be time and resource consuming.

"Scrum simulation with LEGO bricks" is a workshop which attempts to inspire the concepts of Scrum to practitioners within a short time limit [5]. This simulation has been widely adopted and translated into at least fifteen languages. However, a major

H. Unger et al. (Eds.): IC2IT 2018, AISC 769, pp. 317–326, 2019.
https://doi.org/10.1007/978-3-319-93692-5_31

drawback of this workshop, the cost of LEGO, could limit its implementation. This is especially true in organizations with limited budget.

"Plasticine Scrum" is an alternative method of Scrum simulation. It replaces LEGO with plasticine as the main material, reducing costs by 90%. Several adjustments have been made to the simulation in order to suit the different properties of plasticine. The results of a pilot workshop appear to be satisfactory. However, more evidence is needed to verify that the efficiency of this alternative workshop is comparable to the original one.

This paper conducts a comparative analysis between "Scrum simulation with LEGO bricks" and "Plasticine Scrum". Groups of participants are instructed to experience both simulations in parallel. After that, the participants express their opinions as well as having discussion about the advantages and disadvantages of each simulation.

The second section of this paper introduces the Scrum method and describes the two simulations; "Scrum simulation with LEGO bricks" and "Plasticine Scrum". Then, the research methodology is addressed in the third section. The results of the research are illustrated and discussed in the fourth section. Finally, the fifth section concludes the paper.

2 Scrum Simulations

Scrum is a popular lightweight software development which has been widely adopted by many software developers. The Scrum process splits the product into *sprints*. The most important features, which are called *backlogs*, are prioritized and developed first. Each cycle of development involves approximately 1 to 4 weeks and results in a working deliverable. In Scrum, communication, progress tracking and enhancive reviewing are essential elements. All staff need to participate in a daily *stand-up meeting* to review and discuss their progresses. A *burn-down chart*, which illustrates the overall development, is continuously monitored and updated during the process. Then, at the end of each sprint, they openly exchange their lessons learnt and perceptions in a *retrospective* session. Figure 1 illustrates a cycle of Scrum development.

There are three major roles in Scrum. Firstly, the development team is responsible for all implementation of the product. Ideally, they voluntarily select their tasks and also freely exchange their positions when needed. Secondly, the product owners are team members who act and make decisions for the customers. They are also responsible for leading the review of the deliverables. Finally, the Scrum masters are servant leaders who mainly facilitate the Scrum processes. Unlike traditional managers, their major role is not assigning the tasks but rather supporting and motivating the team.

Learning Scrum obviously needs time. The fact that the Scrum process encourages all members to exchange their position indicates that team members need to possess various skills, both technically and non-technically. In order to master this software development model, the practitioners need to gain experience by participating in several cycles of sprints. Successfully gaining this experience could be particularly problematic for people with less technical backgrounds and experience.

Simulation is a promising approach for teaching Scrum to interested persons. Several variations of simulation have been proposed and implemented in both

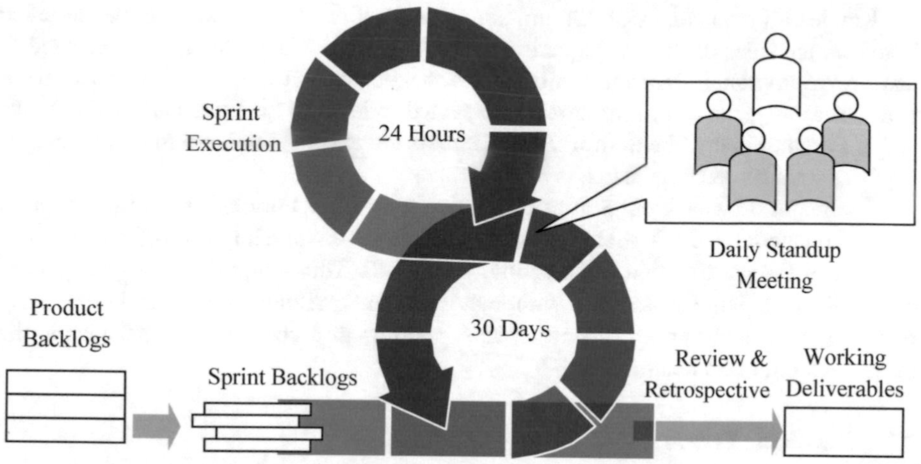

Fig. 1. Scrum [6]

academic and business sectors. However, each simulation has advantages and disadvantages. An appropriate implementation is therefore required in order to maximize their benefits.

2.1 Scrum Simulation with LEGO Bricks

"Scrum simulation with LEGO bricks" is an innovative activity which aims to simulate the entire process of Scrum to practitioners. Instead of developing a real software product, the participants build a city model out of LEGO [5]. In this workshop, one building simulates one product backlog. Since no actual coding is involved in the workshop, it is possible to participate without possessing a technical background. All essential Scrum elements are represented in the simulation and the workshop spans approximately 100 to 120 min.

Three main roles are assigned to the participants. Firstly, the Scrum team members are the main force of the workshop. Each team includes 4–6 participants. Ideally, the workshop should include only two to three teams. Secondly, the role of Scrum master is generally portrayed by the workshop moderator. The main responsibilities of this role include facilitating the Scrum process and encouraging the teams. It is also possible that each Scrum team has their own dedicated Scrum master. In this case, no public Scrum master is needed and the number of the team member increases by one. Finally, the product owner can be also role played by the moderator. The product owner manages and prioritizes backlogs, acts as the voice of the customers and communicates with all Scrum teams.

The simulation begins with an introduction and team organization. Then, the preparation of backlogs and estimation are performed by the players. The backlogs may include simple to advanced buildings such as a bus stop, school, hospital or church, etc. After that, three sprints are implemented. Each of them takes 15 min for planning, sprinting and reviewing, respectively. Finally, the debriefing session takes 15 min.

Key learning outcomes of "Scrum simulation with LEGO bricks" can be classified based on the roles of the participants [7]. Firstly, the product owners are expected to learn to appropriately manage the backlogs, negotiate with the team, and make decisions. Secondly, the scrum masters are expected to learn to facilitate and moderate the use of Scrum. Finally, the Scrum team members are expected to learn to communicate, negotiate and be self-organized.

The major drawback of "Scrum simulation with LEGO bricks" is the high price of the main material, LEGO bricks. In order to organize a workshop for 4–6 participants, the cost of the construction bricks could reach $40. This multiplies in larger groups, especially in a large classroom with many students. Although the bricks are fully reusable, for a large group of participants, such as in a class of 80 participants, the initial cost could be unaffordable.

2.2 Plasticine Scrum

"Plasticine Scrum" is essentially a variation of the "Scrum simulation with LEGO bricks". However, instead of using the famous construction bricks as the main material, this workshop utilizes plasticine as plasticine is much cheaper than LEGO. One kilogram of medium-grade plasticine costs only $2 and is more than sufficient to be used by a group of 4–6 participants. Although not fully reusable, its cost is obviously much cheaper [8].

Due to the difference in characteristics of the main material, the backlogs in "Plasticine Scrum" are changed from components of a city to components of a zoo. The participants are required to build animals and their habitats. The sprint execution period is also slightly expanded to suit the material preparation time.

The three major Scrum stakeholders remain the same. The Scrum master and product owner are role played by the class moderator while other participants portray the role of the Scrum team. Each team consists of 4–6 persons.

"Plasticine Scrum" also begins with the introduction. Then three sprints are performed by the team. Each sprint includes planning, sprint execution and reviewing as in the original. Five more minutes are added to the sprint execution, making each sprint last 20 min. Lastly, a fifteen-minute of debriefing is also represented at the end of each workshop.

An additional feature is proposed in the "Plasticine Scrum". It introduces special backlogs which requires the team to reuse many of their finished products. This is to highlight the importance of the reusability and standardization in software engineering.

3 Research Methodology

In order to comparatively investigate the efficiency of both simulations, two workshops were organized. The first workshop was participated in by 22 postgraduate computer engineering students who enrolled in an advanced software engineering class. Almost all of them were full-time professionals and were very capable of programming. The second workshop involved 18 senior undergraduate computer engineering students on a software project management class. All of the participants were members of the

Department of Computer Engineering, Faculty of Engineering, Chiang Mai University, Thailand. Most of them did not have experience in Scrum development.

In each batch of the workshops, the participants were divided into 4 groups, called A, B, C and D, each of which consisted of 4–6 persons. All groups participated in 6 Scrum sprints, involving three sprints of LEGO and three sprints of plasticine. Groups A and B began with LEGO while groups C and D started with plasticine. The participants were instructed to switch materials at the beginning of the forth sprint and take over the project, including finished and unfinished products developed by their counterparts. In other words, from the fourth sprint onwards, group A and B switched their entire development environment with group C and D, respectively, and continued their colleagues' unfinished project. In this way, all participants were given equal opportunity to experience both simulation modes. The number of sprints and the midway switching were kept secret at the beginning of the workshop.

A minor adjustment was made to the Plasticine Scrum. The five additional minutes of sprint execution was intentionally removed so both groups of participants could finish each sprint at the same time. This resulted in 15 min sprints for both LEGO and plasticine teams. Nonetheless, the plasticine groups were instructed to knead their main material during the introduction session. The entire simulation took approximately 140 min. This included a 30-min introduction and a 20-min debriefing.

During the final two sprints, the participants were given two special backlogs. The LEGO groups were asked to build a cultural center and a deluxe village while the plasticine groups were assigned an elephant village and a circus. Firstly, the cultural center needed several heritage buildings which were developed during the previous sprints as well as new animal mascots. The second special backlog, the deluxe village, included at least six houses with approximately the same size and design. The elephant village, the third special backlog, consisted of several rigid structures such as crowd stand and elephant canvas. Finally, the circus included a ticket booth and various decorations for animals. In this way, both simulation groups were forced to reuse and modify their previous finished products as well as develop features which were not fully compatible with their main materials.

Before the debriefing took place, all participants were given an anonymous questionnaire. This questionnaire comprised of 30 questions, addressing the appropriateness, knowledge, skills, fun and various factors of the workshop. Samples of the questions include "How do you rate the appropriateness of LEGO as the main material?", "How do you rate the overall sanitation of the plasticine material?" and "How do you rate the ease of implementation of LEGO?". The last question of the survey inquires the preference of the participants whether they prefer LEGO, plasticine, or either. Additionally, the participants were also allowed to propose their perceived advantages and disadvantage of both simulation modes.

4 Results and Discussions

After being divided into groups, although all of the students were excited and ready for the workshop, the groups which were assigned with LEGO seemed to be happier with their material. This is likely to be a result of the luxurious feeling of LEGO when compared to plasticine.

During the simulation, as expected, none of the students completed their first sprint without work-in-progress. Then, some of them performed better and successfully delivered their promised features by the end of the second sprint. After that, most of the groups had almost no problem reaching their deadlines. This quick development suggests that both modes of workshop are efficient and have a rather low learning curve. The students were also noticeably satisfied with the simulation.

At the beginning of the fourth sprint, the participants were instructed to switch their materials. This is to allow all participants to experience both materials and to fairly discuss their advantages and disadvantages. Due to this major change in the fourth cycle, the students developed slightly less features compared to their previous sprints. Yet, they all completed their sprint backlogs.

The final sprints of both batches were successful. Although some negotiations were needed during the review, the special backlogs were also deemed accomplished.

Table 1 designates the overall feedback of the simulation. It can be clearly seen that the participants believe that they gained significant knowledge from the workshop. Even the students who claimed to have high knowledge on this development model insisted that he or she gained more understanding after the workshop and increased this score from 4 (before the simulation) to 5 (after the simulation). Two students who admitted that they had not heard about Scrum before the workshop rated their knowledge after the workshop as 3 (average knowledge). A one-way ANOVA also reveals that there is no statistical significance between participants who began the simulation with LEGO or plasticine on this aspect.

Table 1. Participants' feedbacks on their knowledge of Scrum before and after the workshop.

	Before		After	
	Mean	SD	Mean	SD
Knowledge of Scrum before and after the workshop	2.20	0.853	4.10	0.496

Note: SD = Standard deviation

Feedback for the overall appropriateness is not significantly different. Specifically, LEGO is reported to be more appropriate than plasticine in activities regarding backlog management, modification and integration, and testing. On the other hand, the participants believe that plasticine is a more appropriate material during the designing stage.

A further investigation on the differences between the two materials is performed based on paired sample t-test. As displayed in Table 2, at the 95% confidential interval (Sig. < 0.05), three significant differences are found. While the participants find that plasticine is more appropriate during the design, they appear to prefer LEGO during the integration and the testing. This is probably due to the fact that LEGO is much stronger than its alternative. As a result, the LEGO buildings are easier to be flipped, rotated and moved around. On the other hand, plasticine products could be easily damaged during the inspection. Therefore, the participants need to put extra care during this process. In contrast, designing based on plasticine is perceived to be easier since the material can be freely shaped, unlike LEGO which the designers need to imagine the pixelated version of the products.

Table 2. Paired sample t-test of appropriateness of LEGO and plasticine for different development stages.

	Paired differences		t	Sig.
	Mean	SD		
Overall appropriateness	.12821	.95089	.842	.405
Appropriateness for backlog management	.23077	.77668	1.856	.071
Appropriateness for design	**−.28205**	**.82554**	**−2.134**	**.039**
Appropriateness for implementation	.00000	.98710	.000	1.000
Appropriateness for modification and integration	**.50000**	**1.10940**	**2.850**	**.007**
Appropriateness for testing	**.37500**	**.97895**	**2.423**	**.020**

Note: SD = Standard deviation; Sig. = Significance (2-tailed)

Focusing on the simulation factors, paired sample t-tests reveal three statistical significances at the 95% confidence interval (Sig. < 0.05). Not surprisingly, as can be seen from Table 3, the recyclability and cleanliness of LEGO are obviously higher than plasticine. Indeed, LEGO is entirely reusable and do not produce waste during the workshop. This material gained a full score from almost every student. In contrast, plasticine can be only partly recyclable, even when being used carefully. It also obviously has less sanitation than LEGO. However, for freedom of creativity, plasticine is rated significantly higher by the students. This may be a result of the flexibility of the material which can be easily shaped to match the participants' imagination.

Table 3. Paired sample t-test of simulation factors of LEGO and plasticine workshops.

	Paired differences		t	Sig.
	Mean	SD		
Technical skills gained from the workshop	−.05128	.51035	−.628	.534
Ease of use of the material	−.05000	1.37654	−.230	.820
Recyclability of the material	**1.07692**	**1.17842**	**5.707**	**.000**
Cleanliness of the material	**2.02564**	**1.20279**	**10.517**	**.000**
Fun factor of the workshop	.15000	.80224	1.183	.244
Freedom of creativity when using the material	**−.51282**	**.96986**	**−3.302**	**.002**

Note: SD = Standard deviation; Sig. = Significance (2-tailed)

Based on the participants' narrative feedbacks, the most dominant advantage of LEGO is the cleanliness. They suggested that, especially when compared to the other material, LEGO is much cleaner. Ease of implementation is found to be the second most impressive factor in LEGO workshop. The students reported that this material gave them a high level of freedom of choosing and assembling the products as they desired. The third most beneficial preference is its strength. It is obvious that, as a construction set, LEGO and its products are particularly strong. They are unlikely to lose their shape unintentionally during the implementation and inspection. Ease of modification and readiness of use are the fourth and fifth ranked advantages of this main material, respectively. The participants found that modifications of LEGO products are effortless and can be performed quickly. It does not need preparation time as in plasticine simulation. Other preferences of this material include its recyclability, capability of relocation, ease of sizing standardization, and modularity.

In contrast, various challenges from using LEGO in the simulation are raised. The most concerned issues of this material involve the low variety of bricks and unsuitability for developing detailed products. Arguably due to the use of the standard construction set, the students found that the variety of LEGO bricks, i.e. sizes, colors, and shape, is limited. This lack of flexibility makes it difficult to create small details and curvy shapes. This prevents the participants from fulfilling the requirements of their products. This problem could be resolved with more investment. The participants also report that due to the bricks' small size, it is rather difficult to locate the required pieces. This is indeed an exact opposite side of plasticine where any component sizes and shapes can be created from scratch. Other notable drawbacks of using LEGO include its high price and limitation of design. Some participants also reported that they took too much time to locate the desired LEGO parts.

Plasticine, on the other hand, is reported to have major advantages on the ability to create detailed and complex features. The students also found that designing and implementing the design were quick and easy. They were satisfied that they can use this material to fully express their ideas without much needs for explanation. Moreover, plasticine is noted for its exceptionally low cost and availability in stores. Other positive feedback on this material involve fast development time, ability to create various colors, zero lead time for locating materials and ease of forming the product. A student interestingly reported that one of plasticine's advantages is that products are easily demolished and subsequently rebuilt.

Undoubtedly, the major drawback of this material is its low sanitation. Without an adequate protection, plasticine could easily pollute the workshop environment. The long preparation lead time is reported to be another main disadvantage. It is undeniable that the participants need extra preparation time prior to the sprint execution. Kneading the plasticine during the introduction and sprint planning might not be adequate in certain cases. Another key drawback of plasticine is its low strength. As expected, the participants complained that plasticine is not suitable for creating solid structures. Some students also argue that in order to build a plasticine product, the practitioners need to possess certain degree of artistic and imagination skills. Other notable comments include difficulties to relocate the products and difficulties to control the scale of the products.

When being asked to choose a preferred material, mixed feedback was received. 42.50% of the participants chose LEGO as their preferred material while plasticine received 25% of the votes. In addition, 32.50% of the students indicated that both material can be substituted. Interestingly, although LEGO was the most preferred material as expected, the vote for plasticine was not far behind. In fact, the combined votes for plasticine or either were found to be the majority in this study. This intriguingly suggests that the use of plasticine was not unwelcome to the participants.

Table 4 summarizes key learning outcomes from LEGO and plasticine scrum based on The [7]'s material. It can be seen that although several settings are modified, all of the outcomes are achieved. This is due to the fact that the process and Scrum elements of both workshops are the same.

Table 4. Comparison of key learning objectives from LEGO and plasticine workshops.

Outcome	LEGO	Plasticine
Product owner	✓	✓
Management of backlog	✓	✓
Measurement of team velocity	✓	✓
Negotiation with the team	✓	✓
Scrum master	✓	✓
Moderation of the meetings	✓	✓
Facilitation of the process	✓	✓
Mediation of the stakeholders	✓	✓
Coaching for the team	✓	✓
Scrum team	✓	✓
Self-organization	✓	✓
Focused communication	✓	✓
Estimation and measurement	✓	✓
Problem solving	✓	✓

5 Conclusion

Simulation is one of the more popular approaches for teaching software development methodologies in both academic institutes and entrepreneurial organizations. However, due to certain limitations, especially time and budget, the use of simulation could be limited. "Scrum simulation with LEGO bricks" is an efficient workshop which highlights and educates all of the Scrum essentials to the participants in a short period of time. The major problem of this simulation is that its initial material costs could be extremely high, especially when organized for a large group of participants. As a result, an alternative simulation, "Plasticine Scrum", is proposed in order to tackle such limitation. Variations such as requirements and timing are introduced in order to cope with the alternative material. This study investigates the efficiency of both simulation by conducting two batches of controlled workshops with 18 undergraduate and 22 graduate students from the Department of Computer Engineering, Faculty of Engineering, Chiang Mai University, Thailand.

In both batches of simulations, students were divided into groups. All students had chances to work equally using LEGO and plasticine. They were then asked to rate several perceptions towards both main materials after the workshops were completed.

The survey suggests that the efficiency of both materials are not obviously different. Due to their natural characteristics, LEGO and plasticine have their own advantages and disadvantages. While LEGO is more dominant on its cleanliness and recyclability, plasticine wins on costs and creativity perspectives. Interestingly, when it comes to the final choice, although LEGO is found to be the most preferable material for the 42.5% of the participants, a quarter of the students indicate that they prefer plasticine while the rest of them are happy with both materials.

Based on the findings, it seems to be that "Plasticine Scrum" could be an efficient alternative to "Scrum simulation with LEGO bricks". Due to its vastly cheaper cost, plasticine could be an ideal material for a large group of participants.

References

1. Reel, J.S.: Critical success factors in software projects. IEEE Softw. **16**(3), 18–23 (1999)
2. Stern, R., Baumann, C., Lattermann, S.: Key Success Factors for Software Development Within a Systems Engineering Environment (2013)
3. Beck, K., Beedle, M., van Bennekum, A., Cockburn, A., Cunningham, W., Fowler, M.: Manifesto for Agile Software Development. http://agilemanifesto.org
4. Rubin, K.S.: Essential Scrum: A Practical Guide to the Most Popular Agile Process. Addison-Wesley Professional, Upper Saddle River (2012)
5. Krivitsky, A.: Scrum Simulation with LEGO Bricks (2011)
6. Schwaber, K.: Agile Project Management with Scrum. Microsoft Press, Redmond (2004)
7. The Agile42 Team: Scrum LEGO City. http://www.agile42.com/en/training/scrum-lego-city
8. Ramingwong, S., Ramingwong, L.: Plasticine scrum: an alternative solution for simulating scrum software development. LNEE, vol. 339, pp. 851–858 (2015)

Transformation of State Machines for a Microservice-Based Event-Driven Architecture: A Proof-of-Concept

Roland Petrasch(✉)

Faculty of Science and Technology, Thammasat University, Bangkok, Thailand
roland.petrasch@gmail.com

Abstract. The implementation of a state machine for a system or a subsystem in the form of an event-driven architecture (EDA) in a microservice environment requires model transformations. This paper presents a model-to-model transformation for this case based on UML state machines, UML class diagrams for the microservice architecture, a mathematical specification of the transformations and the implementation of the transformation with QVTo. A run-time state or workflow engine as a single-point-of-failure can be avoided with the choreographic and event-driven approach persevering the ability to generate a state machine as an additional system component that provides the advantages of a state or workflow engine features like monitoring, and visualization.

Keywords: Statechart · FSM · UML state machine · Model transformation Microservice · Event-driven architecture · State and workflow engine Choreography

1 Introduction and Related Work

Microservices are decoupled, independent deployable services that can have a positive effect on software quality characteristics like modularity and maintainability. Especially scalability and fault tolerance are interesting in this context: a microservice that is deployed more than once, contributes to the avoidance of a single-point-of-failure that exist for instance when using a workflow engine. Such an engine orchestrates the workflow as a whole; therefore, a failure of the engine leads to a failure of all components and services that take part in the workflow. On the other hand, an event-driven approach follows a choreographic idea: Each microservice publishes or subscribe to events but remains autonomous fulfilling the tasks of the workflow. If one of these microservices fails, it can be replaced easily be another instance of the same microservice type and other services are not affected.

However, the usage of an event-driven architecture (EDA) instead of a workflow or state engine comes at a cost: The state machine or process model (workflow) needs to be transformed into a publish-subscribe logic and an event channel is necessary. This also means that the direct connection between a state machine or a workflow with the run-time system is lost, because only events are exchanged between the services.

H. Unger et al. (Eds.): IC2IT 2018, AISC 769, pp. 327–336, 2019.
https://doi.org/10.1007/978-3-319-93692-5_32

This paper addresses the problems of an event-driven architecture in using an automatic model-to-model transformation and code generation approach. Many code generation approaches and tools for state machines exist, e.g. Rational Statemate [1] or Visual Paradigm [2], but they do not explicitly consider target architectures like microservices and EDA. Microservice patterns (and anti-patterns) including EDA exist, e.g. [3–5], but since they are provided as pattern descriptions (with some code examples), no formalization of transformations or transformation implementation on a concrete model level is provided.

2 State Machines and Workflows

Finite-state machines (FSM, [6]) can be used for the behavioral specification of system elements on different levels, e.g. system/subsystem level, service/component level, module, and function level (for object-oriented system this would be the class and method level). A widespread representation form or notation for FSMs are statecharts introduced by Harel [7]. A deterministic FSM is a directed graph defined as a quintuple

$$StateMachine := (\Sigma, S, s_0, \delta, F) \quad (1)$$

with Σ = the input alphabet, S = set of states, s_0 = initial state, δ = state-transition function (δ: $S \times \Sigma \rightarrow S$), and F = set of final states (subset of S). Figure 1 shows the statechart for an alarm system: After the start, the system is in the state "idle". When a motion is detected, the state changes to "alarm active". From there the alarm can be deactivated (transition back to "idle") or stopped (transition to the final state).

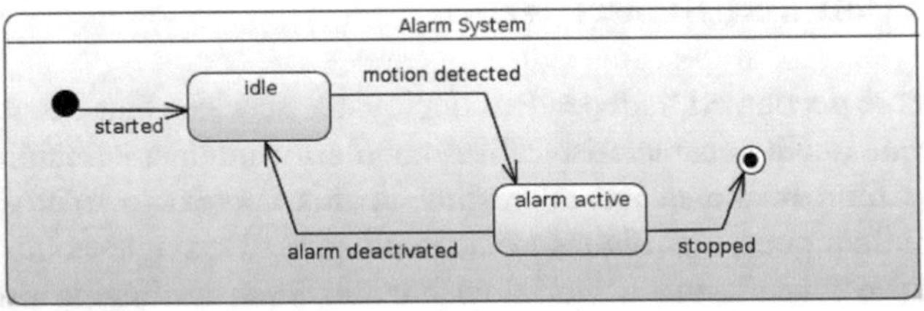

Fig. 1. Simple statechart for an alarm system

UML Statecharts [8] are an extended variant (Extended Hierarchical Automatons) implementing new concepts, e.g. hierarchically nested states and orthogonal regions, combining the characteristics of Mealy machines and Moore machines and addressing the phenomenon known as state and transition explosion. Figure 2 depicts a part of the metamodel for UML state machines ([8]. P.320) that contains the extensions for the transition in the form of the trigger, constraint (guard), and behavior (effect). It also shows the meta-class for the regions.

In comparison to the transition function δ for a classical FSM a transition in UML statecharts is a tuple $t := (sv, tv, \widehat{tg}, g, \propto, \iota, \widehat{tc})$ [9] where $sv \in V(\text{vertex}), tv \in V, \widehat{tg} \subset Trig(\text{trigger}), g \in C(\text{constraint = guard}), \propto \in B(\text{associated behavior = effect}),\ \iota \in R$

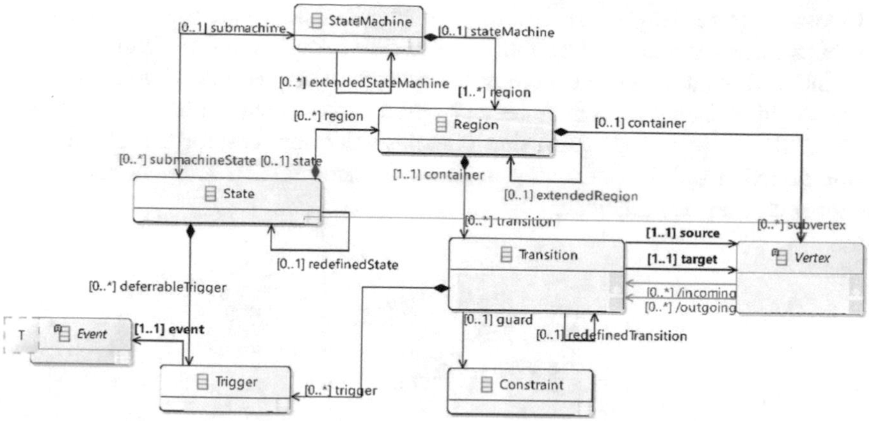

Fig. 2. Metamodel for UML state machines (Eclipse UML 5.0.0 ecore model)

(container of the transition), and $\widehat{tc}$ for a special situation (join/fork pseudo-state connection). The complete mathematical definition for the UML state machines (abstract syntax and formal semantics) is given in [9].

BPM (business process models) also called workflows like BPMN models, UML Activity diagrams or flowcharts also specify behavioral aspects of a system (or its elements) but focus on the flow of logic of a process rather than the states. Therefore, a statechart represents a sequence of states that an element (or object) goes through including the transition (event, guard, action in the case of UML statecharts) and a workflow shows the sequence of actions (or logical steps) of a process. However, the separating line between workflows and statecharts is often crossed, e.g. with State Machine Workflows [10] or UML Activity diagrams (workflow) that are based on Petri-nets ([8], p. 298, 390) which can be described as state-transition systems [11]. Many aspects of workflows that are relevant for the software development in general, and the approach presented here in particular, also apply to statecharts, e.g. the role of a workflow engine (or a state machine) for a software system that implements a microservice architecture. The decision whether to specify the behavior of an object with process models or statecharts depends highly on the domain, the requirements, the object itself, and the behavioral characteristics. The discussion of this aspect would go beyond the scope of this paper.

3 Microservices and Event-Driven Architecture

With a microservice architecture [12, 13] an application is structured "as a collection of loosely coupled services, which implement business capabilities" [14]. It seems that microservices have a lot in common with active software components ("Componentization via Services" [15]) and relatively small services in SOA. On one hand that is true and some of the main characteristics of a microservice are well-known in software engineering and apply also to other software artefacts like modules, e.g. high cohesion and loose coupling, single responsibility, or encapsulation. One the other hand, characteristics like clustered services that help to avoid single point of failure, or "Smart endpoints and dumb pipes" [15] are new concepts in connection with services.

A simplified model for a microservice-based design is shown in Fig. 3 in the form of an UML component diagram that contains microservices for the alarm, for the motion detection and for the facility master data management. The font-end app (user interface) uses an API gateway.

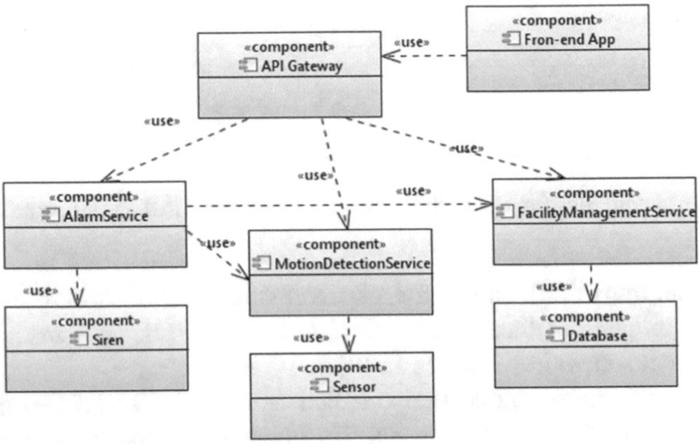

Fig. 3. Simple software design model with microservices using an UML component diagram.

Despite their popularity, microservices are not an end-in-itself, i.e. they do not solve all architectural problems of software systems and there are cases where for example a monolithic software system might make sense, but the recent developments in the theory and practice of microservices are promising, e.g. microservice patterns [13]. However, some aspects are still subject of research and discussions, e.g. the size of a microservice [16]. This leads to another open question concerning microservices: How can a workflow or state machine be designed in microservice architecture that stretches across the boundary of individual microservices? Two, not completely orthogonal approaches exist [17]:

- Orchestration: A central service for controlling or drive the process must be created. This service has the "knowledge" about all steps of the process (or state machine).
- Choreography: The services get informed about its task, but remain relatively autonomous, i.e. they decide themselves how to fulfill the task.

Both microservice collaboration techniques have advantages and disadvantages concerning the software design and implementation, e.g. orchestration leads to a powerful service that can become the single point of failure making the system less fault-tolerant. On the other hand, it is enabling the usage of a state engine similar to a workflow engine, that can manage the behavior in a flexible, central and direct manner. "Direct" in this context means that the state machine does not require any transformation in order to be used in the software system, i.e. the model can be used directly as an input artefact for the run-time-engine.

The choreography approach can be implemented with an Event-Driven Architecture (EDA) [18, 19]. This causes a transformation of the workflow logic: A state machine or a workflow leads to more decentralized and decoupled structure (at least when more than one object is involved in a state or a transition (event, guard condition, or action), because the producer or publisher of an event is not orchestrated by a central instance and is not directly coupled with the consumer of an event that only needs to subscribe to certain events without any "knowledge" of the state machine itself. An active object can be a consumer and also a producer of events. The event channel is realized as a message-oriented middleware or a point-to-point communication. Many UI frameworks with a MVC (model-view-controller) architecture also follows the EDA: UI elements listen (subscribe) to events caused by user interactions and react accordingly (controller).

The following UML sequence diagram (Fig. 4) shows the application of the publish-subscribe pattern (see Publish-Subscribe Channel EIP [20]) with the alarm and motion detection microservices and an event channel as the message-oriented middleware. It demonstrates a state machine of the alarm system (Fig. 1) that has been implemented as microservices (Fig. 3). The alarm service subscribes to the event "motionDetected" and the necessary action is performed when the motion detection service fire the event.

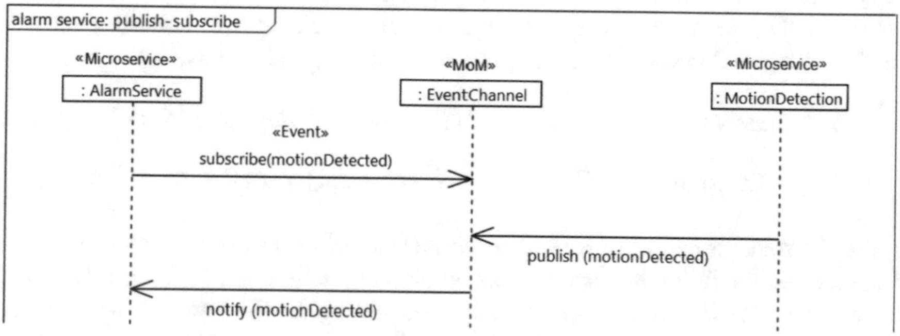

Fig. 4. UML sequence diagram for publish-subscribe pattern application with 2 microservices using an event channel as MoM (message-oriented middleware)

4 Concept for the Transformation of State Machines

For the transformation of the behavioral specification in the form of a state machine into a software design artefact, a formal specification (or concept) is used in order to provide a precise and implementation-independent description. For this purpose, the mathematical notation of set theory as described in [21] is applied. The advantage of following this approach has advantage that it supports "the process of designing model transformations", it leads to a better "understandability of the resulting code", and it facilitates explaining its meanings to others" [21].

Before this notation is introduced, the general idea of model transformations and the transformation of state machines to a software design on a structural level is explained briefly: One or several source models serve as an input for the transformation, e.g. a UML state machine, that is transformed into one or more target models, e.g. a class model, represent the output of the transformation (Fig. 5). Each model needs to be defined by a meta-model (language specification). Therefore, a model-to-model transformation involves at least one meta-model (source and target meta-model can be identical). The target model can be transformed to code (code generation). This approach follows the idea of a model-based or model-driven engineering (MDE).

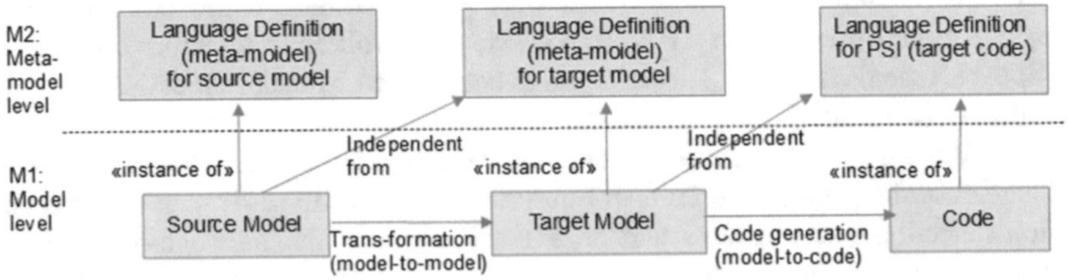

Fig. 5. Model-to-model transformation and model-to-code transformation

According to the UML metamodel (Fig. 2), a state machine Eq. (1), with the set-theory notation is defined as the Cartesian product of states, transitions, initial state, and final states. The powerset is used to express a multiplicity (upper bound) greater than one [21]. A transition consists of triggers, a constraint, and effects.

$$StateMachine \subseteq \mathbb{P}(State) \times \mathbb{P}\,(Transition) \times State \times \mathbb{P}\,(State) \tag{2}$$

$$\text{Transition} \subseteq \times\, \mathbb{P}(Trigger) \times (Constraint) \times \mathbb{P}\,(Behavior) \tag{3}$$

For the transformation of a state machine into an event-driven architecture using microservices, the following example for an alarm system is used (Fig. 6): The system is in an idle state (S_1) until a motion has been detected, which causes the system to change to the state "silent alarm active" (S_2). After 5 min, an audible alarm is activated (S_3) that uses a siren. The alarm can be deactivated so that the system becomes inactive (S_4), and a restart leads to its reactivation (idle state, S_1).

The software design model for the structural viewpoint comprises three microservices (alarm control, motion detection, and timer), several sensors and actuators, and a message broker for the event channel/store (Fig. 7).

During the state machine definition, it is possible to refer to the design classes and methods, e.g. the action "createAlarm" in the transition T_1 references the constructor of the class "AlarmControlService" (Fig. 7). For the transition T_2 the microservice "TimerService" is used, and in T_4 the Signal for the alarm deactivation is managed by the alarm service. The state S4 uses the method "deactivateAlarm" (class "AlarmService") for the entry behavior.

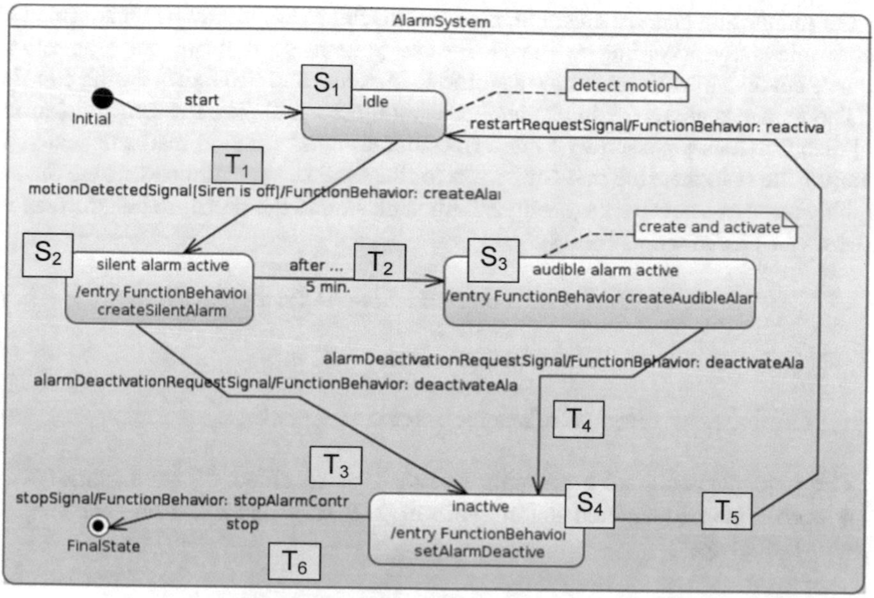

Fig. 6. UML state machine for the alarm system

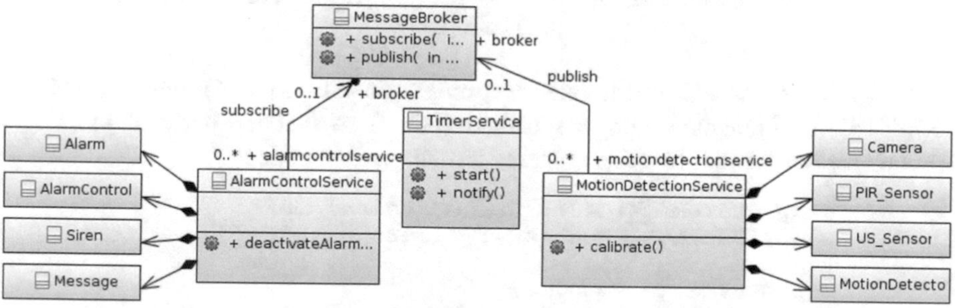

Fig. 7. UML class diagram for the software design of the alarm system

The example makes it easier to understand the necessary outcome of the transformation: Each service needs to subscribe to events or publish events. For the transformation of a state machine into an event-driven architecture using microservices, model elements of states and transitions need to be transformed, e.g.

- states with entry, exit, and do behaviors with method references lead to subscription of the trigger events for the classes of these methods.
- triggers that contain an event with a method reference of a class will publish a corresponding message.
- effects that reference a method will be transformed into subscription like states with entry, exit, and do behaviors.

For monitoring and visualization, the state machine should also be transformed into a class hierarchy according to the GoF state pattern, so that not only the EDA is created, but also the original state machine is implemented and exists during run-time.

The transformations are implemented using QVTo [22]. The set-theory notation in [21] also provides a possibility to describe model transformation mathematically. For example, the state machine transformation for the GoF state pattern part where the state machine leads to a new package with a context class, and the states are transformed into classes, can be defined as follows:

$$StateMachineToGoF : StateMachine \rightarrow Package \times \mathrm{Class} \tag{4}$$

$$StateToClass : State \rightarrow Class \tag{5}$$

$$RegionToPackage : Region \rightarrow Package \tag{6}$$

The implementation of a mapping can also be specified: “StateMachineToGoF” maps each region (quantified union symbol [21]) to a package using the mapping “RegionToPackage”:

$$\begin{aligned} &StateMachine2GoF(sm): \\ &\quad result.name = sm.name, \\ &\quad result.packagedElement += \bigcup_{reg \in sm.regions} \{RegionToPackage(reg)\} \end{aligned} \tag{7}$$

In QVTo, this would lead to three mappings (Eqs. 4, 5, and 6). For “StateMachineToGoF” the implementation is specified (Eq. 7) so that the body of the QTVo implementation for that function contains transformation code.

```
mapping StateMachine :: StateMachineToGoF :
   pkg : Package, context : Class
{
     pkg.name := sm.name;
     pkg.packagedElement += sm.ownedElement[Region] ->
        map RegionToPackage();
...}
mapping State :: StateToClass : Class { ...}
mapping Region :: RegionToPackage : Package { ...}
```

The implementation of the QVTo code based on the mathematical specification is a manual process. The QVTo implementation is used for automatic model-to-mode transformation that leads to the publish-subscribe behavior according to the state machine-EDA-mapping. Figure 8 shows the class diagram after the transformation took place: The microservices and the events are associated via a publish or subscribe relationship.

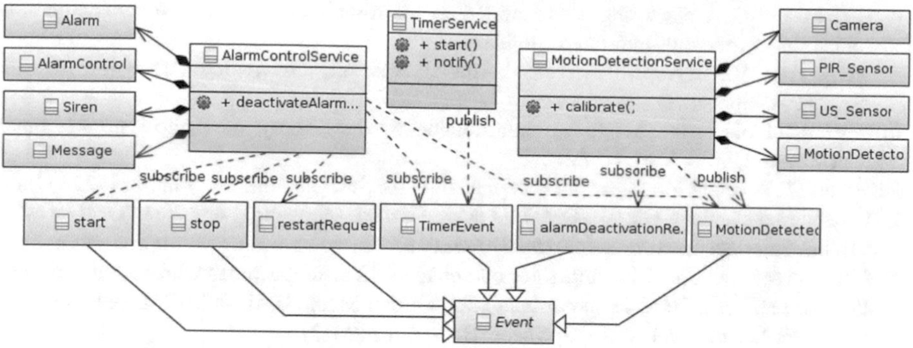

Fig. 8. Design classes for the EDA (detail)

The design model can be used for code generation, e.g. with MOFM2T [23]. This enables the system to be regenerated and redeployed automatically after changes to the state machine had been made.

5 Conclusion

The model-to-model transformation of a state machine for a microservice-based and event-driven architecture (EDA) proved that in principle, it is possible to combine workflow or state machine approaches with EDA. Without transformation and code generation, event management (publish-subscribe logic) has to be implemented manually and cannot be changes easily making the EDA-approach less flexible. On the other hand, the advantages of state or workflow engine that instantiates and manages the state machine directly (without transformation and EDA) leads to possible singe-point-of-failure that is avoided with EDA through choreography instead of orchestration. Another advantage can also be included in this approach: The monitoring, visualization abilities that typically only a state or workflow engine provides, is also available with the transformation approach because it is possible to generate a context class and state classes (according to the GoF state design pattern) that subscribe to all events of the state machine so that a run-time state machine exist. However, not all aspects of state machines in connection with microservices and EDA were discussed, e.g. CQRS [24], orchestration vs choreography [25–27], so that more research is necessary.

References

1. IBM: Rational Statemate. https://www-03.ibm.com/software/products/en/ratistatt
2. Visial Paradigm: How to generate code from state machine diagram. https://www.visual-paradigm.com/tutorials/how-to-generate-code-from-state-machine-diagram.jsp
3. Krause, L.: Microservices: Patterns and Applications: Designing Fine-Grained Services by Applying Patterns (2015)

4. Richardson, C.: Refactoring a Monolith into Microservices. https://www.nginx.com/blog/refactoring-a-monolith-into-microservices/
5. Alagarasan, V.: Seven Microservices Anti-patterns. https://www.infoq.com/articles/seven-uservices-antipatterns
6. Gill, A.: Finite State Machines. Introduction to the Theory of Finite-State Machines. McGraw-Hill, New York (1962)
7. Harel, D.: A visual formalism for complex systems. Sci. Comput. Program. 231–274 (1987)
8. Object Management Group (OMG): OMG Unified Modeling Language (OMG UML), version 2.5, document formal/2015-03-01 (2015)
9. Liu, S., et al.: A formal semantics for complete UML state machines with communications. In: Johnsen, E.B., Petre, L. (eds.) Integrated Formal Methods, IFM 2013. Lecture Notes in Computer Science, vol. 7940. Springer, Heidelberg (2013)
10. Microsoft Developer Network: State Machine Workflows. https://msdn.microsoft.com/en-us/library/ms735945(v=vs.90).aspx
11. Rozenburg, G., Engelfriet, J.: Elementary net systems. In: Reisig, W., Rozenberg, G. (eds.) Lectures on Petri Nets I: Basic Models – Advances in Petri Nets. Lecture Notes in Computer Science, vol. 1491, pp. 12–121. Springer (1998)
12. Newman, S.: Building Microservices: Designing Fine-Grained Systems, 1st edn. O'Reilly Media, Sebastopol (2015)
13. Richardson, C.: http://microservices.io
14. Amundsen, M., McLarty, M., Mitra, R., Nadareishvili, I.: Microservice Architecture - Aligning Principles, Practices, and Culture. O'Reilly Media, Sebastopol (2016)
15. Lewis, J., Fowler, M.: Microservices - a definition of this new architectural term. https://martinfowler.com/articles/microservices.html
16. Morris, B.: How big is a microservice? http://www.ben-morris.com/how-big-is-a-microservice
17. Wolf, O.: Introduction into Microservices. https://specify.io/concepts/microservices#choreography
18. Bruns, R., Dunkel, J.: Event-Driven Architecture - Softwarearchitektur für ereignisgesteuerte Geschäftsprozesse. Xpert.press, Sringer (2010)
19. Fowler, M.: Event logs, State Flows, Microservices…. How do these relate? https://garysmicroservices.wordpress.com/2016/01/06/event-logs-state-flows-microservices-how-do-these-relate
20. Hohpe, G., Woolf, B.: Enterprise Integration Patterns: Designing, Building, and Deploying Messaging Solutions, 1st edn. Addison-Wesley, Boston (2003)
21. Tikhonova, T., Willemse, T.: Documenting and designing QVTo model transformations through mathematics. In: Lorenz, P., Cardoso, J., Maciaszek, L., van Sinderen, M. (eds.) Software Technologies ICSOFT 2015, pp. 344–359. Springer (2015)
22. OMG (Object Management Group): MOF Query/View/Transformation (QVT). Vers. 1.3, doc. formal/16-06-03 (2016)
23. OMG (Object Management Group): MOF Model to Text Transformation Language (MOFM2T). Vers. 1.0, doc. formal/2008-01-16 (2008)
24. Fowler, M.: CQRS. https://martinfowler.com/bliki/CQRS.html
25. Netflix Conductor: A microservices orchestrator. https://medium.com/netflix-techblog/netflix-conductor-a-microservices-orchestrator-2e8d4771bf40
26. Rücker, B.: Why service collaboration needs choreography AND orchestration. https://blog.bernd-ruecker.com/why-service-collaboration-needs-choreography-and-orchestration-239c4f9700fa
27. Webster, C.: From APIs to Microservices: Workflow Orchestration and Choreography Across Healthcare Organizations (2016)

Knowledge Discovery from Thai Research Articles by Solr-Based Faceted Search

Siripinyo Chantamunee[1(✉)], Chun Che Fung[1], Kok Wai Wong[1], and Chaiyun Dumkeaw[2]

[1] School of Engineering and Information System, Murdoch University, Perth, Australia
csiripin@wu.ac.th, {l.fung, k.wong}@murdoch.edu.au
[2] Informatics Innovation Research Center, Walailak University, Nakhon Si Thammarat, Thailand
dchaiyun@wu.ac.th

Abstract. Search engine plays an important role in information retrieval as being the preferred tool by users to locate and manage their desired information. The volume of online data has dramatically increased and this phenomenon of impressive growth leads to the need of efficient systems to deal with issues associated with storage and retrieval. Keyword search is the most popular search paradigm which prompts user to search the entire repository based on a few keywords. From research article collection, using keyword search only may not be enough for researchers to explore academic documents related to their interests from the entire repository. Knowledge discovery tool has recently received much attention in order to compensate the weakness of keyword search usage for academic collection. This paper presents the practical system design and implementation of a knowledge discovery tool in terms of faceted search. This study focused on Thai research articles for use by Thai scholars. The proposed faceted search system was constructed based on the Apache Solr search platform. The methodology of data preparation, knowledge extraction and implementation are discussed in the paper.

Keywords: Knowledge discovery · Faceted search · Thai research articles
Search engine · Knowledge extraction

1 Introduction

Search engine has been extensively developed and widely used. However, the complexity of human language and human information's needs are progressively changed over time. One potential problem is due to issues associated with vocabulary and it is related to keyword search. At present, keyword search is the most popular search paradigm in most retrieval systems. The approach prompts a user for a few keywords and it then searches the entire repository based on the given keywords. Users are requested to supply the few short keywords to represent their intent. These keywords or user queries become an issue due to the fact that users, in practice, may give words which do not match to the existing index terms or the context of the document

H. Unger et al. (Eds.): IC2IT 2018, AISC 769, pp. 337–346, 2019.
https://doi.org/10.1007/978-3-319-93692-5_33

collection. Search engine performance does not only depend on how efficient they store and retrieve the information, but also the relevance of the returned documents. The success of search engine with keyword search is potentially dependent on the quality of keywords supplied to the system.

Query formation seems difficult when users do not know the exact words to be used and retrieval systems may have few supportive features to assist the users. Page ranking is a technique which has been used by Google. It has become popular and it can locate information by considering the number of visits of the webpage link [1]. The technique needs to process on a large amount of information related to page popularity, it therefore may not be suitable for small retrieval systems, or specific collections on topics such as academic, medical, science and technology disciplines. Search in specific collections, in particular, needs the exploration search based upon the knowledge of existing contents. This is unlike general online retrieval systems which focus on current popular content. Page rank therefore seems not appropriate for specific retrieval systems.

Research articles have been generated every minute and thus it needs knowledge discovery tool for researchers to explore this huge volume of documents. Knowledge discovery is a process to find knowledge within the collection domain. Faceted search has been introduced and received much attention for presenting knowledge bounded to returned documents. Once users make their search, the number of attributes called facets will be listed, in most systems, on the left side of the search panel. Facets represents knowledge perspective of the returned results such as research discipline, date of publication, author, place, organization, and other relevant information. Facet itself can be defined as knowledge representation of the collection. Underlying knowledge of returned documents can be extracted and represented in the form of facet attributes. Faceted search gives opportunity for searchers to navigate the arranged facet lists to any specific topic which mostly closes to their interests. Not only one's interest can be found, but also other related documents are displayed so that searchers will be able to associate their works to the nearest topics.

Facet extraction is one important process to achieve faceted search. The diversity of documents' format in the collection becomes the problem for facet extraction especially in non-English documents such as Thai language. This requires pre-processing techniques such as word and sentence segmentation, name entity recognition, and other processes. It requires data preparation prior to the facet extraction process. This research has focused on Thai language search system so that processing techniques for Thai language have to be taken into consideration.

This paper presents and discusses the practical work on knowledge discovery with faceted search from the Thai research articles collection. The system aims to provide efficient knowledge discovery tools for scholars interested in the Thai research articles. In this study, the search and indexing mechanism were implemented on the Apache Solr search platform [12].

The rest of the paper is organised as follows. Section 2 provides the related works and backgrounds. Section 3 introduces the proposed system design and implementation along with the methodology of data preparation, metadata and faceted extraction for Thai documents. Finally, Sect. 4 gives a discussion on the system design and implementation, and it also concludes the paper.

2 Backgrounds

2.1 Knowledge Discovery with Faceted Search

Faceted search can be defined as an exploration search approach aims to provide support for users to explore the underlying knowledge from a collection of documents. It can be consideration as a combination of keyword search and browsing. Retrieval systems need faceted search to compensate the weakness of keyword search regarding the issue of query formation. Ranganathan [7] firstly introduced facet to describe document characteristics by five attributes including energy, personality, matter, space and time. The attributes had been later developed and named to faceted metadata, facet terms, and faceted value [8].

Faceted search provides users with hierarchical structure of knowledge categories or taxonomies covering contents of the search results. Facet attributes or terms are usually represented in hierarchical manner and are placed on the left side of the screen. In recent years, faceted search has been taken into consideration either in research [2, 3], or being used as main features in online retrieval systems [4, 5]. Faceted search generally achieves higher precision and recall than keyword search as previously reported in [2, 6]. It is close to directory navigation or taxonomy-based search, but it is able to represent more coverage knowledge filtered from search results and achieve better classification effectiveness [8]. Figure 1 illustrates an example of faceted search from the IEEE Xplore digital library [9]. The feature enables users repetitively refining their search and exploring resources from the arranged knowledge, and it has also been implemented in other specific systems such as computing literature ACM digital library, biomedical literature PubMed, and E-commerce sites such as Amazon and EBay.

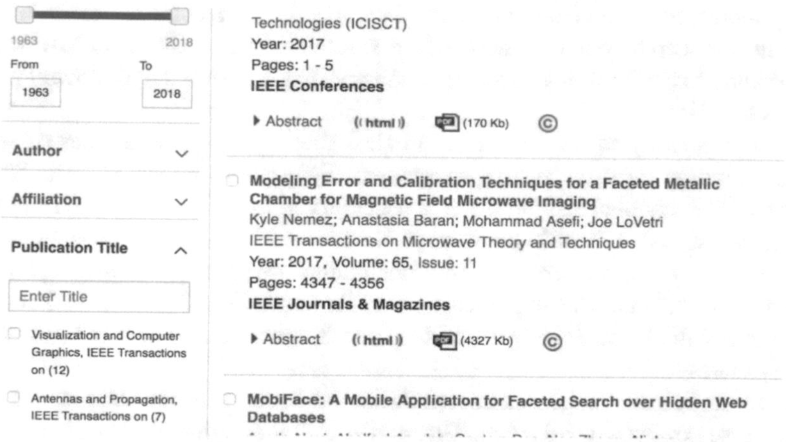

Fig. 1. Faceted search interface of IEEE Xplore digital library where facet attributes and their values are arranged on the left panel bounded to the search results on the right side. The facet attributes include year, author, affiliation, publication title, publisher, supplemental items, and conference location (Source: IEEE Xplore digital library [9])

For academic digital library, faceted search is particularly useful and necessary for retrieval systems to ensure scholars that they can access all the articles related to the knowledge they are interested in. This will advance the benefit of knowledge searching. Not only one particular article will be found, other related articles in the same area will also be listed.

A general framework for faceted search system was introduced by Zheng [8]. The framework consisted of three major processes which are (1) faceted taxonomy generation including works on extracting facet terms from unstructured or structured documents and constructing hierarchical structure, (2) query refining with compound term generation and facet ranking, and (3) results ranking which usually applied traditional scoring techniques from information retrieval systems as illustrated in Fig. 2.

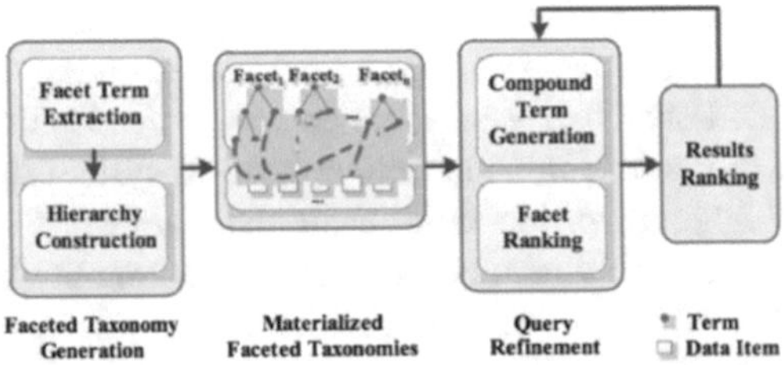

Fig. 2. General framework for faceted search system (Source: Zheng [8])

2.2 Existing Search System for Thai Research Article

Search system on Thai research article collection was first introduced in 2000 by KMUTNB and it has been in operation over 18 years. It currently needs a new system to deal with storage and retrieval issues since Thai research articles have been continually increased. Knowledge discovery is a mandatory feature for the new system especially for search system on academic documents. The challenge is how to extract facet or knowledge from this collection and serves as a knowledge discovery platform for Thai scholars.

In the repository of Thai research articles, there are a large number of research articles including research reports, theses and dissertations collected from Thai universities and educational institutions. It is necessary to have an efficient retrieval system with faceted search to handle knowledge discovery from this large volume of documents. The system has to process Thai documents which requires language analysis technique for Thai language in particular. In order to get a word from textual document, word segmentation technique is a mandatory task since Thai sentences have no space between words.

However, one challenge remains, multilingual documents are commonly found in the Thai research article collection. This is due to the fact that Thai academics do not usually translate English technical words in Thai writing practices. Thai scholars are likely to write Thai and English words together in the same sentence. The system needs to determine which technique will be used when Thai and English words are found in the same documents, even in the same sentence. This still requires work on language processing on multilingual documents. System scalability is another concern as the new system needs to scale up its ability to handle the increasing impact from the huge volume of data.

2.3 Solr-Based Search System

The Apache Solr is a search and indexing platform introduced by open source community in order to provide a scalable, reliable, and fault tolerant system construction tool for online information retrieval [12]. The platform supports faceted search, document filtering and multilingual documents from different languages including English, German, Chinese, Spanish, Mongolian [13], Thai, and etc. Solr indexing system is able to accept a wide range of data formats such as XML, Comma Separated Values (CSV) file, JSON, relational table, and etc. Using the Solr platform, online retrieval systems can be established more convenient and efficiently.

3 System Design and Implementation

3.1 System Architecture for Thai Faceted Search

Faceted search system for Thai research article based on the Solr platform is illustrated by the system architecture shown in Fig. 3. Searchers interact to the system via web interface which prompt users with a search box and search results with facet items were then displayed when searched. The system acquires real-time metadata and full-text files from Thai librarians over the country and, once new document come in, update handler module then performs incremental update to the index system. User query and metadata given to the system are fed to Thai-English text analysis module where the detail will be explained in Sect. 3.5.

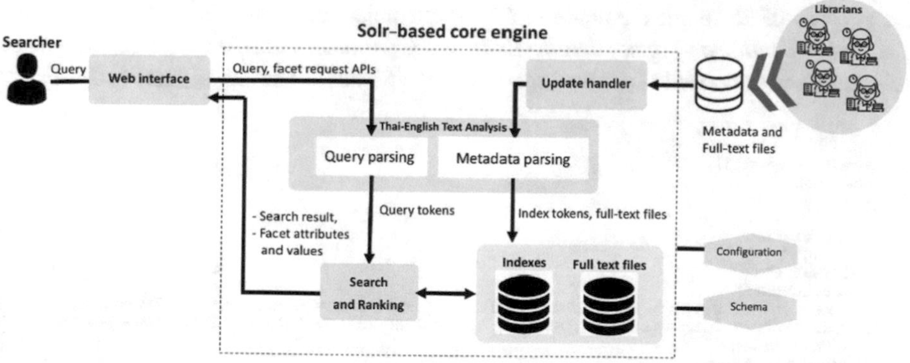

Fig. 3. System architecture for Solr-based Thai faceted search

3.2 Data Collection

At the time of development, the Thai Research Articles Collection had 372,849 Thai full-text academic articles created by 474,638 authors from 171 Thai universities and educational institutes. Table 1 presents a breakdown of the collection classified by material type. The collection included the varieties of material types from various disciplines including science and engineering, education, law, environment science, health science and nursing, management, economic, information technology, and agriculture.

Table 1. Detail of Thai research article collection classified by material type

Material type	Number of documents
Research article	44,241
Thesis	292,130
Research report	36,478

3.3 Data Preparation

Metadata of the documents in the collection was manually extracted by Thai librarians from several Thai educational institutions as the format of entry data was varied based to the cataloguing principle of each organization or librarians themselves. Data cleaning has to be operated on this metadata. At this stage, this issue is not yet to be investigated and this is part of the subsequent development. Data preparation in this study was to check the completeness of data fields which were required for retrieval purpose. Four mandatory fields must exist and it includes: *field creator/author, title, subject, and material type.*

The proposed metadata was constructed in the form of Dublin Core (DC) metadata as it being one of the common metadata standards for many digital libraries. Figure 4 gives an example of the Dublin Core metadata. Simple Dublin Core metadata element set consists of 15 major elements, fields or attributes including *title, creator, subject, description, publisher, contributor, date, type, format, identifier, source, language, relation, coverage, and right management* with 56 sub-elements in total [10, 11].

```
<?xml version="1.0"?>
<metadata
      .
      .
  <dc:title>
    UKOLN
  </dc:title>
  <dc:subject>
    national centre, network information support, library community, awareness, re-
search, information services,public library networking, bibliographic management
  </dc:subject>
  <dc:publisher>
    UKOLN, University of Bath
  </dc:publisher>
</metadata>
```

Fig. 4. Example of the implementation of Dublin Core metadata in XML format (Source: Dublin Core metadata initiative [11])

The usage of Simple Dublin Core could not accommodate all the information required for the proposed system. It is necessary to store more details for the material type 'thesis'. Simple Dublin Core metadata set was then extended from 15 elements to 16 elements and the additional element was named *"thesis"* with 4 sub-elements – *degree name, level, discipline, and grantor.*

3.4 Knowledge or Facet Extraction

Six elements from 16 elements were selected to facet attributes including elements: *type, language, contributor, subject, date_issue*, and *thesis* with sub-elements: *level, discipline, grantor* – eight faceted attributes in total. Two-level facet was implemented, presenting the top level of facet attributes and its values down in the second level. The following code is Solr configuration using by client-side code (using c#) when requesting Solr core engine to create facet attributes and send the facet results back to client to display. *Qp* represents the user-defined objects of query request. *SolrNet* is .NET client for connecting Apache Solr and calls method *SolrFacetFieldQuery* for setting facet attributes.

```
Qp.AddFacets(New SolrNet.SolrFacetFieldQuery("type"))
Qp.AddFacets(New SolrNet.SolrFacetFieldQuery("language"))
Qp.AddFacets(New SolrNet.SolrFacetFieldQuery("contributor"))
Qp.AddFacets(New SolrNet.SolrFacetFieldQuery("subject"))
Qp.AddFacets(New SolrNet.SolrFacetFieldQuery("date_issued"))
Qp.AddFacets(New SolrNet.SolrFacetFieldQuery("thesis_level"))
Qp.AddFacets(New SolrNet.SolrFacetFieldQuery("thesis_descipline"))
Qp.AddFacets(New SolrNet.SolrFacetFieldQuery("thesis_grantor"))
```

3.5 Sorl-Based Index System and Query Search for Thai Collection

Single core of Solr-based faceted search system was constructed based on Solr search platform version 6.0.0 [12]. The system obtained real-time metadata from librarians to create index terms usage by search and ranking module. Regarding the nature of the documents as being Thai-English documents, schema configuration for text analysis used class *solr.ThaiTokenizerFactory* for processing Thai word segmentation, *solr.StopFilterFactory* for removing Thai and English stop words, and *solr.LowerCaseFilterFactory* for setting all English words to lower case. It is noted that there is no need of word stemming for Thai language. The same configuration was done with user query.

Solr configuration (file *manage-schema*) to define a specific field type (in the example, named *text_tdc*) to process Thai textual sentences is displayed as following.

```
<fieldType name="text_tdc" class="solr.TextField">
   <analyzer type="index">
      <tokenizer class="solr.ThaiTokenizerFactory"/>
      <filter class="solr.StopFilterFactory" ignoreCase="true" words="stopwords.txt" />
      <filter class="solr.LowerCaseFilterFactory"/>
   </analyzer>
   <analyzer type="query">
      <tokenizer class="solr.ThaiTokenizerFactory"/>
      <filter class="solr.StopFilterFactory" ignoreCase="true" words="stopwords.txt" />
      <filter class="solr.LowerCaseFilterFactory"/>
   </analyzer>
</fieldType>
```

3.6 System Implementation

The developed Solr-based faceted search system operates on an 64-bit Windows server with Intel CPU Xeon 1.87 GHz, 12 GB ram, and 150 GB Hard disk space. The system was implemented by using Microsoft .NET framework as the development tool. The interface is displayed in Fig. 5. For instance, 7,481 documents related to given user query *"ป่าไม้"* ("forest" in English) were returned with a list of facet or knowledge attributes and their values aligned on the left panel. Once query *"ป่าไม้"* was given, the system executed Thai word segmentation which resulted into two words – *"ป่า"* (forest in English) and *"ไม้"* (wood in English) so that there were three words to search through the indexes which were *"ป่าไม้"*, *"ป่า"* and *"ไม้"*. As shown in the search result in Fig. 5, the documents related to keyword *"ป่า"* were also returned as seen in the example as the first document.

In the search result panel, the interface displayed the detail of each individual document including document title, material type, author and name of contributor in Thai language. Eight facet attributes and their values were displayed either in Thai or English language regarding to the metadata given from librarians which sometimes they could be the same meaning. For example, for educational level, some librarians input "Master's degree" instead of "ปริญญาโท" which they were the same meaning. This is due to the face that data cleaning has not been performed on Thai and English words conversion. This is similar to as the words "ปริญญาเอก" and "Doctoral Degree". This also occurred with other facet attributes. Figure 6 displays an example of faceted attribute and their values for user query "ป่าไม้" ("forest" in English). In particular, one interesting knowledge is about research discipline and document subjects where searchers could possibly match their interests toward this filtered knowledge list.

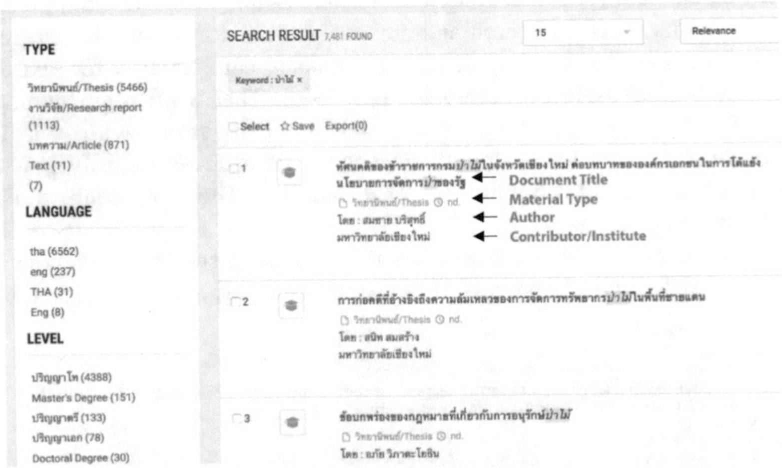

Fig. 5. Implementation of faceted search interface for Thai research article collection with the example of user query/keyword *"ป่าไม้"* or "forest" in English. The number in parentheses beside facet attributes indicates the number of documents found by each facet attribute.

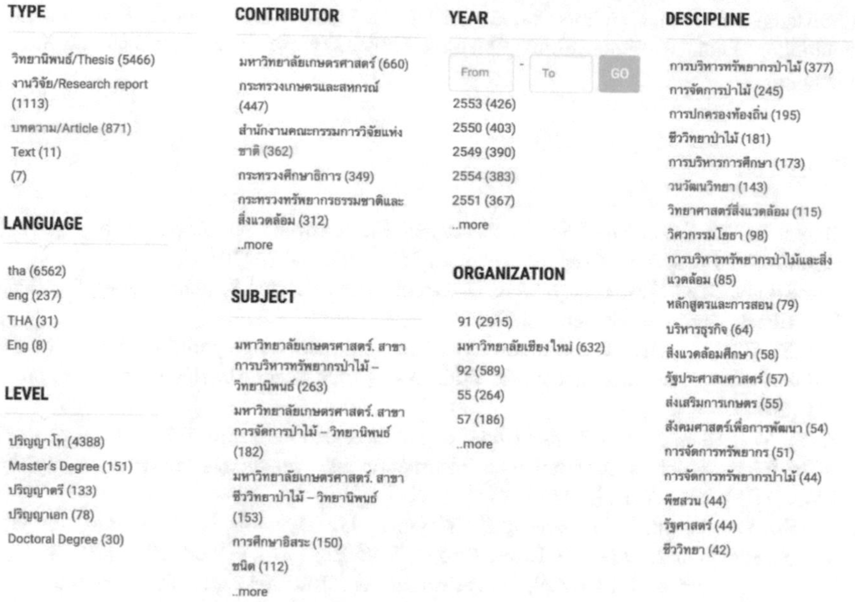

Fig. 6. Example of knowledge discovered from Thai research collection in form of faceted attributes extracted from the returned documents with user query/keyword *"ป่าไม้"* (*"forest"* in English)

4 Discussion and Conclusion

This paper presents work on knowledge discovery on Thai research article repository in the form of faceted search. The detail of practical system design and implementation for faceted search is explained together with data preparation, the knowledge extraction from eight facets and two-level faceted search user interface implementation. Solr search platform was used in this study to construct the proposed faceted search system.

Nevertheless, faceted search for Thai research article still has many issues. Firstly, librarians have to put in a lot of effort to manually extract the metadata from all the articles. It takes a considerable amount of time for an article to be ready for use. Secondly, single level of facet values can only be created in this study. Some levels of knowledge from search results may possibly be missing, not expanded and limited. For example, knowledge on "plants" could be expanded to "plant mutation", "plant quarantine", "plant conservation", "tree", "dye plant" and "plant protoplast". The knowledge about *"tree"* could be extensively expanded to another level which is *"tree planting"*. Ideally, all the knowledge should be properly covered and organised. Users will then have insights as regard to the kind of knowledge representing the search results. Thirdly, the problem of metadata format mismatching needs to be addressed. These issues shall be addressed in the planned future work.

Acknowledgement. The authors would like to acknowledge Suradech Petcharanon and Phamphiphop Phunnaputtimok from Walailak University for their assistance on metadata preparation.

References

1. Gupta, R., Shah, A., Thakkar, A., Makvana, K.: A survey on various web page ranking algorithms'. Int. J. Adv. Comput. Technol. **5**(1), 2046–2052 (2016)
2. Sacco, G.M., Tzitzikas, Y.: Dynamic Taxonomies and Faceted Search: Theory, Practice, and Experience, Springer, Heidelberg (2009)
3. Liu, X., Zhai, C., Han, W., Gungor, O.: Numerical facet range partition: evaluation metric and methods. In: International World Wide Web Conferences Steering Committee, pp. 662–671 (2017)
4. Kong, W., Allan, J.: Precision-oriented query facet extraction. In: Proceedings of the 25th ACM International on Conference on Information and Knowledge Management, pp. 1433–1442. ACM, New York (2016)
5. Vandic, D., Aanen, S., Frasincar, F., Kaymak, U.: Dynamic facet ordering for faceted product search engines. IEEE Trans. Knowl. Data Eng. **29**(5), 1004–1016 (2017)
6. Yogev, S., Roitman, H., Carmel, D., Zwerdling, N.: Towards expressive exploratory search over entity-relationship data. In: Proceedings of the 21st International Conference on World Wide Web, pp. 83–92. ACM, New York (2012)
7. Ranganathan, S.: Elements of Library Classification. South Asia Books, Bombay, New York (1991)
8. Zheng, B., Zhang, W., Feng, X.F.B.: A survey of faceted search. J. Web Eng. **12**(1&2), 041–064 (2013)
9. IEEE Xplore digital library. http://ieeexplore.ieee.org/Xplore/home.jsp
10. Mathieu, C.: Practical application of the Dublin core standard for enterprise metadata management. Bull. Assoc. Inf. Sci. Technol. **43**(2), 29–34 (2017)
11. Dublin core metadata initiative. http://dublincore.org/documents/dces
12. Shahi, D.: Apache Solr: an introduction. In: Apache Solr. Apress, Berkeley (2015)
13. Ma, L., Bao, W., Bao, W., Yuan, W., Huang, T., Zhao, X.: A Mongolian information retrieval system based on Solr. In: International Conference on Measuring Technology and Mechatronics Automation, pp. 335–338 (2017)

Author Index

H. Unger et al. (Eds.): IC2IT 2018, AISC 769, pp. 347–348, 2019.
https://doi.org/10.1007/978-3-319-93692-5

Printed in the United States
By Bookmasters